P9-CFD-943

A Short History of the Universe

A Short History of the Universe

Joseph Silk

SCIENTIFIC
AMERICAN
LIBRARY

A Division of HPHLP
New York

ISBN 0-7167-5048-1

Silk, Joseph
A short history of the universe / Joseph Silk.
p. cm.
"A Scientific American Library book."
Includes index.
ISBN 0-7167-5048-1
1. Cosmology. 2. Astrophysics. I. Title
QB981.S554 1994 94-21771
523.1—dc20 CIP

ISSN 1040-3213

Printed in the United States of America

Scientific American Library
A division of HPHLP
New York

Distributed by W. H. Freeman and Company
41 Madison Avenue, New York, New York 10010
20 Beaumont Street, Oxford OX1 2NQ, England

1 2 3 4 5 6 7 8 9 0 KP 9 8 7 6 5 4

This book is number 53 of a series.

Contents

To Phil

Preface

Flashes of light illuminate the dark sky. Fireballs briefly glow in a rainbow of colors and fade away. Soon the Pacific fog rolls in, covering the San Francisco Bay in a dense gray blanket that quenches any light or sound. Perhaps the early universe was once such a fireworks display on a much vaster scale, as outbursts of star formation heralded the creation of the first galaxies. Astronomers peer hopefully through interstellar obscurity, far in distance and back in time, to catch a glimpse of any pyrotechnics that may be visible with the aid of the largest telescopes available, on desolate mountain peaks and in earth orbit. Not to be outdone, physicists labor in underground caverns, constructing their own elaborate versions of telescopes to search for traces of exotic particles that carry secrets from the beginning of time.

The study of cosmic evolution comprises a unique amalgam of disciplines. Participants include astronomers who peer at the heavens, mathematicians who devise elaborate theories of the fundamental nature of space and matter, physicists who construct complex machines to help unravel the innermost nature of elementary particles, and philosophers of science who search for the ultimate answers and, indeed, the ultimate questions. In telling the saga of the origin of the universe, I have interwoven these disciplines to construct a more coherent and unified framework. There is a name for the emerging interdisciplinary blend of these fields, particle astrophysics, that captures the essence of the various approaches to understanding the cosmos. Whether we will succeed in our pursuit I would not venture to guess. But whatever the outcome, the process of discovery will be a source of enlightenment, of joy, of enrichment, of passion, of growth, of wisdom, of vision—not merely to the scientists involved but to the entire community of humans who have ever looked up at the night sky and pondered the meaning of their existence in this vast universe.

I first felt the allure of cosmology when, as an undergraduate in search of distraction from tedious degree courses at Cambridge, I attended a series of lectures by Dennis Sciama. Bowled over by the passion in Sciama's approach, I determined that my fate lay in the stars. An eclectic interlude followed as I worked in the Manchester laboratory of radio astronomer and space physicist Roger Jennison, who guided me toward my graduate studies at Harvard. There I came under the spell of David Layzer. An unorthodox cosmologist of remarkably broad interests, Layzer inspired me to acquire perspective and insight in tackling many problems, no matter how seemingly irrelevant. During a marvelous summer at Woods Hole, George Field provided the impetus that launched me toward studying the evolution of cosmic structure, a voyage of mystery that has never ceased to amaze me, nor has it yet ended. These were the primary influences on my early career, and to them I am deeply grateful. A first and most formative postdoctoral year was spent at Fred Hoyle's newly established Institute of Theoretical Astronomy, as it then was known, where intense discussions with Martin Rees, Hubert Reeves, Gary Steigman, and Michael Werner, to name but a few, helped further broaden my perspectives. In later years, many other colleagues, too numerous to single out, have shared in my research interests, and these too I gratefully acknowledge.

This book began as a series of lecture notes for a course that I taught at Berkeley. I am grateful to successive generations of students whose questions helped me hone the ideas that I describe here, and to whom I hope I have transmitted some of my enthusiasm about cosmic matters. I wish to acknowledge several colleagues for critically commenting on the manuscript: the late William Kaufmann, Jay Pasachoff, Martin Rees, and Douglas Scott. My editor at the Scientific American Library, Susan Moran, is to be commended for her devotion to overseeing this project. I thank Travis Amos for his diligent pursuit of photographs and Carol Frost for creating the index. I am grateful to Jerry Lyons and to his successor at the Scientific American Library, Jonathan Cobb, for having persuaded me of the merits of this endeavor. I am indebted to many colleagues for having provided illustrative and topical material, in many cases directly from their computers or telescopes, including Ed Bertschinger, Roger Blandford, Francois Bouchet, Ted Bunn, Jack Burns, Avishai Dekel, Alan Dressler, George Efstathiou, Gus Evrard, Carlos Frenk, Margaret Geller, Richard Griffiths, Lars Hernquist, John Huchra, Nick Kaiser, Anthony Lasenby, David Malin, Don Mathewson, Yannick Mellier, Chris Mihos, Mike Norman, John Peacock, Jim Peebles, Bruce Peterson, George Smoot, Gordon Squires, Albert Stebbins, Michael Strauss, Frank Summers, Alex Szalay, Brent Tully, Martin White, and Simon White. I hope this book reflects the boundless enthusiasm of the graduate students and postdoctoral researchers whom I have been privileged to help guide.

Joseph Silk
Berkeley, California
July 1, 1994

A Short History of the Universe

Prologue

The universe began in a violent explosion that occurred about 15 billion years ago: this is the modern hypothesis that has replaced the myths of classical Greece and Rome, of ancient China and India. We feel certain that our theories have more truth than the beliefs of our ancestors, yet are we so much smarter than they were? Perhaps a thousand years in the future, the big bang theory will itself be regarded as a twentieth-century myth.

I am an optimist, however, who finds our current paradigm so compelling that I can only imagine it will eventually be subsumed into a greater theory, without losing its essential features. This conviction provides justification enough to describe the archaeology of the universe. By probing fossil fluctuations in the distribution of matter on the one hand and fully formed galaxies on the other, the oldest stars and the largest structures, one can reconstruct almost the entire history of cosmic evolution.

The story has been told before, but there are two good reasons for a fresh approach. The study of the cosmos was recently revitalized with the discovery of seed fluctuations from which structures later emerged. This was the crucial missing link needed to reconcile the big bang theory with the large-scale distribution of matter around us, and it injected a note of reality into the science of cosmology. From a cosmic perspective, one might say that pregalactic material was discovered that is yielding intimate clues about an origin in the remote past. This is a timely moment to describe our view of cosmic evolution.

An equally compelling reason is that there exists no truly accessible and modern description of the big bang theory, with its many ramifications arising from the domains of astronomy and of the physics of elementary particles. One of the perennial fascinations of the science of cosmology is that people, both lay cosmologists as well as the professionals, view it as having the potential to answer the "ultimate questions" about our place in the universe, the creation and existence of the universe, and indeed the existence of God. It is by no means coincidental that the big bang epic has excited the attention of theologians and philosophers as well as astronomers, mathematicians, and physicists.

Some of these thinkers have viewed the theory as providing confirmation of religious views of creation. The science historian and mathematical physicist E. T. Whitaker declared in 1942 that "when by purely scientific methods we trace the development of the material universe backwards in time, we arrive ultimately at a critical state of affairs beyond which the laws of nature, as we know them, cannot have operated: a Creation in fact. Physics and astronomy can lead us through the paths to the beginning of things, and show that there must have been a Creation." In 1951, Pope Pius XII, under the influence of Whitaker, went the additional step. He averred in an address to the Pontifical Academy of Sciences that "thus with concreteness which is characteristic of physical proofs, it [science] has confirmed . . . the well-founded deduction as to the epoch [some five billion years ago] when the cosmos came forth from the hands of the Creator. Hence, creation took place in time. Therefore, there is a Creator. Therefore, God exists!"

On hearing these words, one can well imagine that the President of the Pontifical Academy, eminent cosmologist and cofounder of the big bang theory Abbé Georges Lemaître, must have stirred uneasily. To compare the primeval explosion from which the universe emerged to the miracle of creation must have seemed to leave him, a proponent of the Primeval Atom phase that preceded the big bang, on somewhat uncertain and heretical ground. Lemaître insisted that physics would suffice to describe the beginning of the universe: "Cosmogony is atomic physics on a large scale."

The big bang was not an easy pill to swallow, for scientists and theologians alike. Eminent astrophysicist and science popularizer Arthur Eddington was "unwilling to accept the implied discontinuity in the divine nature." Others went further. The pioneering cosmologist E. A. Milne concluded in his magnum opus *Relativity, Gravitation and World Structure*, published in 1935, that "the system to which we have likened the universe is an intelligible system. It contains no irrationalities save the one supreme irrationality of creation—an irrationality indeed to physics, but not necessarily to metaphysics. . . . Theoretical cosmology is but the starting point for deeper philosophical enquiries."

Some scientists conceded the battle for understanding how the universe began to the theologians, who after all had been wrestling with it for centuries. Astronomer Robert Jastrow described the cosmologists' dilemma thus, in a quote beloved by theologians: "It seems as though science will never be able to raise the curtain on the mystery of creation. For the scientist who has lived by his faith in the power of reason, the story ends like a bad dream. He has scaled the mountains of ignorance; he is about to conquer the highest peak; as he pulls himself over the final rock, he is greeted by a band of theologians who have been sitting there for centuries."

In contrast, some eminent scientists have no recourse to any deity in constructing a suitable cosmology. One convenient way out is the assertion that time itself was created at the moment of the big bang. This is not a very radical idea, for St. Augustine wrote in the fifth century, "The world and time had both one beginning. The world was made, not in time, but simultaneously with time." This was a remarkably prescient notion: to physicist Steven Weinberg, "it is at least logically possible that there was a beginning, and that time itself has no meaning before that moment."

However, as the mathematical physicist Stephen Hawking points out, a proper formulation of this concept of the beginning of time, as well as that of space, must await a quantum theory of gravity, should it be forthcoming. In this case, "there would be no boundary to space-time and so there would be no need to specify the behavior at the boundary. There would be no singularity at which the laws of science broke down and no edge of space-time at which one would have to appeal to God or some new law to set the boundary conditions for space-time. One could say: 'The boundary condition of the universe is that it has no boundary.' The universe would be completely self-contained and not affected by anything outside itself. It would neither be created or destroyed. It would just be." In other words, the universe is the way it is because the universe was the way it was.

Eloquent expression of this cosmic agnosticism was prophetically penned in 1920 by, again, Arthur Eddington: "We have found that where science has progressed the furthest, the mind has but regained from nature that which the mind has to put into nature. We have found a strange footprint on the shores of the unknown. We have devised profound theories, one after another, to account for its origin. At last, we have succeeded in reconstructing the creature that made the footprint. And Lo! it is our own."

Discoveries and breakthroughs in cosmology proceed at a breathtaking pace. It is increasingly difficult for the theologians to keep up. At the same time, cosmologists have not been deterred from dabbling in occasional theological metaphors. The shower of images reached a crescendo in 1992 with the epochal discovery of ripples in the cosmic microwave background. Newspapers around the world, less discriminating perhaps than the scientists anticipated, jumped on the cosmic connection. The most notorious examples, 40 years after Pius XII's endorsement of the new cosmology, compared the long-sought fluctuations to various attributes of God. These vary from "His face," "His handwriting," and "His mind" to mere relics such as the "Holy Grail."

To properly appreciate the significance of such statements, it would be helpful to have laid out for one the workings of modern cosmology at an accessible level. This book is devoted to such a goal. I hope that the following chapters will be sufficiently transparent to the many lay cosmologists among us that such connections can be more fully appreciated, although, I hasten to add, not necessarily justified.

An infrared view, at a wavelength of 2 micrometers, of the center of the Milky Way, in Sagittarius.

1

Building Blocks of the Cosmos

Throughout the ages, cosmologists have imposed their world views on the fundamental nature of the universe. Modern astronomers have proved no exception. As the extragalactic universe began to be explored in the twentieth century, cosmologists felt impelled to incorporate their concepts of how the universe should be. In 1931, Ernest Barnes, a mathematician and theologian who was also the Anglican Bishop of Birmingham, asserted, among other things, that an infinite universe was abhorrent to nature. Many years later, physicist John Archibald Wheeler, father of the black hole, echoed this sentiment in arguing for a finite universe. Others, beginning with Albert Einstein, adopted the Platonic ideal that the universe should have no center and should be perfectly isotropic, with no preferred direction. Only recently have we been able to test these assumptions, as some of the most fundamental issues in cosmology have been reassessed in the light of modern astronomical data.

Cosmological Principles

It is inevitable that an astronomer studies objects remote in time as well as in space. Light travels a distance of 300,000 kilometers in one second, or ten thousand billion kilometers in a year. The nearest star, Proxima Centauri, is 4.2 light-years from us, so we see it as it was about four years ago. The nearest galaxy comparable to our own Milky Way, the Andromeda galaxy, is two million light-years distant. We are seeing that galaxy, which is visible in a dark sky to the naked eye, as it was when Homo sapiens had not yet evolved. A large telescope is a time machine that can take us part way to creation. With a modern telescope, we can examine regions from which light emanated more than five billion years ago, before our sun had even formed. To a cosmologist, the issue of creation is unavoidable.

Our nearest neighbor galaxy, Andromeda, of comparable size to the Milky Way.

There are three possibilities that one may envisage for the creation of the universe.

1. The beginning was a singular state, not describable by physical science. A skeptic might ask, what did God do before He created the universe? The apocryphal answer is that He was preparing Hell for people who might ask such questions (attributed to St. Augustine).
2. The beginning was the most simple and permanent state imaginable, containing within itself the seeds of future evolution. This is the modern view.
3. There was no creation, and the universe is unchanging and of infinite age.

We can try to distinguish between the latter two possibilities, the only two options on which scientific tools can be brought to bear. To test this approach to cosmology, one searches for the correct physical laws that describe the initial state of the universe.

The scientist commences his study of the universe by assuming that the laws of physics which are locally measured in the laboratory apply more generally throughout the cosmos. In this spirit, cosmology, the science of studies of the universe, is developed by extrapolating locally verified laws of physics to remote locations in space and time. Objects and events at these locations can be probed with modern astronomical techniques. If these probes were to prove that, contrary to our expectations, the physics we learn in high school is not universally applicable, the scientist then would proceed to explore a more complex physics that may reduce to generalizations of local physics. In a theory of the universe, simplicity is sought on sufficiently large scales. The successful theories in physics and mathematics are invariably the simplest, with the least number of arbitrary degrees of freedom. Postulating that Atlas held up the heavens (where did he come from? why didn't he get bored? or sleepy?) requires many more ad hoc assumptions than the modern view of the celestial sphere. This is of course the view that the orbits of the planets in the gravity field of the sun suffice to stop them falling onto the earth like so many shooting stars.

Such considerations about the simplicity of a successful theory are incorporated into a simple principle that serves as a guide for building a model of the universe. This cosmological principle states that the universe, on the average, looks the same from any point; that is to say, it is isotropic. The principle is motivated by the Copernican argument that the earth is not in a central, preferred position. If the universe is locally isotropic, as viewed from any point, it must also be uniform in space. So the cosmological principle states that the universe is approximately isotropic and homogeneous, as viewed by any observer at rest. This particular version of the cosmological principle is the foundation stone of modern cosmology, and has been borne out by modern observations.

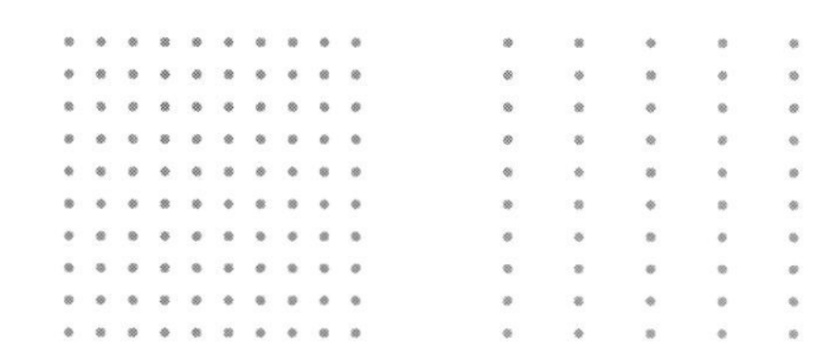

If the universe is homogeneous, there are equal numbers of galaxies in comparable volumes. If the universe is isotropic, the number of galaxies is the same in all directions. The universe actually is homogeneous and isotropic (left diagram), although it could have been homogeneous, but not isotropic (right diagram).

A stronger version, the perfect cosmological principle, goes further: the universe appears the same from all points and at all times. In other words, there can have been no evolution: the universe must always have been in the same state, at least as averaged over long times. In this sense, the perfect cosmological principle contrasts with the weaker version, which allows the possibility of very different past and future states of the universe. The perfect cosmological principle spawned the theory of the steady state universe described later.

Unlike other branches of science, cosmology is unique in that there is only one universe available for study. We cannot tweak one parameter, juggle another, and end up with a different system on which to experiment.

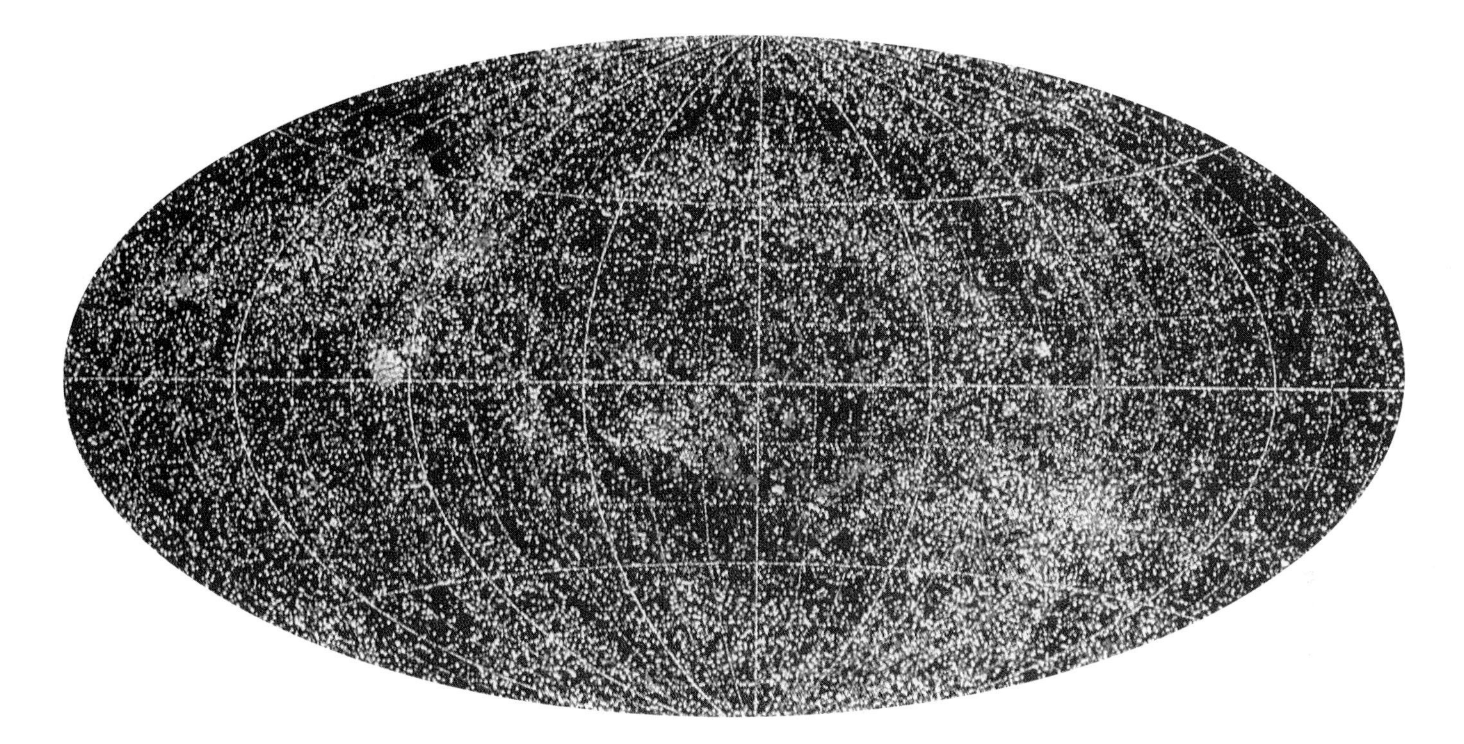

The sky as viewed by the x-ray telescope aboard the ROSAT satellite is nearly isotropic. Approximately 50,000 x-ray sources are shown. The color coding is such that blue measures high x-ray energies and red low x-ray energies, over the 0.1 to 3 keV range of the telescope detector. The near isotropy shows that most of the sources must be very distant galaxies and quasars.

We can never know how unique is our universe, for we have no other universe with which to compare it. The universe denotes everything that is or ever will be observable, so we can never hope to glimpse another universe.

Nevertheless, we can imagine other possible universes. One could have a universe containing no galaxies, no stars, and no planets. Needless to say, man could not exist in such a universe. The very fact that our species has evolved on the planet Earth sets significant constraints on the possible ways our universe has evolved. Indeed, some cosmologists think that such an anthropomorphic approach may be the only way we can ever tackle such questions as, why does space have three dimensions, or why does the proton have a mass that is much larger (precisely 1836 times larger) than the electron, or why is the neutron just 0.14 percent heavier than the proton? If none were the case, we certainly would not be here.

One can take the argument further. Perhaps our actual existence requires the universe to have had three space dimensions and the proton mass to be 1836 electron masses. This conclusion is called the anthropic cosmological principle: namely, that the universe must be congenial to the origin and development of intelligent life. Of course, it is not an explanation, and the anthropic principle is devoid of any physical significance. Rather it limits the possibilities. There could be a host of radically different universes that we need not worry about.

The anthropic cosmological principle argues that the universe must have been constructed so as to have led to the development of intelligence. This version of the cosmological principle begs the question of how likely is the development of life, and indeed the observable universe, from suitably random initial conditions. At least in principle, this is an issue resolvable by physics rather than by fiat. Inflationary cosmology, described in Chapter 5, purports to do just this.

Cosmology comprises the vast and the infinitesimal. In measuring the ingredients of the astronomical universe, astronomers deal with a range of magnitudes that stretches the limits of human conception. I begin with a look at this range of magnitudes to set the scene for the cosmological drama that will subsequently unfold.

Orders of Magnitude

Astronomers have a language of their own, that at times makes it difficult for those outside the field. This language involves specific terms: who but an astronomer would ever use a parsec, a distance of 3.26 light-years, to measure the unit of distance? In addition, the language of astronomy takes a heuristic approach to numbers. It is enough in many instances to specify a number to two significant figures or even one. Factors of 10 are called "orders of magnitude": it is important to have the correct order of magnitude, but often that suffices.

There is a proliferation of factors of 10 that is endemic to astronomy. We know that the mass of the sun is 2×10^{33} grams, a large number even in tons (2×10^{27}). Such large exponents may seem meaningless, but, if one errs by an order of magnitude, the star is a dwarf or a giant, and very different in its observable properties. To reduce the size of the exponent, it is common to use the mass of the sun ($M_\odot$) as the unit of mass in astronomy. Our Milky Way galaxy weighs in at about 100 billion, or 10^{11}, $M_\odot$, in terms of the visible matter. We may be off by 10 billion $M_\odot$, in either direction, but we have to settle for an accuracy of 10 percent. Rounding off to one, sometimes two, significant figures is normal in astronomy.

Of course, there are examples where astronomical data is of exquisite accuracy. Our universe is bathed in a cold sea of microwave radiation; the temperature of this cosmic background radiation is measured to be 2.73 degrees Kelvin, or −270.43 degrees Celsius. But specifying it as 3 degrees Kelvin usually is adequate. The problem invariably encountered is that the remote stars and galaxies yield so few photons for our detectors that we indeed have the sparsest information in most cases. Astronomers are data starved, despite the flood of information from new telescopes and satellite experiments, simply because the objects being studied are generally so dim.

The range of orders of magnitude needed to describe the physical universe is huge. For example, how large a number can one imagine as having any physical significance? Let us consider the number of atoms in the observable universe, as a simple illustration of orders of magnitude. One gram of hydrogen, the predominant form of matter, contains about 10^{24} atoms (in order of magnitude). Hence the sun, with a mass of 2×10^{33} grams, contains 10^{57} atoms (in order of magnitude), and the Milky Way, containing some 100 billion suns, about 10^{68} atoms (in order of magnitude). On a truly dark night, the observer can see with the naked eye one galaxy, Andromeda, which is the counterpart in stellar content to our own Milky Way. With the aid of the largest telescopes, the sky is found to be teeming with galaxies. As many as 200,000 per square degree can be counted. (For comparison, the sun covers about one-quarter of a square degree.) Since the entire sky contains 41,000 square degrees, there are consequently nearly 10 billion galaxies in the observable universe, or about 10^{78} atoms (in order of magnitude). This is the largest meaningful number in the universe.

To find the smallest number in the universe, we look for the smallest object. An atom has a size of the order of one-hundred millionth of a centimeter, yet is far from the smallest object in the universe. The nucleus of a hydrogen atom is about 10^{-13} centimeter across. But the single proton forming the hydrogen nucleus is known to be composite, to consist of smaller entities called quarks. Truly fundamental particles such as quarks have pointlike interactions. A pointlike particle has no measurable diameter; instead, a measure of the effective dimension of such a particle is given by the uncertainty that quantum theory assigns to its position. One can never pinpoint with complete precision the location of a particle. The greater the mass of a particle, the smaller is the quantum uncertainty in location, a measure of length that is called the Compton wavelength. The smallest Compton wavelength is for the most massive particle that is permitted by theory to exist without collapsing to a black hole. Such a particle has a mass of 10^{19} proton masses and a Compton wavelength corresponding to a dimension of about 10^{-32} centimeter. This, one can claim, is the smallest length scale that can exist (if such particles exist) in the universe.

The universe beloved of astronomers consists primarily of stars and galaxies. These luminous objects provide a measure of the universe, and are the key to unlocking its secrets. For example, it is the brightest stars that can be seen from afar as cosmic beacons, and that are vital to the reckoning of extragalactic distances. Also important are the dark clouds of gas and dust, whose measurement is a clue to the amount of matter in the universe and to the mysteries of star formation. It is impossible to understand how cosmologists study the universe without knowing something about these objects.

Clouds in Space

A star begins life as an interstellar cloud of gas and dust. The interstellar gas consists of cold clouds of a variety of sizes embedded in a more uniform, warmer gas layer that pervades our galaxy. The clouds range in size from a parsec to a hundred parsecs, and contain between a few solar masses and a few million solar masses of gas. The denser clouds are predominantly dark clouds of *molecular* hydrogen (H_2); the more diffuse clouds are *atomic* hydrogen (HI). A third form of hydrogen, *ionized* hydrogen (HII), surrounds massive stars, and is in brightly glowing nebulae.

Spectroscopy is one of the astronomer's most important tools for studying gas clouds, and the universe in general. Every atom has its characteristic spectral signature: it absorbs or emits radiation at a specific series of wavelengths. An instrument called a spectrograph enables astronomers to obtain the spectral signatures of the atoms inside a gas cloud. This instrument breaks light down into its constituent wavelengths, to give a spectrum. Excited atoms in the cloud returning to their unexcited states emit radiation that shows up in the spectrum as bright "emission lines" at the signature wavelengths.

Emission lines are only visible when a cloud is viewed in the absence of a bright background light source. Alternatively, if a gas cloud is observed against a background of hotter radiation from some other source, its atoms will absorb the background radiation at their characteristic wavelengths, creating gaps or "absorption lines" in the spectrum of the radiation. Spectra produced by the atmospheres of stars silhouetted against the opaque hot stellar cores have unlocked the secrets of the stars, and the detailed composition of these glowing bodies has been inferred. The atoms move to and fro in the hot atmosphere of a star, and as a result stellar absorption lines are characteristically broadened.

Interstellar clouds were discovered about 50 years ago, when narrow absorption lines were first detected in stellar spectra. The stars being studied were occasionally variable or binary stars, with spectra whose absorp-

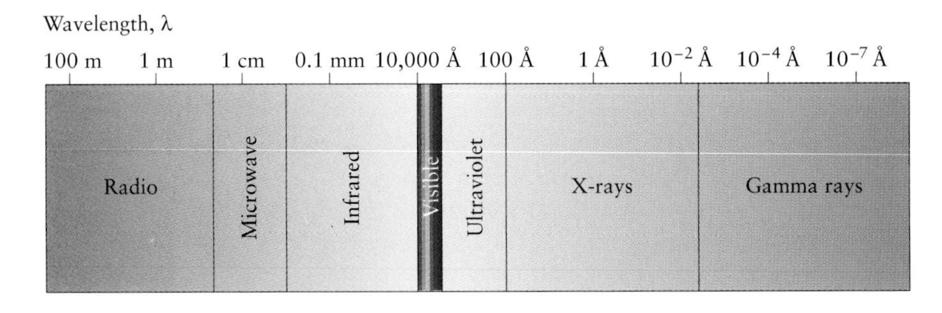

Radiation comes in a variety of wavelengths, which make up the electromagnetic spectrum. Wavelengths decrease continuously from left to right as one goes from radio waves to microwaves, infrared waves, visible light, and ultraviolet light, x-rays and gamma rays.

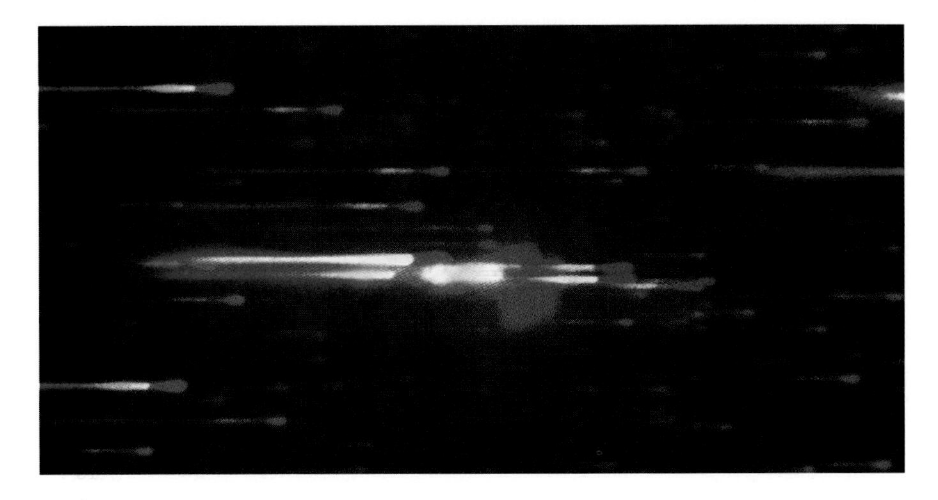

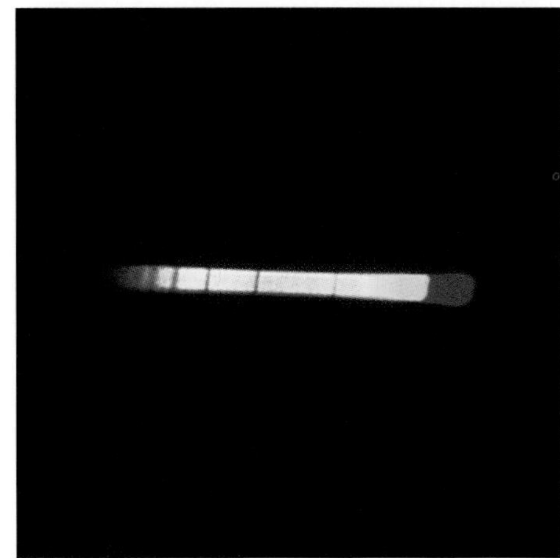

Left: The emission spectrum of the Orion nebula, obtained by refracting light from the nebula through a prism. Right: Dark absorption lines are clearly visible in this objective prism spectrum of the star Sirius.

tion lines should have disappeared as the star dimmed or became hidden behind its binary partner; yet some absorption lines, much narrower than the usual stellar lines, were unchanging in strength or wavelength. Their narrow width showed that they had originated in a cold gas located between us and the star at a temperature of less than 100 degrees Kelvin, much colder than typical stellar temperatures of at least a few thousand degrees. The traditional spectrographs could analyze only visible light, so the lines seen were produced by sodium, calcium, and a few other relatively rare elements that have spectral lines in the visible part of the spectrum. These elements were soon identified as being trace constituents in clouds of atomic hydrogen.

The atomic hydrogen in these clouds was discovered in the 1950s, with the development of radio astronomy. An atom of hydrogen may be visualized as containing two tiny magnets, associated with the electron and the proton. The magnets may be either pointing in opposite directions or aligned in the same direction. These two states have slightly different energies, and a transition from one state to the other results in the emission

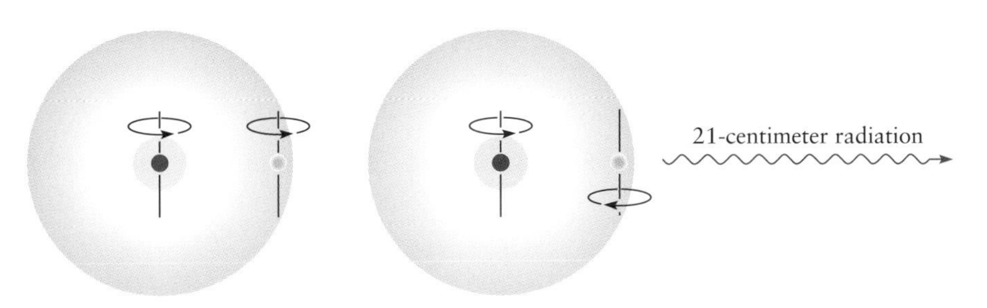

The 21-centimeter radiation from hydrogen atoms is created by the spin flip of the electron. When the electron spins in the same direction as the proton, it can spontaneously jump to a lower energy state in which it spins in the opposite direction, with the emission of radiation at a wavelength of 21 centimeters. The reverse, flip-back, occurs when the atom absorbs radiation at that wavelength.

or absorption of a photon with a wavelength of 21 centimeters, in the radio portion of the spectrum. Despite the very low probability of such a transition occurring (it takes about 10 million years for the magnet to flip), the immense numbers of atoms in an interstellar cloud generate enough photons to be observable with radio telescopes. With the advent of instruments able to analyze the ultraviolet spectrum in the 1960s, the abundant heavy elements, such as carbon, nitrogen, and oxygen, were also discovered in interstellar clouds.

A dark cloud of gas and dust, where stars are being formed. Rays from the newly formed stars are exciting the gas, causing the edge of the cloud to glow in the light of the hydrogen alpha line.

The galaxy contains about 2×10^9 $M_\odot$ in the form of atomic hydrogen (HI), in clouds typically of diameter 10 parsecs, temperature 100 degrees Kelvin, and density only about 10 atoms per cubic centimeter. The principal absorption line produced by atomic hydrogen is in the far ultraviolet spectral region at 1216 angstroms, where 1 angstrom is 10^{-8} cen-

timeter. Called the Lyman alpha absorption line, it was only detected when the first rocket-borne telescopes were lofted high enough above the atmosphere to detect far ultraviolet radiation from space, after ultraviolet astronomy was developed.

Since the atomic clouds were known to contain dust, it was conjectured that the dust in a sufficiently dense cloud would shield the cloud center from ultraviolet light. Light at these energetic wavelengths breaks molecular bonds; when shielded from its effects, two hydrogen atoms are able to join to create molecular hydrogen (H_2). Thus molecular hydrogen should be the dominant form of hydrogen in a dense cloud. This conjecture was confirmed with the first detections of H_2 molecules by the recording of their ultraviolet absorption lines.

Most interstellar molecules, however, are detectable not in the ultraviolet but at microwave frequencies. Clouds of molecular hydrogen, despite the chill of interstellar space, are warm enough, at a few tens of degrees Kelvin, to excite common molecules such as water and ammonia into higher levels of rotational energy. There are quantized energy levels for rotation of compact molecules, similar to energy levels seen in atoms, but with much less energy. After being excited to higher energy levels, these molecules emit at microwave frequencies as they return to the lowest energy state. These low-energy rotational transitions lead to photons of much lower frequency (at radio and microwave frequencies) than the photons produced by electronic transitions in atoms (typically visible and ultraviolet). It required the development of *microwave* astronomy in the 1970s before the densest interstellar clouds were studied and shown to contain an extraordinary variety of molecular species.

Astronomers searching for spectral lines at microwave frequencies have discovered hundreds of different molecules, including methyl alcohol, ammonia, water, formaldehyde, and some relatively complex molecules. The most common molecule, after H_2, is carbon monoxide (CO),which has a strong spectral line at a wavelength of 2.6 millimeters. Only at densities above 100 atoms per cubic centimeter is H_2 the dominant form of hydrogen. For molecules to predominate, the clouds must be exceptionally dark and shielded by dust from the diffuse interstellar radiation field of starlight. In these molecular clouds, typical hydrogen densities are 1000 atoms per cubic centimeter and typical temperatures are about 30 degrees Kelvin.

Although the temperatures are low, the thermal energy of the atoms supplies enough kinetic energy to keep a cloud's atoms in a state of continuous jostling. These thermal motions exert a pressure force that tends to make the cloud expand, and thereby counter the opposing tendency of the cloud's own gravity. The life of an interstellar cloud is a continuous struggle of the sustaining force exerted by thermal pressure against the relentless grip of gravity that drives the cloud toward collapse. As clouds

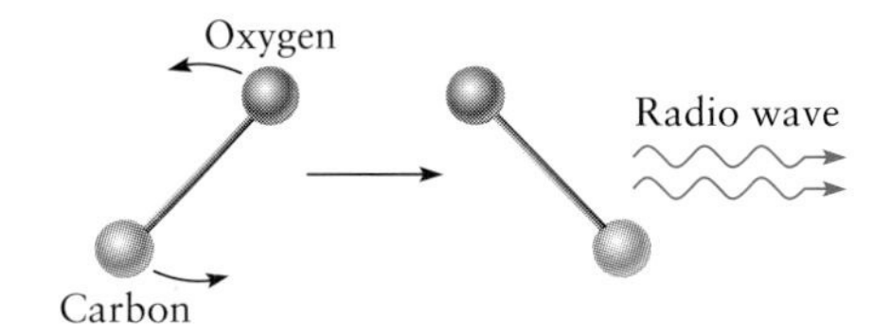

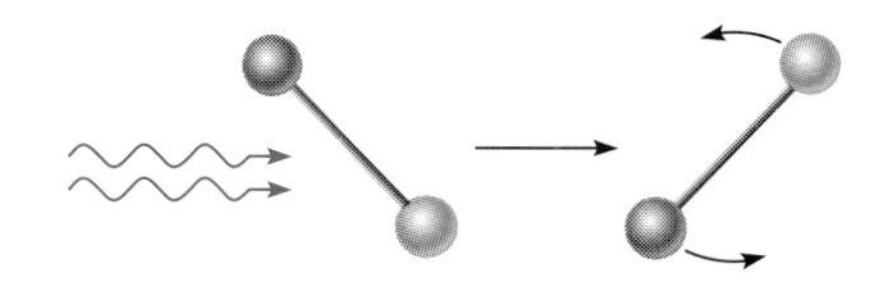

A carbon monoxide molecule emits radiation at a wavelength of 2.6 millimeters when the rotating molecule jumps to a state of lower angular momentum. The reverse process, absorption of radiation at this wavelength, happens when rotation speeds up.

orbit the galaxy, however, they accrete mass. They also radiate and cool. Hence gravity eventually wins the battle. Once thermal energy cannot withstand the force of gravity, clouds begin to undergo gravitational contraction. Gravitational collapse soon ensues, and eventually the cloud becomes so dense and so opaque to molecular radiation that its core heats up. A gravity-powered *protostar* forms, or, most likely, a cluster of protostars. The protostar phase of the sun demarcated the formation of the solar system. Gravity power, however, has a finite duration. The protostar phase is long (more than 30 million years) for stars of mass less than the sun, but relatively short for the most massive stars.

Powering the Stars

Suppose gravity was the only energy source available to power the sun. As the forming protosun contracted out of the diffuse cloud of interstellar gas, the center of the sun would heat up as the cloud became compressed. The gravitational energy supply of the sun is $E = GM^2/R$, where G is Newton's constant of gravitation, M is the sun's mass, and R is its radius. With $M = 2 \times 10^{33}$ grams and $R = 7 \times 10^{10}$ centimeters, the sun's gravitational energy supply is 4×10^{48} ergs. The luminosity L of the sun is its power usage—the rate at which it radiates energy, or 4×10^{33} ergs per second. Hence the lifetime of the sun, if powered only by gravitational contraction, would be E/L, or about 30 million years. This is far less than the age of the oldest rocks in the solar system, which are about five billion years old. Nuclear energy provides the solution to this apparent contradiction.

Transmutation of the elements by nuclear fusion, our modern alchemy, is the energy source that powers a star. Hydrogen, the most abundant element in the universe, is gradually converted into helium under the intense heat and pressure at the center of the sun. An atom of helium has a mass seven-tenths of a percent less than four hydrogen atoms: this mass difference emerges as almost pure energy in the form of gamma rays, positrons, neutrinos, and the kinetic energy of motion of these particles. The radioactive energy released in the center of the sun diffuses out and is degraded into harmless yellow light by the time it leaves the solar atmosphere. Eventually, however, a star will exhaust its supply of hydrogen fuel.

To estimate the lifetime of our sun, suppose that its core (about 10 percent of its mass) is available for nuclear fuel. According to Einstein's famous law, mass can in principle be converted into energy to yield c^2 ergs per gram, where c is the speed of light. This reasoning assumes 100 percent efficiency of conversion, only realizable if matter annihilates with antimatter. Stars certainly contain no antimatter. Their energy source is nuclear fusion, which has a conversion efficiency of about seven-tenths of a percent. This means that the energy reservoir of a star is about seven-tenths of a percent of the mass of the core if converted completely into helium. The core of the sun, where the temperature is high enough for nuclear reactions to occur, is about one-tenth of the mass of the sun. The total available energy reservoir of the sun therefore amounts to 0.0007 Mc^2. Here 0.1 $M_\odot = 2 \times 10^{32}$ grams is the mass of the core and we multiply by $0.007c^2$ to obtain the equivalent energy. This calculation gives us a nuclear energy supply of 1.4×10^{51} ergs. The fuel is enough, when used at the rate of 4×10^{33} ergs per second, to last about 10 billion years. We conclude that the sun is about halfway through its consumption of hydrogen fuel.

One manifestation of the nuclear reactions going on inside the core of the sun is the generation of weakly interacting particles called neutrinos. Every thermonuclear reactor produces neutrinos: they are a telltale signature that is unavoidable. The nuclear fusion theory of solar energy predicts that neutrinos are emitted in large numbers from the center of the sun. These weakly interacting particles travel at the speed of light and escape from the sun directly.

Solar neutrinos have been detected on the earth by a remarkable experiment that monitors 100,000 gallons of cleaning fluid, carbon tetrachloride (CCl_4), in a vat at the bottom of the Homestake gold mine, nearly a mile underground in South Dakota. The deep site was chosen to reduce interference from cosmic rays. The common isotope of chlorine absorbs a neutrino to form a nucleus of a rare, radioactive isotope of argon:

$$\mathrm{Cl}^{37} + \text{neutrino} \longrightarrow \mathrm{Ar}^{37} + \text{electron.}$$

Every two months or so, the cleaning fluid is flushed, filtered, and carefully checked for the minute amount of radioactive argon. About one atom of radioactive argon is predicted to form per day as a consequence of the absorption of solar neutrinos. The bimonthly monitoring usually finds a few atoms of argon, and by this means, neutrinos have been confirmed to be produced in the sun. However, the experiment only measures about a third of the predicted flux. Either our model for the temperature in the center of the sun is not quite correct, or there is new physics about neutrinos awaiting discovery.

That neutrinos have actually been detected from the center of the sun is an astounding result. It provides absolute proof of an ongoing nuclear reactor, some 96 million miles from the earth, that is powering the sun. However, the chlorine experiment, which has been running for 20 years, is only sensitive to the most energetic of the neutrinos from the sun. Other experiments are now underway to probe the full range of solar neutrinos. Two of these (in Italy and in Russia) use gallium as the detector fluid to study a reaction in which a radioactive nucleus of a germanium isotope is

An underground neutrino "telescope" in New Mexico. It contains some 200 tons of liquid scintillator surrounded by 1200 phototubes, which search for flashes of light from neutrinos that interact with the detector liquid.

produced when a neutrino strikes an atom of gallium. A third experiment in Japan detects the light flashes produced by fast electron recoils that are induced by neutrinos scattering off water. All of these experiments are presently counting neutrinos from the sun, and are located under a kilometer-equivalent of rock, either in a deep mine or below a mountain, to suppress spurious neutrino signals introduced by cosmic rays.

The Aging of the Stars

A protostar slowly contracts and heats up. Once the temperature in the core of the protostar rises above about one million degrees Kelvin, nuclear reactions commence, and a star has been formed. At this stage, the inward tug of gravity balances the outward force exerted by the higher central pressure, and the star is in hydrostatic equilibrium. The temperature and pressure are so high that the hydrogen atoms are fully ionized to give free electrons and protons, and nuclear fusion operates to overcome the repulsion between protons to liberate energy. The nuclear reactions provide a stable heat source.

As a star exhausts its supply of hydrogen, its core contracts and heats up, enabling the repulsion between helium nuclei (each with twice the charge of hydrogen) to be overcome so that helium fusion can begin. Unfortunately, the fusion of two helium atoms, $2He^4 \longrightarrow Be^8$, forms an unstable isotope of beryllium that quickly breaks down (Be^9 is the stable form of beryllium). The solution to how helium can fuse to heavier elements was found by two theorists. First, Edwin Salpeter pointed out in 1953 that He^4 and Be^8 shared a common property (a similar level of energy when the nuclei were excited), which implied that two helium nuclei had a greatly enhanced probability of merging to form the Be^8 nucleus. Consequently, the beryllium could be produced just as quickly as it self-destructed. But this was not enough beryllium to make the next most abundant element, carbon. Within a year, however, Fred Hoyle predicted that there should be an energy level in common between beryllium and the common isotope of carbon C^{12}, again when both nuclei were excited. The shared energy level greatly increased the probability of beryllium capturing another helium nucleus, allowing a reaction (called the triple alpha process) to occur in which a grand total of three helium nuclei merge to form a nucleus of carbon. Beryllium is now bypassed entirely as an intermediary stage. The boost in the probability of such captures is rather like enhancing the ability of a player to catch a baseball by giving him or her a baseball glove. Only a year after Hoyle's prediction, the existence of the excited state of C^{12} was confirmed in an experiment. The production of carbon in stars is the secret of life: the carbon in our bodies formed aeons ago from the triple alpha process inside supergiant stars that are now long dead.

As the core begins to burn helium, its luminosity dramatically increases. The outer layers of the star swell up like a balloon, and the star becomes a red giant. Our sun, for example, is destined to become a red giant in about five billion years, and the earth will then be enveloped inside the fiery stellar atmosphere. Helium burns at about 100 million degrees Kelvin to form carbon, and even heavier elements are formed by massive stars in their later stages of evolution. Essentially all of the heavy elements are formed inside stars by *nucleosynthesis*.

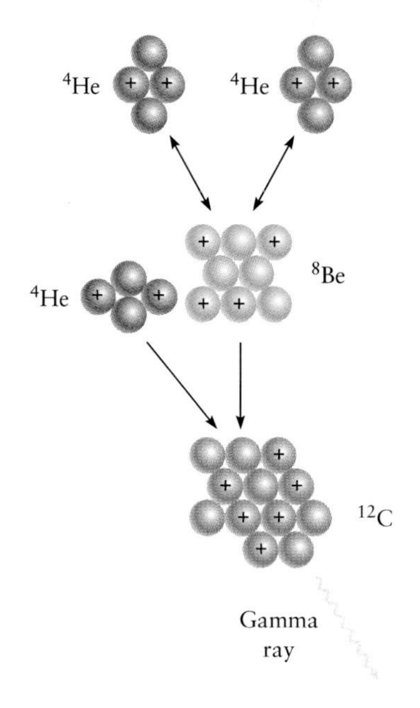

Three helium nuclei combine into a nucleus of carbon, after first forming beryllium-8 in an intermediate step. Helium nuclei are known as alpha particles, and hence the name triple alpha process is given to this reaction, which occurs in the cores of aging stars.

The Nearest Stars

The sun is a run-of-the-mill star, somewhat above the average in its mass and its luminosity, but a pallid counterpart of the brightest stars. Some stars are a few times the mass of the sun, some are as massive as 100 solar masses, but the typical star near us is about one-third the mass of the sun. Stars occasionally come in pairs, or binary systems, and the mutual pull of gravity determines the orbital motions of the stars. This mutual dance allows astronomers to directly measure the stellar masses of binary stars.

The masses of other stars are measured indirectly by observing their luminosities and colors. The luminosity of a star depends sensitively on its mass: double the mass of the star, and its luminosity is larger by a factor of 10. The more luminous the star, the hotter it usually is. A star is rather like an almost perfect furnace, or blackbody. The hotter a blackbody, the shorter the wavelength of the typical radiation that it emits; the cooler the blackbody, the longer the wavelength of the radiation. Hence hot blackbodies are blue, cool blackbodies are red. In general, the color of the emitted light is a measure of the temperature of the blackbody. Astronomers measure the surface temperature of a star from its color, or in a more sophisticated approach, by obtaining a spectrum of the light. Because stars, to a first approximation, behave like ideal radiators of light, we can infer the size of a star from its color and luminosity: the luminous, hot, blue stars are giants; the dim, cool, red stars are dwarfs.

The nearby stars include a broad mixture of colors and of luminosities. We learn about these stars by studying nearby examples, which are sufficiently bright that astronomers can obtain detailed information on composition, as well as, in some cases, size and even mass. The sample of nearby stars includes some hundreds of other stars in addition to the closest star, our sun. We find that a few stars are red giant stars, a thousand times the size of the sun, while others are white dwarfs, one-thousandth the size of the sun, in both cases comparable in mass to the sun. Most stars are neither giants nor dwarfs. The greatly distended or shrunken stars are probably in a very advanced stage of evolution compared to that of the sun.

A pattern emerges: most of the stars lie on a straight line, called the main sequence, in a diagram that plots stellar luminosity against surface temperature. The diagram is so useful, and so ubiquitous in astronomy, that it has a name, the Hertzsprung-Russell diagram, after astronomers Ejnar Hertzsprung and Henry Norris Russell. On the main sequence the stellar temperature increases progressively as the luminosity increases. We interpret the main sequence as the locus of stars of differing masses, all of which are still burning hydrogen. It takes a long time before the hydrogen is exhausted, some 10 billion years in the case of the sun, and so a star's position on the hydrogen-burning main sequence changes very little until the hydrogen is burnt. The sun is just a yellowish main sequence star. Were it even 10 parsecs distant from us, it would be barely visible to the naked eye as a faint point of light.

In addition, there are outliers, stars far from the main sequence. These stars have extreme luminosities, both high and low, and are correspondingly giants or dwarfs. There are red giants and blue giants, red dwarfs and white dwarfs. A star like the sun is destined to become a red giant as it burns helium. Because this phase of stellar evolution is short-lived, taking a hundred million years or so, red giants are rare relative to main sequence stars. A white dwarf is the final destiny of a star like the sun, when all of its nuclear fuel is exhausted.

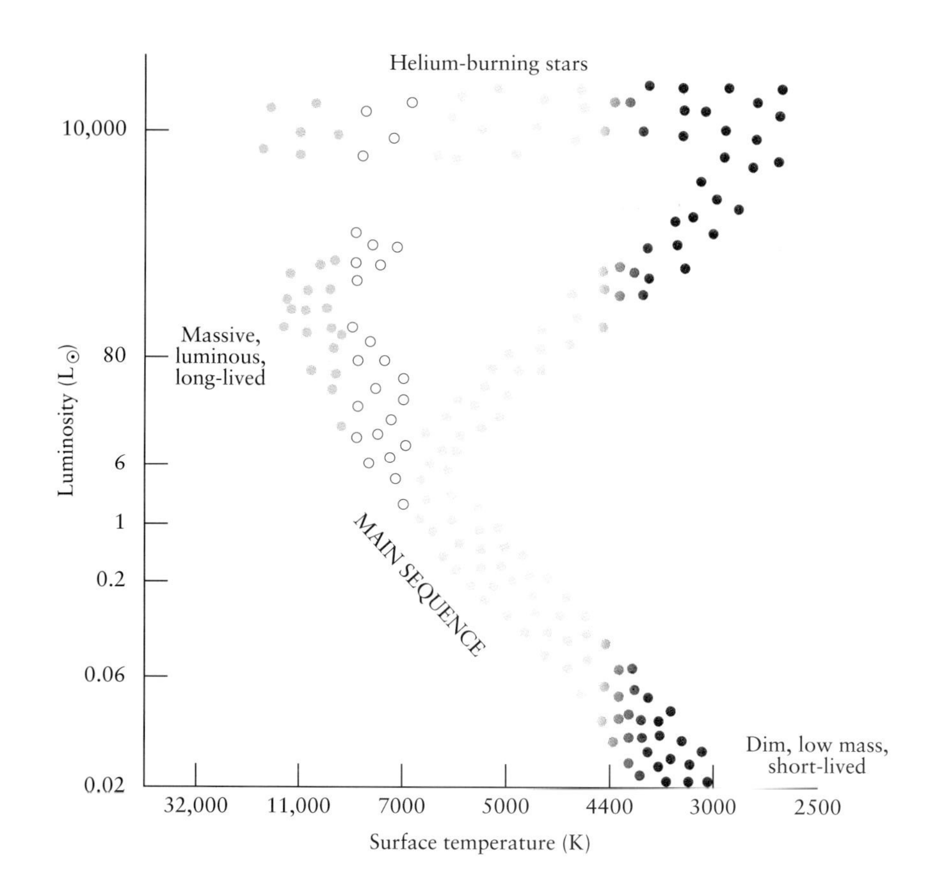

The Hertzsprung-Russell diagram plots stellar luminosity (vertical axis) against stellar surface temperature (horizontal axis). Hydrogen-burning stars define a "main sequence," spanning dim, low-mass, short-lived stars and bright, massive, long-lived stars.

The Galaxies

Stars are rarely alone. They form in clusters and looser groupings, which, in aggregate, constitute a galaxy. Galaxies themselves come in distinct varieties, mostly spiral or elliptical. The spiral galaxies are the birthplaces of most of the stars in the universe.

Spiral galaxies were originally known as spiral *nebulae*, because at first they were seen simply as fuzzy objects that could not be resolved into stars. Early in this century, Edwin Hubble was able to infer that even the nearest spirals were at very large distances from us. The closest spiral is the Andromeda galaxy, which Walter Baade eventually determined to be at a distance of 750 kiloparsecs, or two million light-years. It was clear that the spiral nebulae were "island universes," stellar systems of size comparable to the Milky Way.

In the first decades of the twentieth century, it became apparent that galaxies could be classified by shape. Some revealed a spiral structure, others were amorphous in appearance. Hubble developed a classification of galaxies, based on their morphology, that is still in use today. Spiral galaxies all have spiral arms, but are classified according to the appearance of the arms and the central bulge. Type *Sa* has tightly wound arms and a prominent spheroid or bulge. Type *Sb* has looser arms and a smaller bulge. Type *Sc* has very loose arms and a small bulge. More interstellar gas and dust are found as one goes from Sa to Sc, and more stars are forming. Most bright galaxies are spirals. The Milky Way is thought to be an Sb type, as is Andromeda. Both have luminosities equivalent to about 10 billion suns.

Many spirals have a bar-shaped bulge, the ends of which coincide with the beginnings of the spiral arms. These galaxies are also classified with letters *a*, *b*, and *c*, according to tightness of arms and size of bulge, and are designated SBa, SBb, and SBc, to distinguish them as barred spirals. All spirals are found to be rotating.

Elliptical galaxies are pure spheroid, and possess no spiral arms and no disk, nor any significant amount of interstellar gas or dust. They are categorized according to shape. E0 galaxies are circular in appearance (spherical), and types E1 through E6 are progressively more elliptical (flattened). They are found to contain only old stars, of mass similar to or less than the sun.

The largest galaxies known are ellipticals. Giant ellipticals such as Messier 87 are 10 times more luminous and about 10 times as massive as the Milky Way galaxy. Most galaxies are dwarfs, and the majority of dwarf galaxies are ellipticals, including our nearest neighbor galaxies, with luminosities of only 10^7 or 10^8 suns. Luminous ellipticals are found not to rotate significantly. Most large elliptical galaxies are found in the cores of rich galaxy clusters.

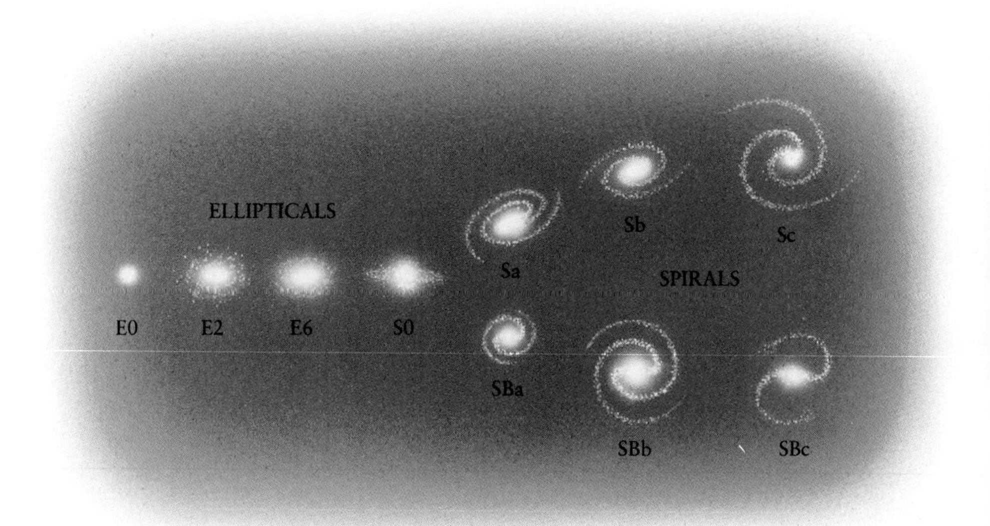

The "tuning fork" diagram depicts and compares the various shapes of galaxies.

A barred spiral galaxy, Messier 83.

Also called lenticular, or lens-shaped, S0 galaxies are an intermediate type of galaxy that has a disk and a bulge, but no interstellar gas or dust. Consequently, S0 galaxies have no young stars or spiral arms. In stellar population, they are identical to elliptical galaxies. S0 galaxies are mostly found in clusters of galaxies.

Irregular galaxies include everything not classified as elliptical, spiral, or S0. Such galaxies typically contain a considerable amount of gas and dust, with vigorous ongoing star formation. The Large and Small Magellanic Clouds are nearby irregular galaxies. Irregulars are galaxies of low luminosity, typically a tenth that of the Milky Way. Irregular galaxies amount to a few percent of all galaxies. Instead of being classified as spiral, elliptical, or irregular, a galaxy can be classified as *peculiar* if it has a particularly unusual shape or other strange feature. Such galaxies might be ring-shaped systems, galaxies with multiple nuclei, peanut-shaped systems, or galaxies with bright, starlike nuclei or tidal tails.

The Darkness of the Night Sky

Observations, even of the most elementary kind, have provided critical clues to unraveling the puzzles of cosmology. Perhaps the most striking observation is one that comes naturally to any human being, from the dawn of time until today, who looks up at the sky on a dark night. Olbers' paradox is "Why is the sky dark at night?" The German astronomer Heinrich Olbers (and before him, others) posed this question in 1823.

Olbers assumed that both the average frequency of stars (and galaxies) in space and their luminosity are approximately constant throughout space and over time. Consider any large shell of matter of specified thickness Δr with the earth at its center. The volume of the shell is the surface area of the shell multiplied by its thickness. The light emitted by the shell is equal to the volume of the shell multiplied by the number of stars per unit volume n_* and the luminosity of a star L. The amount of light reaching the earth from the shell can be calculated, since it is just equal to the light emitted by the shell divided by its area $4\pi r^2$, namely $n_* L \Delta r$. It is found that this amount does not depend on the distance of the shell from the earth, but only on its thickness. As we add up the contributions of more and more distant concentric shells (each of equal thickness), the light measured at the center seems to increase without limit. This is not quite right, since light from a distant star is intercepted by an intervening star. Still, any line of sight must sooner or later run into a star, and so we would expect the sky to be about as brilliant as a star's surface. This conclusion applies at any arbitrary point, and hence it applies everywhere.

Despite the compelling logic of this argument, it is obvious that apart from the Milky Way, part of our own galaxy, the night sky is remarkably dark. Edgar Allan Poe, in a noteworthy convergence of literary imagination and scientific reasoning, was fully aware of this paradox when he wrote in 1848 that "were the succession of stars endless, then the background of the sky would present us a uniform luminosity." Poe advocated a finite universe, and his arguments, although rudimentary, echoed the later discussions of cosmologists in the twentieth century. It was apparent to the English astronomer John Herschel, also in 1848, that Olbers' paradox is not resolved by allowing for starlight to be blocked by interstellar dust, since such dust absorbs and radiates energy.

Possible resolutions are (1) the universe is young, so stars have only been shining for about 10 billion years, or (2) the universe is of infinite age but expanding so as to avoid a state of thermodynamic equilibrium. If stars have only been shining for a finite time, then there has not been time enough for their light to accumulate from an infinite distance, or more precisely from farther than 10 billion light-years. On the other hand, if the universe is infinitely old but expanding, then expansion "cools off" the universe,

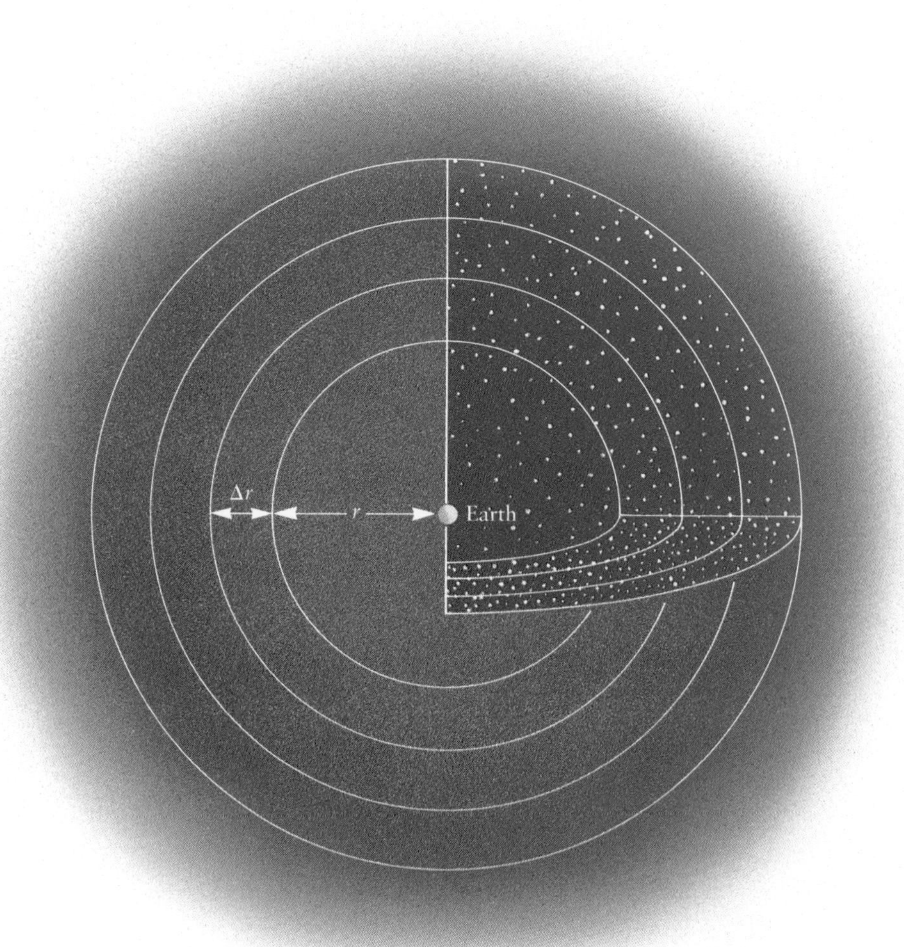

The night sky must include light from the stars in successive spherical shells, into which we can imagine dividing the universe. In an infinite universe the night sky should be as bright as the sun. Consider any large shell of matter of specified thickness Δr with the earth at its center. The amount of light reaching the earth from the shell depends only on its thickness, not its distance. As we add the contributions from ever more distant shells, the amount of light coming to us should increase without limit.

due to the Doppler shift (Chapter 2), so that the energy of photons that are received from a receding source is reduced. Of course, the universe may be both young *and* expanding, but only the second hypothesis requires expansion. Either possibility allows the universe to be infinite in extent. A finite age is not necessarily equivalent to a finite universe, which would provide a third alternative.

The story of Olbers' paradox illustrates that a telescope wasn't needed to reach conclusions about cosmology that have withstood the test of time. The darkness of the night sky has a significance that continues to be crucial to our ideas about the universe, for the speculations that resolved Olbers' paradox turn out to contain elements of truth. Modern observations reveal, as we shall now see, that the universe is expanding and is young.

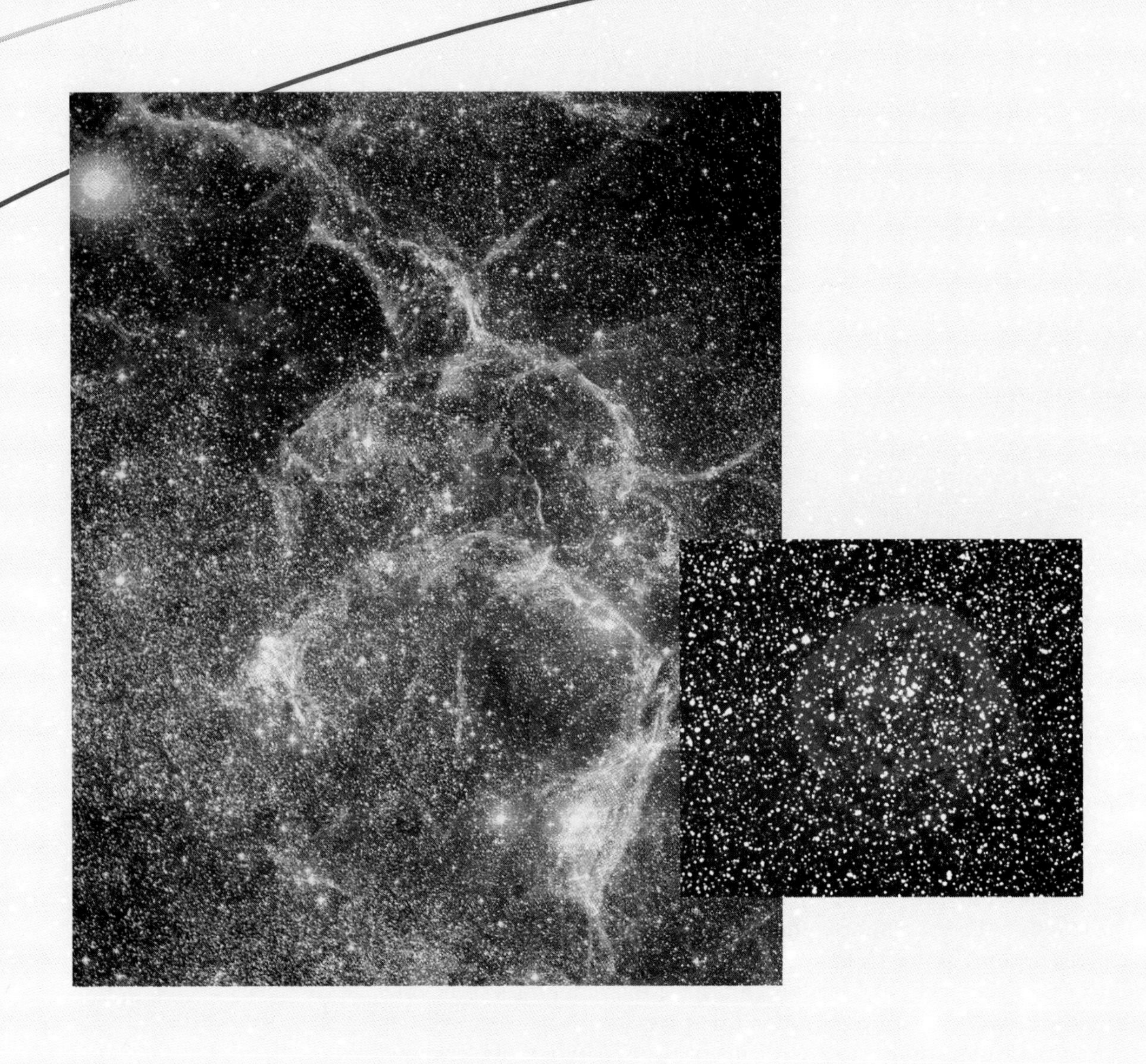

Top: *The Vela supernova remnant, relic of a star that exploded about 11,000 years ago.* ***Bottom:*** *A bubble of hot gas is driven by winds from massive stars in the Large Magellanic Cloud. The hydrogen atoms glow red in the light of the hydrogen alpha spectral line.*

2

The Expansion of the Universe

The twentieth-century revolution in cosmology was the discovery of the expansion of the universe. Prior to the 1920s, it was almost universally believed that the universe was static, centered on our Milky Way galaxy. This world view received a major challenge when systematic motions of recession were measured for spiral nebulae, and was ultimately toppled when in 1929 Edwin Hubble announced the redshift-distance law, which heralded what is now interpreted as the expansion of the universe. From this single result, modern cosmology emerged.

The Doppler Shift

The expansion of the universe was discovered as a consequence of the Doppler effect, the same effect that gives away a speeding motorist caught in a radar trap. The principle is simply that when a source of light or electromagnetic radiation at other wavelengths is moving relative to an observer, the distance between successive wave crests (or the wavelength of the light) is measured by the observer to differ from that measured when the same source is at rest. For motion of approach, the wavelength is reduced and we say that the light is shifted to the blue; for motion of recession, the wavelength is increased, and we say that the light is shifted to the red. The speed of the passing motorist is measured by the shift in frequency of the radar signal.

Consider a distant galaxy emitting ν wave crests of radiation per second. Suppose the galaxy is moving away at speed v. The time between wave crests is $1/\nu$, and the galaxy moves a distance v/ν during this time. The observer measures wave crests separated not by a time $1/\nu$ but by $1/\nu$ plus the additional time for light to traverse the distance v/ν. That time is v/ν divided by c, the speed of light. So the wavelength is increased by the frac-

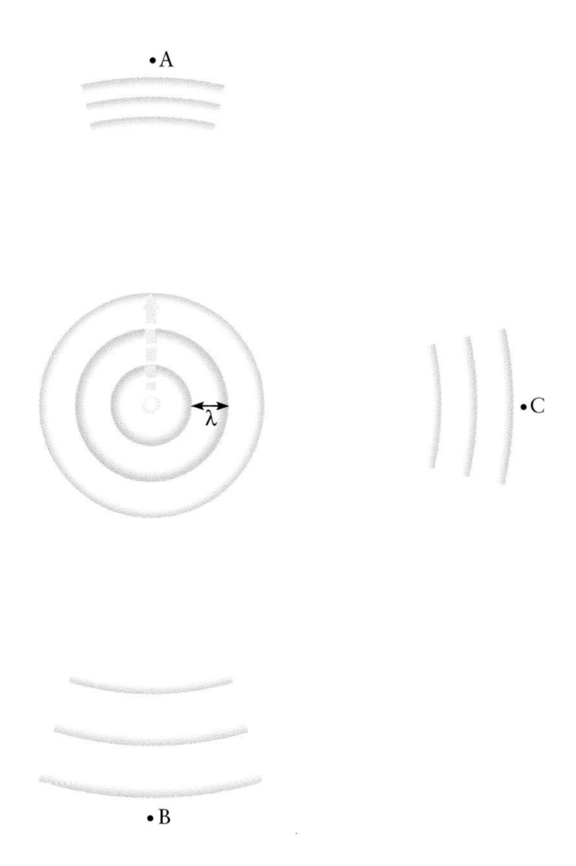

The Doppler shift. The circles and arcs are wavefronts emanating from a star moving in the direction of A and away from B. An observer at A will measure the distance between wave crests (the wavelength) to be shorter, while an observer at B will measure the wavelength to be longer. The observer at C will see almost no difference.

tional amount v/c, and we define this increase to be equal to *redshift* z. The convention is to measure v away from the observer, so that a velocity of approach corresponds to negative v or to a *blueshift* (a reduction in wavelength of emission). Frequency (ν) and wavelength (λ) are related by $\nu = c/\lambda$, so as the redshifted wavelength is increased, the frequency ν is correspondingly reduced. The expression for redshift must be generalized for velocities v that are not small compared to the speed of light (to $z = (v/c)(1 - v^2/c^2)^{-1/2}$). One finds that the redshift (or blueshift) of light becomes infinite as v approaches c.

Light from distant stars or galaxies provides the clues to measuring the velocities of distant objects. The same spectral lines that reveal the compositions of stars can also reveal their motions, and those of galaxies as well. A galaxy is simply the superposition of billions of stars: its spectrum shows distinct lines that are produced by those stars. The to-and-fro motions of stars within a galaxy broaden the spectral lines relative to those of an individual star. In addition, the entire spectrum is shifted to the red or to the blue if the galaxy is approaching or receding from us.

The characteristic pattern of emission lines from a star at rest (redshift $z = 0$) systematically shifts to longer wavelengths as the star recedes from the observer at successively faster velocities. The examples depicted correspond (center) to a 10 percent shift, or a redshift z of 0.1, obtained from a star receding at about 10 percent of the speed of light, and (bottom) to a 110 percent shift at a redshift z of 1.1, obtained from a star receding at about 75 percent of the speed of light.

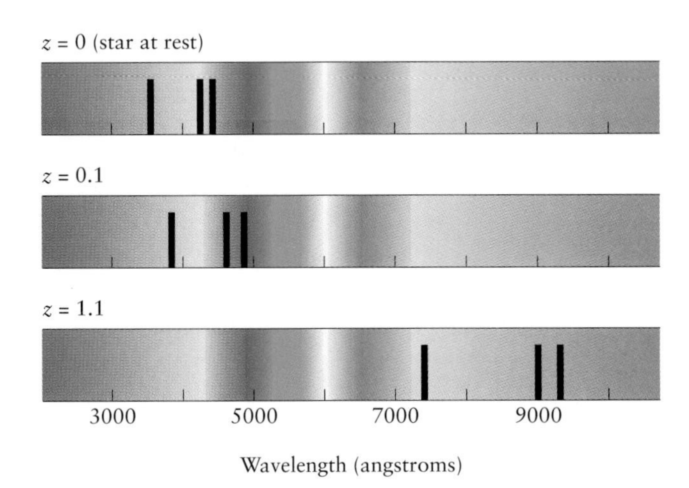

Between 1910 and 1920, the astronomer Vesto Slipher, who worked at the Lowell Observatory, obtained the spectra of many extragalactic nebulae. He found that the majority of these objects, now known to be nearby galaxies, systematically showed redshifts. Not all were redshifted, the Andromeda galaxy being a prominent exception. To make any further progress in cosmology, it was crucial to ascertain the distances to these galaxies. Were they all dwarfs, receding from the Milky Way, itself placed strategically at the center of the universe? Or was our galaxy an insignificant part of a grander scheme? Without a convincing estimate of distance to these nebulae, no choice could be made between these radical alternatives.

With the construction of telescopes able to resolve the most luminous stars in galaxies came the breakthrough that first enabled astronomers to obtain extragalactic distances. Peculiar stars known as Cepheid variables provided the necessary bridge between the Milky Way and nearby galaxies.

The Distances to the Nearest Stars

The closest star is Proxima Centauri, at a distance of 4.2 light-years. We derive its distance precisely, thanks to parallax. What astronomers call parallax is the change in position of a star on the sky as the earth moves from one side of its orbit to the other. Over six months, as the earth moves around the sun, the apparent position of Proxima Centauri relative to far more distant stars changes by 1.5 arc-seconds, an amount that is directly proportional to the inverse of its distance from us. A distance of 1 parsec is defined to correspond to a change of 1 arc-second, or $\frac{1}{3600}$ of a degree, so that 1 parsec is about 3 light-years.

Before the first stellar parallax was measured in 1838, astronomers debated whether the stars were unprecedentedly distant objects or mere acolytes of the solar system. Parallax has truly taken astronomy out of the solar system. The more distant the star, the smaller its parallax deflection angle. One has to turn to a telescope in space to measure angular positions much smaller than 0.01 arc-second. With the exquisite precision of the Hipparcos satellite, designed to obtain parallaxes for thousands of bright stars, one can now directly and systematically measure parallax distances to one-thousandth of an arc-second, or 1000 parsecs. This is a substantial fraction of the scale of our Milky Way galaxy. Armed with accurate observations, astronomers can begin to unravel the secrets of the stars. The crucial weapons in the astronomer's armory are Cepheid variable stars.

At a certain point in their evolution, stars become unstable. The instability is not catastrophic. It does not involve the entire star, as happens at the endpoint of stellar evolution when the nuclear fuel supply is exhausted, but rather only the outer layers.

The exhaustion of the hydrogen will trigger a phase when the stellar core shrinks and heats up to burn helium as a new fuel, while the atmosphere of the helium-burning star greatly expands. At a later stage during this giant phase, electrons in the stellar atmosphere become detached from some of the helium atoms, and the helium becomes partially ionized. This partial ionization is responsible for a large increase in *opacity;* that is, the ionized helium prevents radiation from freely escaping from the star. Any slight tendency to compression in response to the star's gravity heats up the outer layers, but the radiation cannot freely escape, and its pressure causes the star to overexpand. Then the star must contract again, leading to a series of oscillations that continue until the star heats up further and the helium is mostly ionized. The outer envelope of the star pulsates, with a period that may range between days and months, depending on the precise mass and evolutionary status of the star.

One common type of variable star is called a Cepheid variable, after the prototype star δ Cephei. Each Cepheid has its own precise period and a definite average luminosity. The brighter the Cepheid, the longer the period. This relation between period and luminosity was originally recognized in 1912 by Henrietta Leavitt. In her study of Cepheids, she faced the problem that the apparent luminosity of a star changes with its distance. Since at that time only distances to the nearest stars could be determined by parallax, how could she determine the distances to her Cepheids? She overcame this difficulty by choosing to study Cepheids in the Magellanic clouds, the nearest galaxies to us. These Cepheids were all at a similar distance from us: hence the absolute luminosity of any star would have been pro-

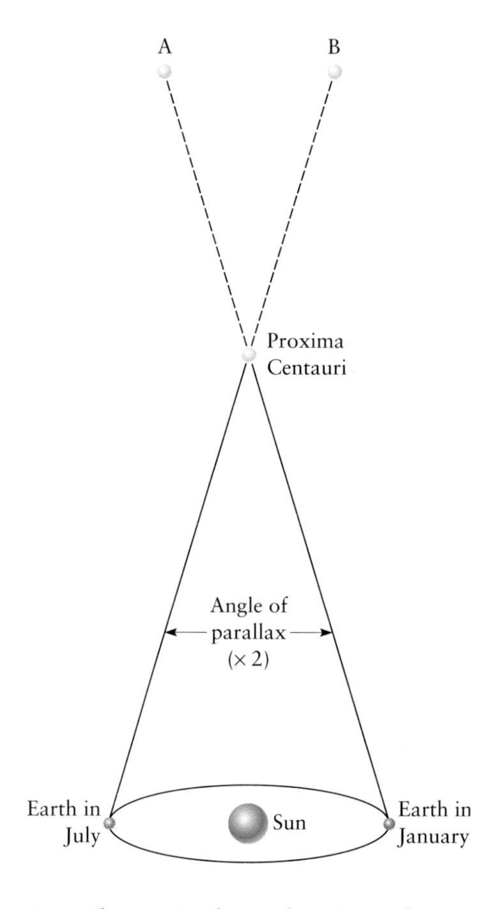

A nearby star is observed at 6-month intervals. The shift in its apparent position from A to B between January and July gives twice the star's parallax. The distance to the star is equal to the mean radius of the earth's orbit divided by the angle of parallax. The star Proxima Centauri has a parallax of 0.8 arc-second, and its distance is therefore 1.3 parsecs (or about 4 light-years).

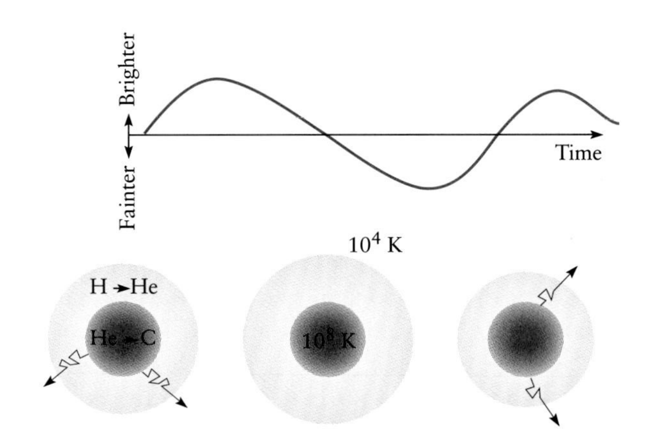

A pulsating star brightens and dims as it goes through successive stages of expansion and contraction. Partially ionized helium in the outer envelope prevents the escape of radiation: the star dims and pressure from the radiation makes the outer envelope swell. The swelling is followed by a stage of contraction, when radiation can once more escape, and the star brightens.

portional to its apparent luminosity. The relationship was calibrated against Cepheids in nearby star clusters, whose distance could be measured directly by their parallax.

A breakthrough came in 1923, when the astronomer Edwin Hubble, working at the Mount Wilson Observatory, identified Cepheids in the Andromeda galaxy, Messier 31. He used the relation that Henrietta Leavitt had found between the period and luminosity of Cepheids to deduce the distance to the Cepheids in Andromeda. Hubble concluded that this spiral nebula was a galaxy in its own right, like the Milky Way, and about a million light-years distant. In fact, Hubble's distance estimate was too small. The Cepheid puzzle was finally deciphered by a German refugee, Walter Baade, working with the 100-inch telescope atop Mount Wilson when Los Angeles was blacked out during the Second World War. Baade recognized that there were two distinct types of Cepheids, and, using the brighter and therefore more distant Cepheids analogous to those studied by Leavitt, derived the modern value of two million light-years for the distance to M31.

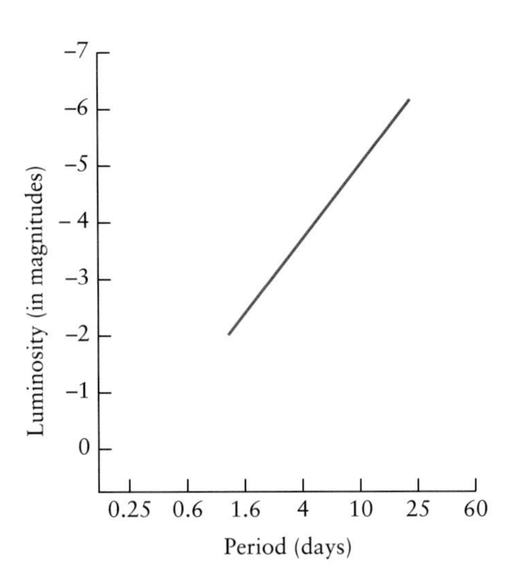

Plotting his estimates of luminosity against period, the astronomer Harlow Shapley fitted all Cepheids onto one curve, which showed luminosity increasing for the longer-period variables. The magnitude scale signifies that brightness increases by a factor of 2.5 for each smaller number (the smaller, or more negative, the magnitude, the brighter the star).

The Redshift-Distance Relation

Once equipped with a description of Cepheid stars in Milky Way clusters at known distance, Hubble compared these variable stars to cepheids in nearby galaxies. He went on to derive distances to many other galaxies, and deduced that despite the fact that Andromeda itself was approaching us at 50 kilometers per second, the galaxies were on the average moving away from us at a velocity v that increased proportionally to their distance. That velocity could be found by multiplying the distance R by a single constant of proportionality H_0, so that $v = H_0R$. This is said to be a *linear* relation between velocity and distance. Here H_0 is Hubble's constant, usually specified as a ratio of velocity to distance. Its modern value is between 50 and 100 kilometers per second per megaparsec (1 megaparsec = 1 mil-

Henrietta Leavitt, discoverer in 1912 of the period-luminosity relation for Cepheid variable stars in the Magellanic Clouds, satellite companions of the Milky Way.

Edwin Hubble, discoverer in 1929 of the expansion of the universe.

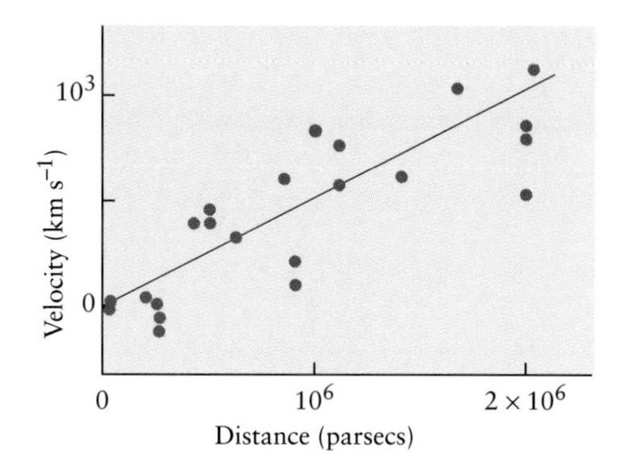

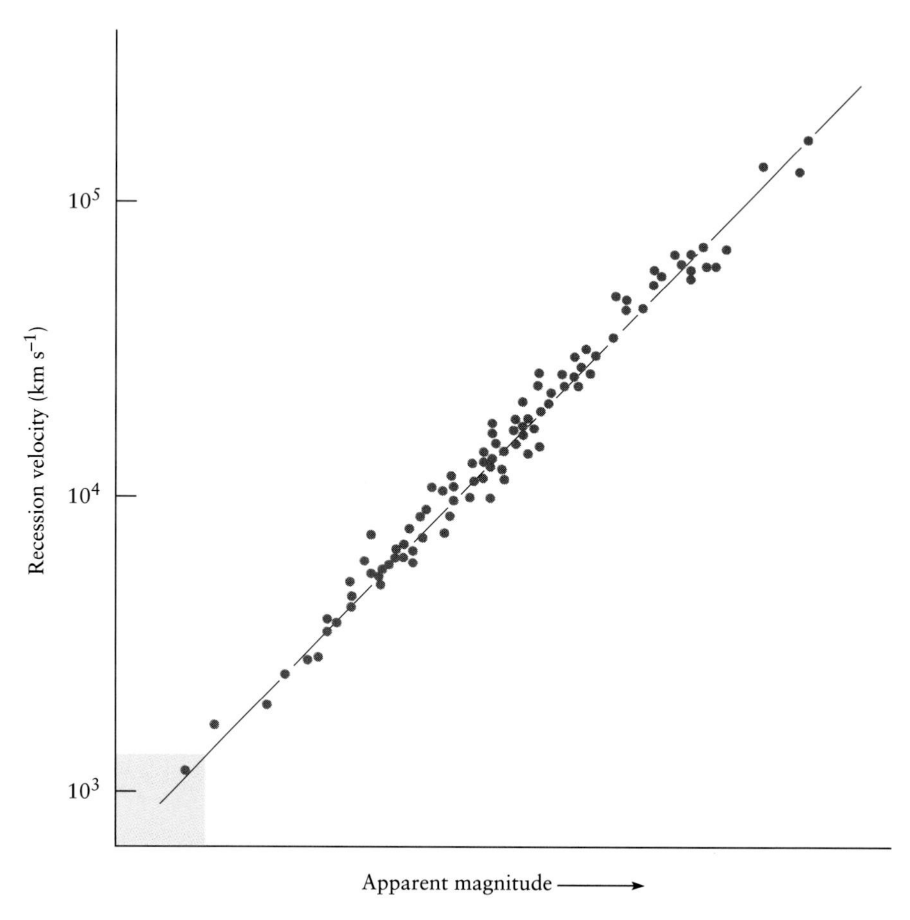

Edwin Hubble made this plot of galaxy velocities versus distances in the 1930s (above left). He only sampled a nearby volume of space, as far as the Virgo cluster. This so-called Hubble diagram was greatly extended by Alan Sandage and his collaborators (right), who compared the recession velocities of the brightest galaxies in galaxy clusters with distances to the clusters as inferred from the apparent magnitudes of these galaxies. A galaxy cluster is so luminous that it can be recognized at a large distance from us. Both plots show that recession velocity increases proportionately to distance.

lion parsecs). Hubble's use of erroneous distances had originally led him in 1929 to deduce a much larger value, of about 500 kilometers per second per megaparsec, for H_0. These distances were subsequently corrected by his own work and the work of Baade and others.

Hubble's result is precisely what one expects in an expanding universe governed by the cosmological principle. Any law of expansion but the linear one results in some direction being preferred over another. Hence the cosmological principle would be violated. Hubble's law does not itself prove isotropy of the universe, however, since it is only directly verified to a fraction of the observable extent of the universe. A modern interpretation of the redshifts of distant galaxies is not simply that of Doppler shifts, as in a gigantic explosion. Rather, the expansion of the universe is the expansion of space, and does not involve the motion of objects across that space, a concept that Hubble himself was never able to fully accept.

We can think of $1/H_0$ as a measure of the duration of the expansion phase. If the galaxies have not decelerated, they would have been touching one another a time $1/H_0$ ago. We shall see later that the presence of matter guarantees some deceleration, so the true age may be somewhat less than $1/H_0$. Nevertheless, $1/H_0$ yields an approximate maximum value for the age of the universe, which is therefore between 10 and 20 billion years (for $H_0 = 100$ or 50 kilometers per second per megaparsec, respectively). For more than 30 years, astronomers have been unable to reduce the uncertainty in H_0 to less than a factor of 2, despite intensive efforts.

Unlocking the Extragalactic Distance Scale

The key step in pinpointing the Hubble constant is to leap from Cepheids in galaxies at 1 to 3 megaparsecs distance to other, more luminous, distance indicators in galaxies at up to 100 megaparsecs distance. Individual Cepheids cannot be resolved at such great distances, even with the Hubble Space Telescope. By taking a giant step outward into the universe, one can bypass any local deviation from Hubble flow due to the gravitational at-

The isotropy of the expanding universe: space expands equally in all directions; matter doesn't expand. Held together by gravitational attraction, galaxies remain the same size although the distance between them increases.

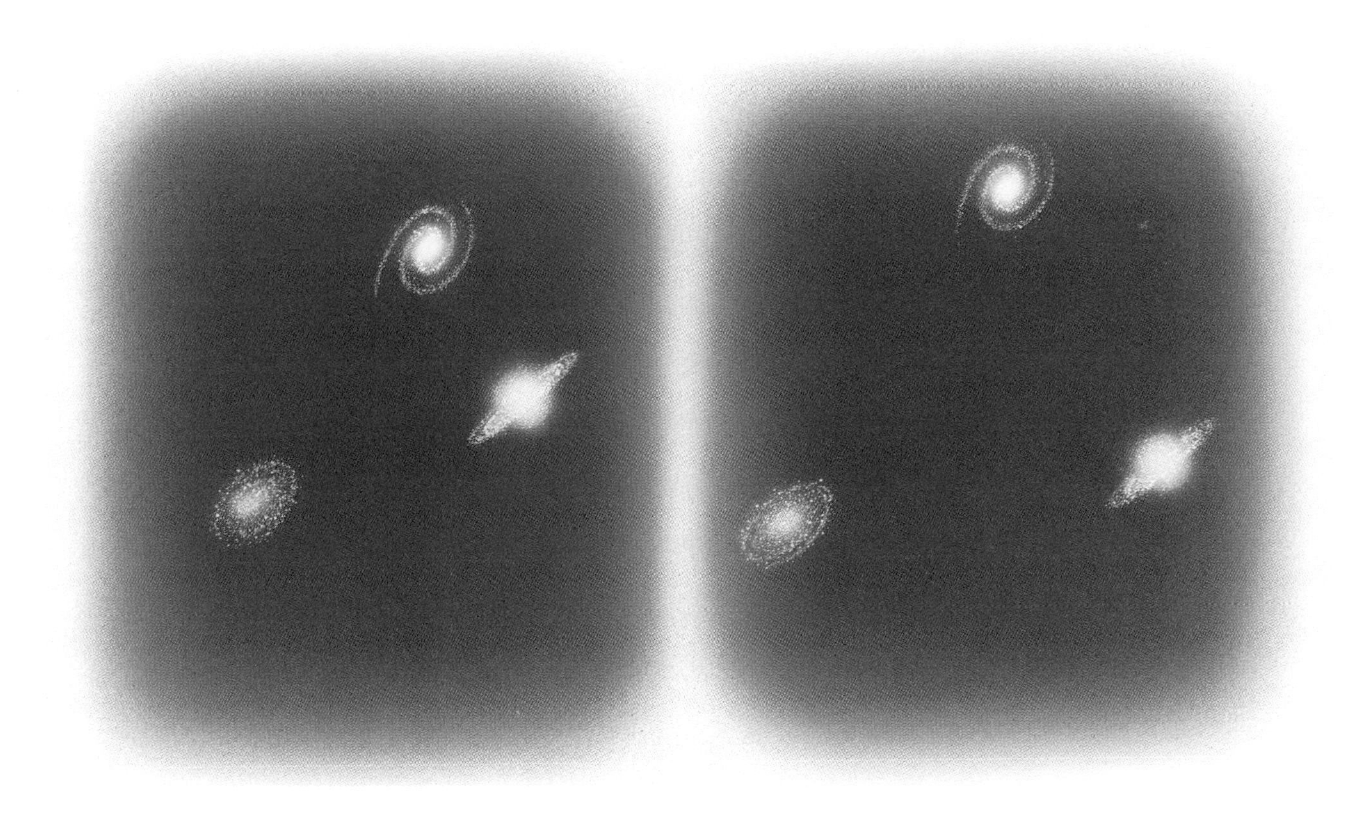

traction of our own local supercluster of galaxies, centered in the Virgo cluster, which is receding from us at a velocity of 1100 kilometers per second. We need to reach out to a recession velocity of several thousand kilometers per second to have a secure distance measurement that is not biased by distortions from the overall Hubble flow.

Standard Candles and Malmquist Bias

The basic problem is to find a good "standard candle" that is the same in distant galaxies as in nearby galaxies. Such an object would be the cosmologist's meter stick, providing a metric measure of distance. Standard candles seen at any distance immediately give a measure of that distance, provided that the distant standard candle is identical to its local counterparts. The problem that almost inevitably arises is that no cosmic standard candle is perfect: all suffer to greater or lesser extent from a common malady that has received the name of Malmquist bias. Simply put, there is no standard candle for which we can be confident that the distant and local counterparts are identical. Standard candles have an intrinsic spread in their properties, and there is a natural tendency to include only the brighter, distant standard candles, thereby leading to a bias in the observer's metric measure. Several techniques have been used to measure distance, but all have problems in their implementation that have hitherto prevented us from obtaining a precise value for Hubble's constant.

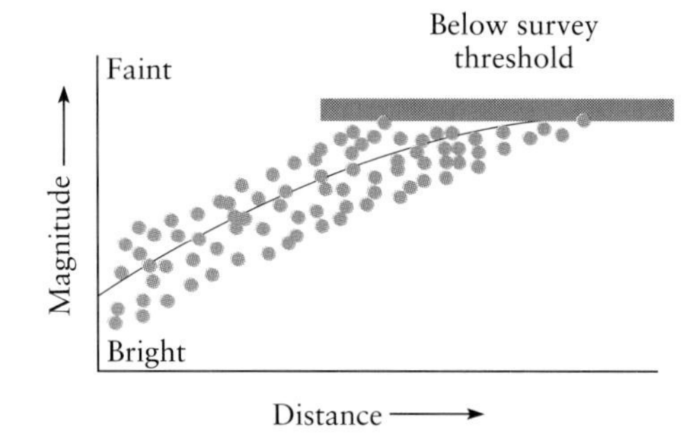

There is a spread in measurements of magnitudes of galaxies at a given distance from us, due in part to measurement error and in part to the dispersion in intrinsic properties of the galaxies; it occurs no matter how good a sample of "standard candle" galaxies is selected. As a consequence of this spread, any survey that cannot measure galaxies of brightness below a certain threshold will include only the intrinsically brightest galaxies at large distance from us, whereas at a small distance one could be sampling both intrinsically bright and dim galaxies. This imbalance leads one to infer an incorrect slope (and therefore Hubble constant) from the magnitude-distance relation.

An early approach to finding distance used as standard candles the brightest stars that could be resolved in a galaxy. These bright stars are helium-burning supergiants. The hope was that the luminosity of a helium-burning star is invariant, depending on stellar properties rather than on the type of galaxy. In that case, the fainter the star, the greater its distance. It turned out, however, that the abundance of metals in the star influences its peak luminosity. Since there is a spread in metals from star to star, the dispersion in stellar luminosities presents a serious problem. In distant galaxies, the fainter stars may not be detectable. One might systematically be measuring brighter stars than in nearby galaxies, and thereby systematically be biasing the inferred distance.

A similar problem arose when the diameters of the largest regions of ionized hydrogen (HII) in spiral galaxies were used as standard candles. The diameters of a well-defined class of objects can be used as standard measuring sticks, in a way analogous to the luminosities of standard candles. HII regions seemed at first to make good measuring sticks: these clouds of gas surrounding massive, hot stars are visible at large distances since the far ultraviolet radiation from the stars ionizes the gas and makes it glow. The problem is that larger HII regions are more easily detectable, especially in fainter, more distant galaxies. Consequently, when the diameters of the

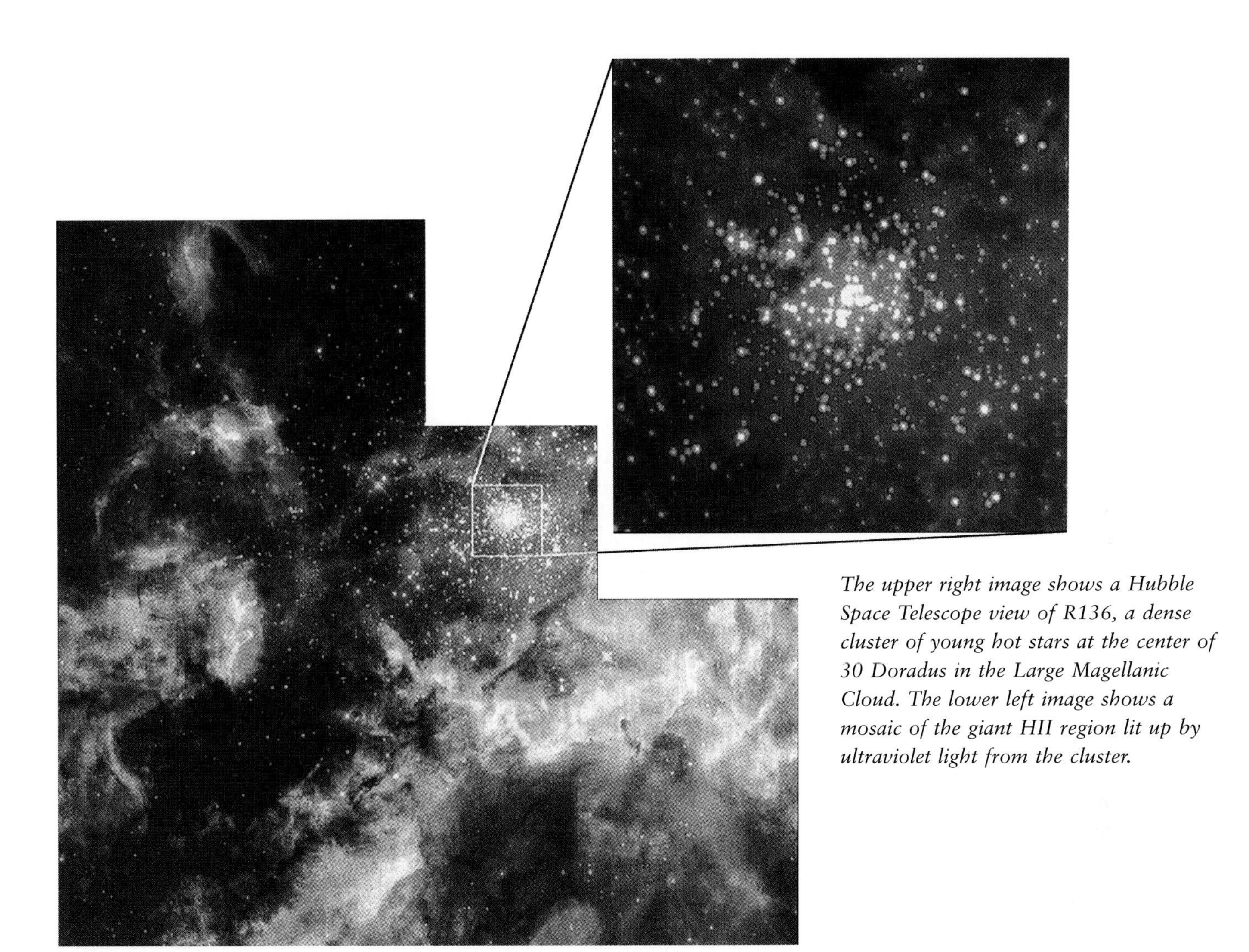

The upper right image shows a Hubble Space Telescope view of R136, a dense cluster of young hot stars at the center of 30 Doradus in the Large Magellanic Cloud. The lower left image shows a mosaic of the giant HII region lit up by ultraviolet light from the cluster.

largest of these regions were measured, one would systematically detect larger HII regions in the more distant galaxies. The result was a bias in the distance determination. This is an example of the Malmquist bias, which is inevitable for any standard candle that has an intrinsic dispersion. Unfortunately, all standard candles that have been used have some dispersion. Although a spread in intrinsic properties is unavoidable for any standard candle, the aim is to find one with the least dispersion. The effect of the Malmquist bias is to underestimate the distances of distant galaxies, leading to too large a Hubble constant.

Planetary Nebulae

For another standard candle, we turn to the planetary nebulae. When a low-mass star has no more nuclear fuel available, the core collapses and the star enters its final state, when it resembles a glowing white-hot lump of highly compressed coal. The star has become a slowly cooling white dwarf of carbon and oxygen. As the core collapses, the surrounding gases are blown off in a gentle explosion, visible as a planetary nebula. The nebula is an HII region of ionized hydrogen gas that has been excited by the hot white dwarf.

In using these nebulae as distance tools, we measure the luminosity of the nebulae, which in each case is determined by the white dwarf mass. One can measure a large number of planetary nebulae in a given galaxy. The oldest are expected to be the most luminous, because their central stars are the most massive and therefore have the most thermal energy. There are two reasons that the oldest white dwarfs have the largest masses. First, their precursor stars were, on the average, more massive than those of the younger white dwarfs. Second, older stars are more metal-poor, and consequently mass loss is less effective during the evolutionary stages after the star leaves the main sequence.

The Helix planetary nebula is a shell of glowing gas ejected by a dying star, whose relic core is visible in the center as a hot white dwarf star.

A similar distribution of planetary nebula luminosities is seen in many galaxies. We conclude that there is presumably a corresponding range of white dwarf masses, whose upper limit is set by the age of the oldest planetary nebulae. This age in turn is determined by the age of the galaxy. If galaxies have similar ages, a not unreasonable expectation, one would then find a similar spread of planetary nebula luminosities in different galaxies. One can then use the collection of planetary nebulae in a galaxy as a standard candle. Scientists have calibrated the mass-luminosity relationship in planetary nebulae using such nebulae in nearby galaxies of known distance. They can then extract the distances to more remote galaxies where the planetary nebulae have been observed.

Surface Brightness Fluctuations

Ideally, we would like to use the brightest stars in a galaxy as standard candles, since these are helium-burning supergiants whose luminosities are determined by their core masses. The distribution of luminosities is again related by the nuclear physics to the limiting mass of a stellar core at the endpoint of its evolution, namely the maximum white dwarf mass. Unfortunately, one cannot resolve the individual bright stars in any but the very closest galaxies, but a modern technique is to measure the fluctuations in surface brightness that are cumulatively due to the brightest stars. This method has been applied to elliptical galaxies as remote as those in the Virgo cluster.

The method is a replay of the standard candle approach. If one has a standard candle that is calibrated to galaxies of known distance, one has a reliable distance estimator. Unfortunately, the galaxies that can provide the necessary calibration for this approach are currently limited to the closest systems: the bulge of the Andromeda galaxy and its companion dwarf elliptical M32. One has a reliable distance only if the stellar characteristics of these two galaxies correspond to those of more distant elliptical galaxies. Clearly, there is some cause to be concerned about possible systematic differences between the nearby calibrators and the remote ellipticals in Virgo and even beyond. The calibrator galaxies are not even ellipticals of comparable luminosity.

The Tully-Fisher Correlation

A novel means of measuring galaxy distances is provided by combining measurements of radiation from both the optical and radio parts of the spectrum. The method relies on an intrinsic relationship between two distinct properties of galaxies. One depends on distance, and one does not.

The first property, the luminosity L of a spiral galaxy, is measured at optical or infrared wavelengths. The measured brightness depends on the distance to the galaxy. To obtain the second property, astronomers examine the gaseous medium in the space between stars. The mass of the interstellar medium of the galaxy consists mostly of hydrogen atoms that emit photons in the radio range, which are detected at the 21-centimeter line in the spectrum of neutral hydrogen. The width of that same 21-centimeter line is a measure of the second property of interest, the galaxy's velocity of rotation. The Doppler shift ensures that the orbital velocities of interstellar gas clouds cause the line profile to broaden, so that its width measures the speed of galactic rotation. One generally finds that the rotation velocity of a galaxy increases with galactic radius to a peak value, V_{rot}. The two properties, luminosity and peak rotation velocity, have been observed to be related: the luminosity is directly proportional to the fourth power of the peak rotation velocity ($L \propto V_{rot}^4$, where $\propto$ means "is proportional to").

The rotation velocity V_{rot} is independent of distance to the galaxy, but the measured light flux is of course distance-dependent. If the same relation holds for all spiral galaxies, the type of galaxy containing a substantial interstellar medium, measurement of V_{rot} from the 21-centimeter linewidth together with galaxy brightness yields the distance to a galaxy. Once this relation is calibrated for nearby galaxies, it can be applied to more remote galaxies to infer their distances. Astronomers apply the relation to spiral galaxies within a narrow range of spiral shapes, to minimize any possible systematic variation with galaxy type. This method of course assumes that the correlation between galaxy luminosity and peak rotation velocity, named after its discoverers Brent Tully and Richard Fisher, is the same for all spiral galaxies of a given type.

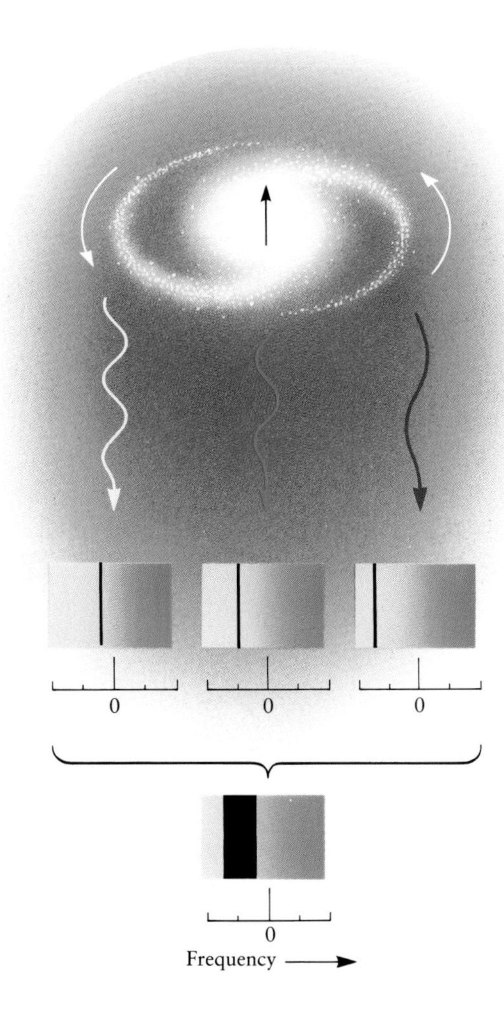

The light from the center of a galaxy that is moving away from the earth is redshifted. One arm of the rotating galaxy will not be moving away from the earth as quickly as the galactic center, so its light will be less redshifted. The other arm will be moving away more quickly, so its light will be even more redshifted. The different redshifts blend into one broad band, whose width depends on the velocity of galactic rotation.

Again, there is an intrinsic dispersion in the Tully-Fisher correlation that leads to a Malmquist bias. Although observers manage to correct for this bias, there are concerns that the relation may vary systematically depending on whether galaxies are in loose groups, in dense clusters, or randomly located. Astronomers do not really have a secure physical basis for the Tully-Fisher correlation, and hence there is reason to be cautious about applying it in diverse environments. The number of calibrator galaxies is presently too sparse to adequately test this source of uncertainty.

Supernovae as Standard Candles

A massive star, like its low-mass counterpart, becomes a giant and even a supergiant when its core helium supply is exhausted. However, a more dramatic final fate lies in wait. The higher gravity force guarantees that every last ounce of thermonuclear energy is extracted from the core, which at the cessation of nuclear fusion is a solar mass of iron. Iron is the most stable element, the ultimate slag heap of the universe. No more nuclear energy can be made available by compressing and fusing it. The core must collapse as the energy supply inexorably runs down.

The result is a neutron star. The iron nuclei break up and the protons and electrons combine to form neutrons as neutrinos carry away the excess heat produced by the compression. A neutron star is a giant atomic nucleus, full of neutrons at nuclear density, a density so high that one teaspoonful of neutron star material weighs a billion tons. The release of neutrinos helps blast off the outer layer of the core. Most of the star explodes as a supernova, ejecting carbon, oxygen, iron and other elements into the interstellar medium. There, these elements mix with hydrogen, and are eventually recycled into a new generation of stars.

A supernova is brighter than a billion suns. When stars die, they brighten so much that individual stars can be detected even in distant galaxies. That suggests that supernovae might be good distance indicators. Individual stars have an advantage as distance indicators over statistical ensembles of stars, since one has a more direct handle on the particular distance calibrator if one can resolve it directly.

The most obvious distance indicators are not the supernovae generated by massive stars as just described, but those occasionally generated by low-mass stars. White dwarfs that have a close companion star sometimes accrete mass from the companion and become unstable; they then collapse and implode as the core first forms iron. Under immense pressure, the iron decomposes into neutrons, protons, and neutrinos. A large amount of energy is suddenly released, some of which is carried off by the neutrinos. The entire star explodes as a supernova. This type of supernova, generally

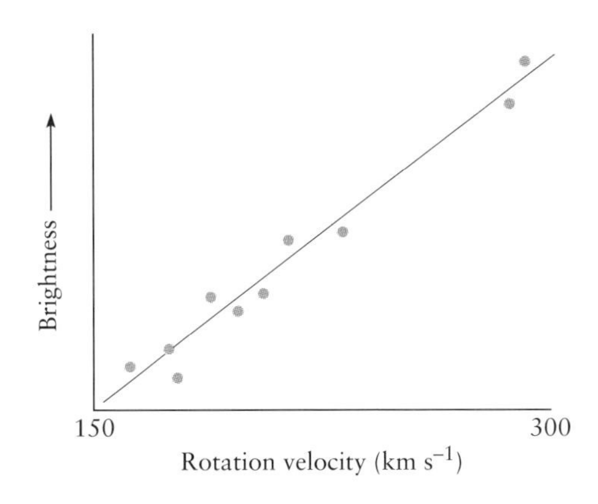

The Tully-Fisher relation. The optical (or infrared) brightness for a sample of spiral galaxies is plotted against their maximum velocity of rotation. The logarithms of the velocities of the two quantities are found to be approximately proportional. The slope of the line that connects the dots, each one of which represents a different galaxy, implies that the luminosity of a spiral galaxy is proportional to the fourth power of its maximum rotation velocity.

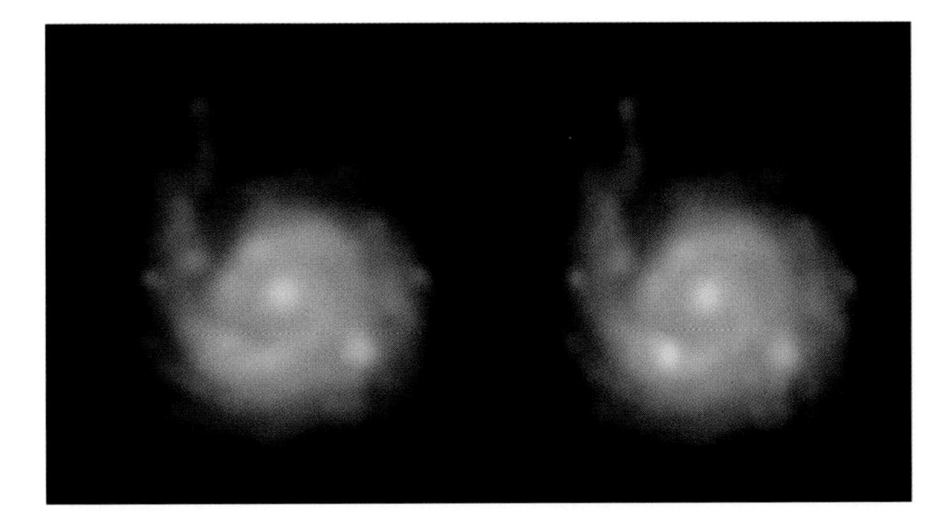

A supernova in the galaxy NGC 3310. The picture at the left was taken in 1987; that at the right within of a week of the outburst of supernova 1991N in April 1991. The supernova (lower left) is practically bright enough to outshine the nucleus of the galaxy, which contains the light from a billion stars.

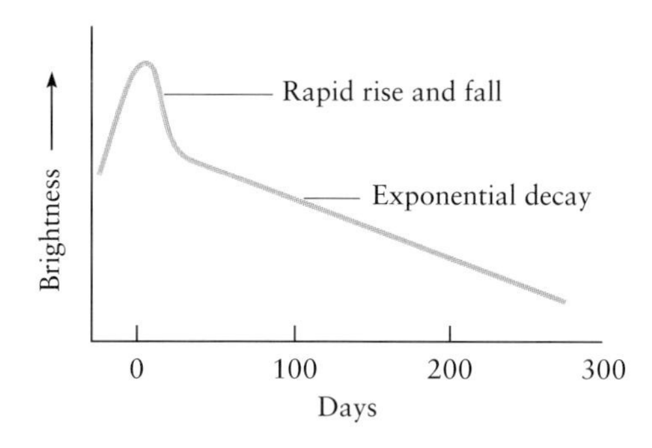

The light curve of a supernova displays a characteristic pattern, especially for supernovae characterized as Type I. The brightness of the exploding star rapidly peaks and declines during the first month, but then enters into a prolonged period of exponential decay. In this latter phase, the time for the brightness to decline by a factor of 2 is approximately 60 days.

involving low-mass stars, is designated Type I to distinguish it from the usual fate of high-mass stars that explode as Type II supernovae. The critical mass above which a white dwarf becomes unstable is 1.4 solar masses, and was first derived by the Indian-American astrophysicist Subrahmanyan Chandrasekhar. Theory suggests that the energy released by the collapse should not vary significantly for different Type I supernovae. As a consequence, supernovae of this type should start off equally bright and then gradually dim at the same rate.

Type I supernovae have been observed in nearby galaxies of known distance, and it does indeed appear that not only does the visible light decay at identical rates (the supernovae have identical "light curves"), but the total light emitted is also the same, to a good approximation. The light is caused by energy released from the radioactive decay of Ni^{56}, a radioactive isotope of nickel produced in great and rather precise quantity during the collapse of the outer layer of the white dwarf core. The unstable nickel isotope decays to iron, Fe^{56}, and the iron is ejected as the core explodes with a final burst of neutrinos. The spectra of the supernovae show that the ejecta consist predominantly of iron and other heavy elements.

Supernovae of Type I are easily detectable in galaxies that are at a distance of 100 megaparsecs or more. If the distant supernovae are identical to the nearby ones, they can then be used as "measuring rods" to infer the distance to the remote galaxies. The difficulty is that one can never be sure that the supernovae of Type I are all identical, at least until one has a sufficiently large collection of nearby examples.

What Is the Hubble Constant?

The various techniques just described may be considered to be independent measures of H_0 that cover similar distance ranges. They are mostly in reasonable agreement. In fact, the Tully-Fisher method, the measurement of planetary nebula luminosities, and the measurement of surface brightness fluctuations all give a similar value of H_0, namely about 80 kilometers per second per megaparsec. However, the supernova method, in particular the use of Type I supernovae as standard candles, favors a value of H_0 equal to 50 kilometers per second per megaparsec. All of these approaches have margins of error smaller than the difference between the two values of H_0.

Direct Measurements

New techniques have been developed to measure H_0 directly, rather than indirectly using a standard candle. The Sunyaev-Zel'dovich method, first proposed by Russian cosmologists Rashid Sunyaev and the late Yaakov

Zel'dovich, uses clusters of galaxies containing a hot, diffuse, gaseous intergalactic medium. The gas emits x-rays, and the x-ray luminosity is proportional to the cluster volume. The gas also scatters photons of the cosmic microwave background, effectively casting a shadow on the microwave background at radio wavelengths. Actually, the scattered background photons gain energy, so that the cluster is dimmer when observed at low frequencies but brighter at high frequencies, relative to the microwave background. The magnitude of this effect is proportional to the length of the path traveled by the photon through the cluster. The x-ray luminosity and the microwave decrement are compared to obtain a direct measure of the size of the region containing hot gas. Since one knows the redshift and can calculate the size from Hubble's law if a value is adopted for Hubble's constant, a value is obtained for Hubble's constant by direct measurement.

A second technique is to observe variable quasars that lie on the far side of an intervening galaxy. Light from the quasar is slightly deflected by the galaxy's gravity field. In effect, the intervening galaxy acts as a gravitational lens that creates multiple images of the background quasar. These images vary as the quasar varies, but with a phase lag that is due to the difference in light path for the different images. The path difference depends on the distance scale, and hence provides a direct measure of Hubble's constant.

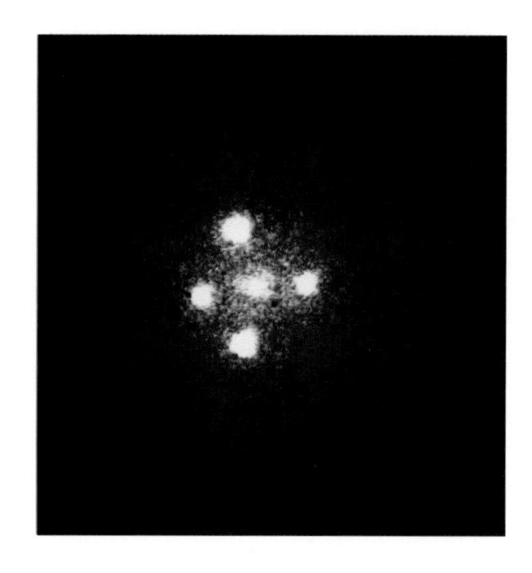

The galaxy at the center of this image is lensing a background quasar and splitting its light into four surrounding images. Since the path length to the quasar is slightly different for each image, any variation in the quasar's light will be echoed in the quasar images at different time delays. Astronomers can thereby extract a distance scale measurement and, hence, a direct measure of the Hubble constant.

Both of these direct methods report preliminary results that favor a low value for the Hubble constant, about 40 kilometers per second per megaparsec. The direct methods are not completely secure, however. For example, if a cluster were elongated along the line of sight, a misleadingly large Sunyaev-Zel'dovich decrement would be measured. If the intracluster gas were clumpy, too high an x-ray luminosity would be inferred.

Supernovae as Expanding Atmospheres

There is a third direct technique for measuring distance. The supernovae associated with the death of massive stars are Type II supernovae. While they are less luminous (by a magnitude at maximum light) than Type I supernovae and their light curves are more dissimilar from one another, they have another characteristic that makes them good distance indicators: they are predominantly composed of hydrogen. Hence their spectral lines are sufficiently simple that the Doppler shift can be measured to give the explosion's speed of expansion. From the expansion speed and the time elapsed during the explosion, we can obtain a measure of the radius of the exploding star, typically a blue supergiant. If the star were an ideal radiator, measurement of its temperature combined with its radius would imply its luminosity, and hence its distance. In practice, a Type II supernova is not a perfect furnace. However, this "expanding atmosphere" method still

Astronomical distances are measured by a variety of techniques, but their accuracy diminishes as larger and larger distances are probed. There are several more indirect distance measures applicable over scales of kiloparsecs to hundreds of megaparsecs. Of the indirect measures, brightest cluster galaxies are the most luminous of the standard candles, and applied to the greatest distances. Direct distance measures commence with parallax and are being developed for much more distant objects.

THE COSMOLOGICAL DISTANCE LADDER

- Parallax (nearby stars)
- Main sequence fitting
- Cepheids

Indirect distance measures

- Supernovae
- Planetary nebulae
- HII regions
- Surface brightness fluctuations
- Tully-Fisher relation
- Brightest cluster galaxies

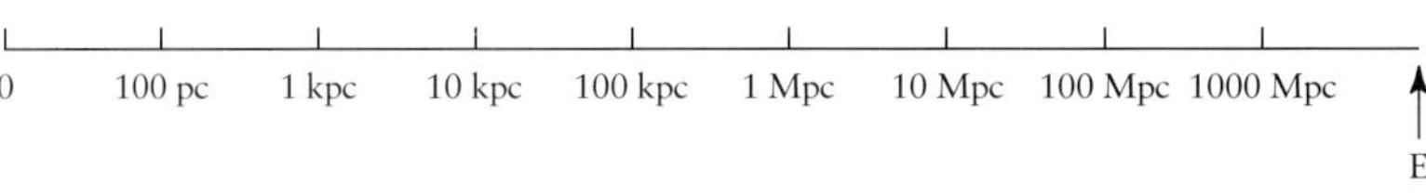

Direct distance measures

- Expanding supernovae
- Sunyaev-Zel'dovich method
- Gravitational lensing of quasars

operates to give a direct measure of the distance to the supernova. It has been applied out to a distance of some tens of megaparsecs. A value for the Hubble constant is found of about 70 kilometers per second per megaparsec. Clearly, the true value must await improved data.

Astronomers anticipate that the refurbished Hubble Space Telescope will make major inroads toward resolving the uncertainty in Hubble's constant, because its improved image resolution will enable it to monitor Cepheid variables in galaxies as distant as 20 megaparsecs. In the longer term, the largest ground-based telescopes will be fitted with special devices to circumvent the effects of atmospheric blurring, and this advance will improve the application of several of the other techniques for determining the distance scale to even more remote galaxies. For the moment, however, we must live with an uncertainty of about a factor of 2 in Hubble's constant, which lies between 40 and 80 kilometers per second per megaparsec.

Steady State Cosmology

Hubble's discovery of the expansion of the universe had led to a crucial inference: if there has been no acceleration or deceleration, all matter must have been piled up at the beginning of the expansion, at a time R/v or $1/H_0$

ago. H_0 is Hubble's constant and was found to be $H_0 = 500$ kilometers per second per megaparsec in Hubble's original work. This means that, with Hubble's distance scale, $1/H_0 =$ two billion years was to be interpreted as an upper limit on the age of the universe.

One may compare this age with the age of the solar system's oldest rocks, as determined by applying the radioactive dating technique. This technique measures the amount by which long-lived radioactive isotopes have decayed. For example, the common isotope of uranium with mass 238 (U^{238}) is unstable, with a half-life of four billion years. It decays into an isotope of lead of mass 205 (Pb^{205}), which is produced only by radioactive decay of uranium. The older the rock, the greater the ratio of Pb^{205} to U^{238}. The present abundance of the lead isotope is measured for different rock and meteorite samples, and from these abundances the age of the rocks is estimated. We infer an age of 4.6 billion years for the oldest known rocks in the solar system, found in meteorites and in lunar samples. The solar system appears to have an age more than twice as great as Hubble gave to the universe.

We can check these numbers by looking at the ages of the oldest stars. Globular clusters are ancient assemblages of stars found outside the disk of the galaxy. The stars in such a cluster were formed at the same time, but die off sequentially, the massive stars dying first. By finding the highest mass of the surviving stars, we can tell the age of globular clusters, the oldest stars in our galaxy. The inferred age is about 14 billion years. Evidently, there is a discrepancy. How could the universe be younger than the oldest stars it contains? In Hubble's day, even the earth seemed older than the universe.

This age discrepancy gave birth to a new theory of cosmology. The steady state universe was conceived by Herman Bondi, Thomas Gold, and Fred Hoyle in 1949. According to Hoyle's account of the evening, these three astrophysicists had attended a film in Cambridge that featured a series of ghost stories. A peculiarity of this film was that its last scene was the same as its first. The movie was continuous, without end. Inspired by the film, it was Gold who first argued that the universe could also be in a continuous time-loop, with no beginning or end. Thus was the steady state cosmology born. A central feature of this theory is that it postulates the creation of matter out of vacuum in order to keep the density of the universe constant and thus satisfy the perfect cosmological principle. Atoms, which eventually aggregate into galaxies, appear out of empty space to replace those that have moved away as space expands. A fundamental concept cherished by all physicists, the conservation of matter and energy, was abruptly abandoned.

The steady state theory was aimed at abolishing the need for an absolute beginning. It has been dubbed "the greatest trick ever given scien-

The globular star cluster 47 Tucanae, at a distance of about 4 kiloparsecs. Clusters such as this contain the oldest stars in our Milky Way galaxy; and are used to establish a minimum age for the universe, if we allow one billion years for the cluster to form, of about 15 billion years.

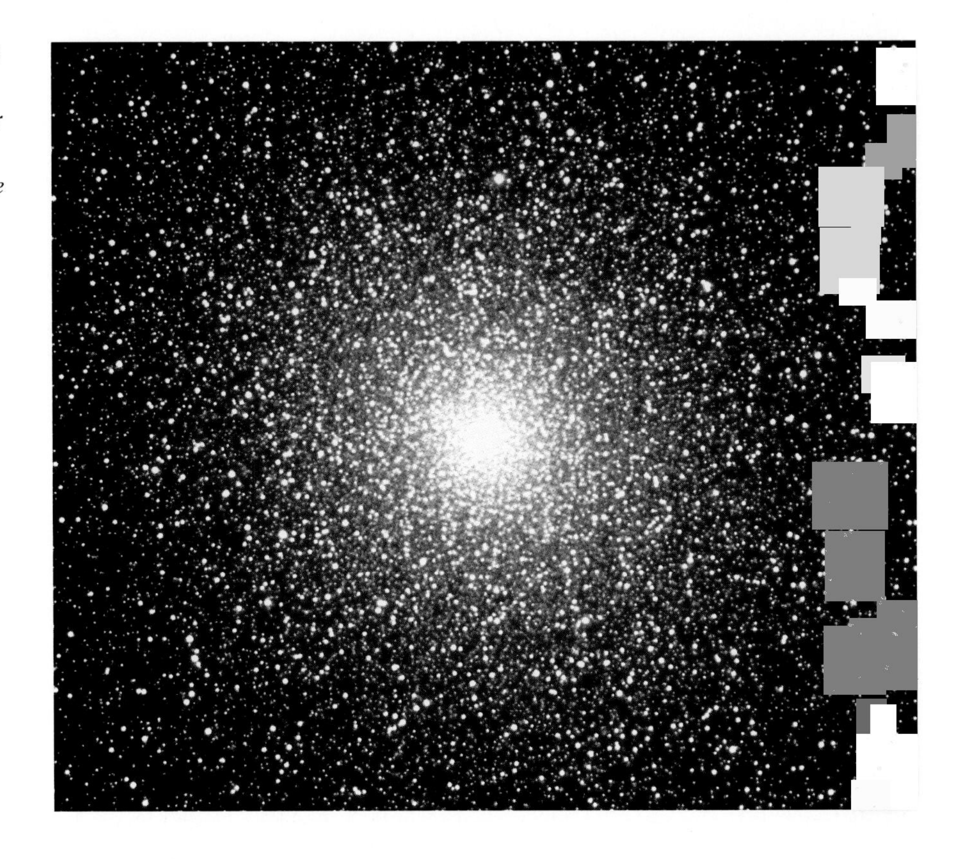

tific veneer" by the Benedictine writer on theology and cosmology Stanley Jaki. However, a strong scientific motivation underlay its development. The steady state postulate was an attempt to resolve the apparent timescale problem.

Steady state cosmology made several key predictions.

1. *There was and continues to be the creation of about one hydrogen atom per cubic meter every 10 billion years.* Matter is assumed to be created out of the vacuum, radically violating the law of conservation of mass and energy.
2. *No evolution at great distance could have occurred.* The steady state hypothesis predicts that, in the mean, we should see the same density of galaxies far away as we observe nearby. No change in density should have occurred over the billions of years to which we can see back by observing distant galaxies.

If the first prediction is true, and hydrogen atoms are created out of the vacuum, one expects antimatter to also be produced. Protons and their antimatter equivalents, antiprotons, annihilate each other when they meet, producing gamma rays. Such occasional annihilations should lead to a diffuse gamma ray background, yet the universe is not glowing with the gamma rays from annihilations. Moreover, it was apparent that galaxies could not be composed of matter and antimatter in equal parts without causing somewhat of a cosmic catastrophe.

Another solution came to mind. Even if the law of conservation of mass and energy is assumed to break down, one needs to avoid violating another fundamental law, namely the law of conservation of electric charge. Hence another possible form for newly created matter is neutrons. These would decay and leave behind a hot, x-ray–emitting gas pervading the universe. Like the expected cosmic gamma rays, however, the x-rays are not seen. To meet these objections, Fred Hoyle and Jayant Narlikar soon modified their theory to postulate that creation only occurred in exceptionally dense cores. These are identified with the nuclei of galaxies, and with the exotic and exceedingly luminous objects called quasars, which occur in great numbers in the early universe and are thought to be associated with the early evolution of galaxies.

Is there evidence to disprove the second prediction as well? Some galaxies are powerful sources of radio signals, and can be detected out to a very great distance. These radio sources have been counted to test the hypothesis that the density is unchanging with epoch. Were the universe denser in the past, as the big bang theory predicts, then after allowing for the amount of volume surveyed, one should measure a substantial increase in the number of faint, distant sources relative to the nearer, brighter sources.

A typical survey counts only those radio sources above a certain limiting radio brightness, determined by the size of the radio telescope, the telescope sensitivity, and the duration of the observation. Suppose that $N(>f)$ is the number of radio galaxies observed brighter than a measured intensity f, which for a radio source of luminosity L at distance d is given by the ratio of the luminosity L of the source to the surface area of a shell at that distance: $f = L/4\pi d^2$. Solving for d, one obtains $d = (L/4\pi f)^{1/2}$. So the distance to which one can see in a brightness-limited survey of sources with identical L is proportional to the square root of the ratio of L to f.

Now the total number of radio sources measured in a survey of the entire sky is simply the product of the density of radio sources and the volume surveyed. The steady state hypothesis predicts that the number density of radio sources is constant. Hence the number density of sources observed brighter than the survey limit f should be proportional to the volume surveyed, or d^3:

$$N(>f) \propto d^3 \propto (L/f)^{3/2},$$

where $\propto$ signifies "is proportional to."

A radio map—made with the Very Large Array radio telescope in Socorro, New Mexico—of the elliptical galaxy NGC 1265 in the Virgo cluster. The radio lobes dwarf the optically visible part of the galaxy, and are swept out behind it in gigantic trailing tails by the pressure of the intracluster gas through which the galaxy is moving. Radio galaxies such as this were found in great number in the early days of radio astronomy, but with imprecise positions and no structural detail. Nonetheless, simply counting these sources led to important conclusions for cosmology.

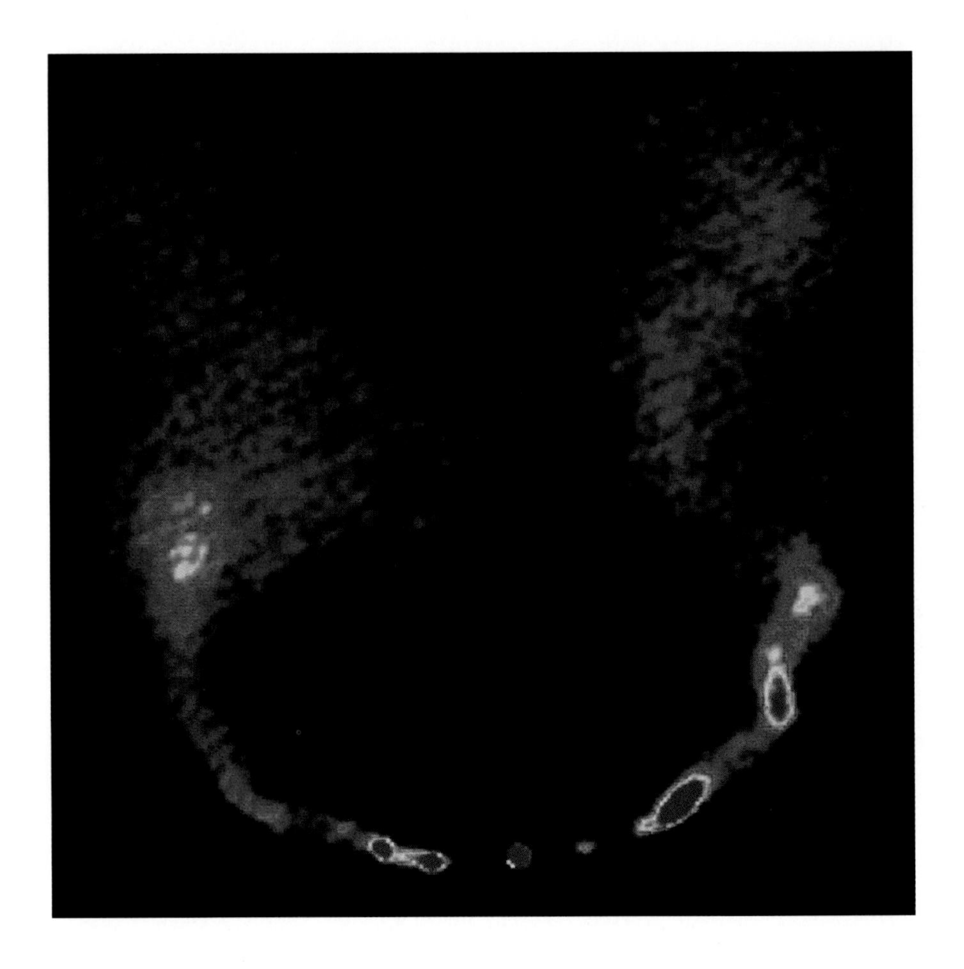

This result predicts that as the sensitivity of the survey is increased, that is to say, as the limiting brightness f is lowered, the number of sources detectable should rise sharply. In a nonevolving universe, as is appropriate for the steady state model, the predicted number should increase proportionally to the inverse three-halves power of source brightness ($f^{-3/2}$). Unfortunately for the steady state theory, observations first made by Martin Ryle and John Bolton in the 1950s revealed a much stronger increase in source counts. Proponents of the steady state model in the 1950s argued that we might be living in a region of the universe where there is simply a local deficiency of radio sources. However, the radio sources were shown subsequently to be primarily radio galaxies and quasars that are several billions of megaparsecs away from us, demonstrating that evolution must be occurring over a timescale of order 10 billion years. Luminous radio-emitting galaxies were far more frequent in the past than they are seen to be today.

The discrepancy between the universal expansion age, on the one hand, and meteoritic and stellar ages on the other hand, was only removed in the 1950s, when a more accurate value for H_0 emerged. The best modern value of 75 kilometers per second per megaparsec gives the age of the universe as $1/H_0 = 15$ billion years.

The final blow to the steady state theory came with the discovery of the cosmic microwave background in 1964. This sea of radiation bathing all of space, discussed in the following chapter, is direct evidence of radiation originating in a dense hot phase of the universe, as predicted by the big bang theory. To explain such radiation in a steady state model, one would have to postulate the universal presence of millimeter-sized dust grains that would absorb an intense radiation field produced by many exceptionally luminous galaxies and reradiate it as microwave photons. This interpretation of the cosmic microwave background is so contrived and requires so many special assumptions that it is regarded by most cosmologists as highly implausible.

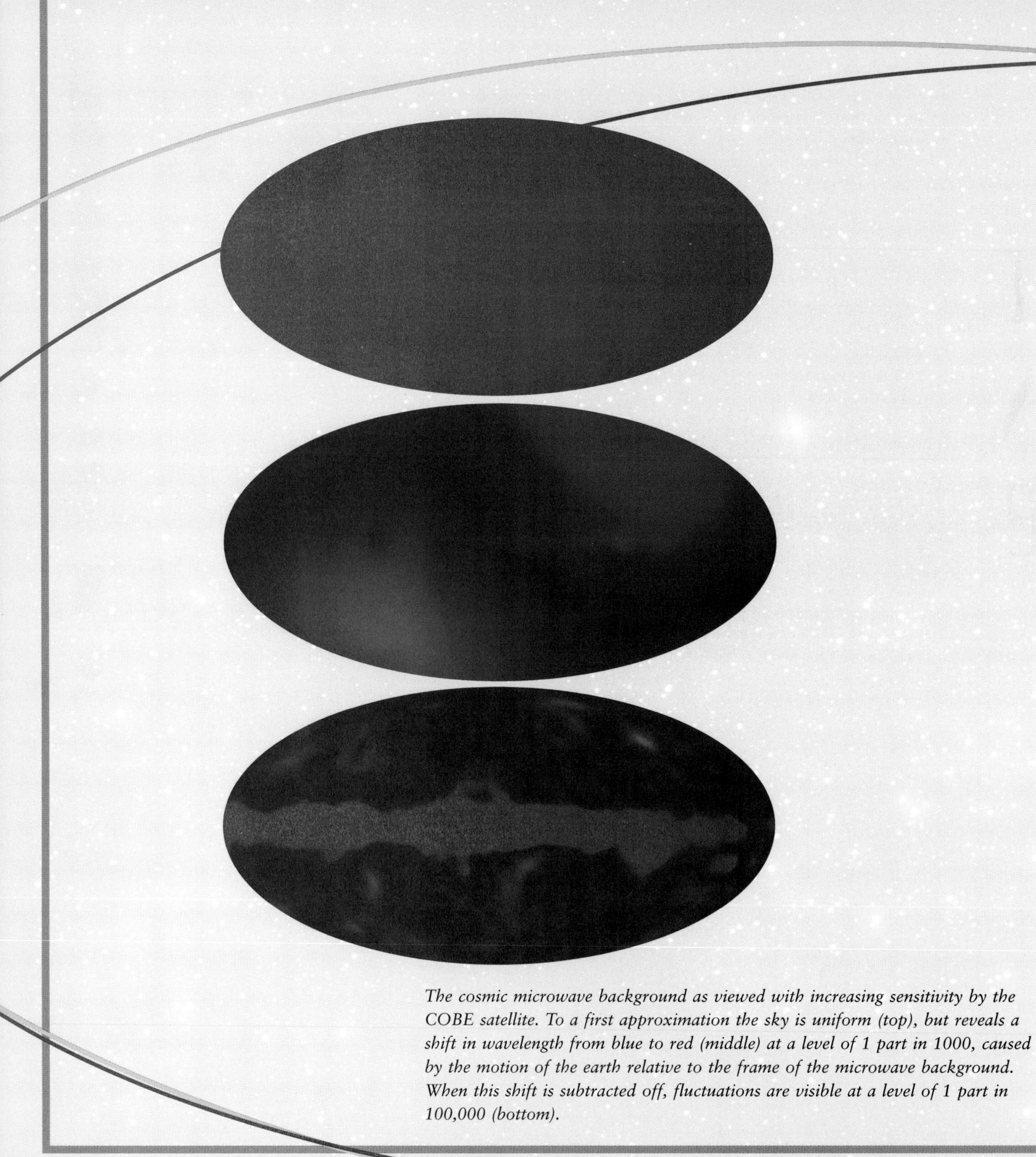

The cosmic microwave background as viewed with increasing sensitivity by the COBE satellite. To a first approximation the sky is uniform (top), but reveals a shift in wavelength from blue to red (middle) at a level of 1 part in 1000, caused by the motion of the earth relative to the frame of the microwave background. When this shift is subtracted off, fluctuations are visible at a level of 1 part in 100,000 (bottom).

3

The Cosmic Microwave Background

Alexander Friedmann, who predicted the big bang cosmology in 1922.

In 1922, a Russian meteorologist and mathematician, Alexander Friedmann, made a discovery that was to resound throughout the century. He realized something that Albert Einstein had overlooked, and initially refused to accept: the universe could be expanding. Einstein had applied the cosmological principle and simplified the equations of gravitation in his new general relativity theory to obtain a cosmological model of the universe that appeared to be static. He even invented a new force, that of cosmic repulsion, to prevent the universe from collapsing on itself due to the attractive tendency of its own gravity. Friedmann discovered that Einstein had made an elementary mathematical error, and overlooked a new solution to his equations, which allowed the universe to expand, with no need for any new forces. Einstein eventually acknowledged that the introduction of cosmic repulsion was one of the greatest mistakes he had ever made.

Working independently in 1927, the Belgian cosmologist Georges Lemaître rediscovered the expansion of the universe and went a step further. He suggested that the redshifts of the galaxies provided evidence for the expanding universe. His insight that the redshift should be proportional to the distance to a galaxy gave a physical meaning to redshifts, something that had been eluding the famous cosmologists of the time, who were focused on a static universe. We saw how in 1929 Edwin Hubble provided the observational underpinning, in the form of galaxy distances, that enabled him to establish the redshift-distance law as an empirical fact. Curiously, Hubble never seemed to fully accept that the linear increase in redshift with distance was evidence for the expansion of the universe. Subsequently, cosmologists have been almost unanimous in taking Hubble's law as the basis for developing the theory of the expanding universe.

The expanding universe theory was difficult for many to accept. It implied that the universe began at a finite time in the past from an extremely

In a radio program in 1950, Fred Hoyle, arch-proponent of
tate universe, which avoided such a singular past, derisively re-
e rival expanding universe theory as the "big bang." Despite
priateness of the notion of sound, the name stuck. The "big ex-
ight have been preferable, but it lacked the resonance. The big
postulates an origin in time, some 10 to 20 billion years ago,
ntire observable universe emerged from a singular state of ex-
ty. The great confirmation of the big bang theory was the dis-
ne relic radiation field from the origin of the universe.

Georges Lemaître, who predicted in 1927 that the universe might be expanding.

The Discovery

In 1964, two young radio astronomers, Arno Penzias and Robert Wilson, discovered a puzzling source of radio noise. The two scientists were working at Bell Laboratories in New Jersey, with an ultrasensitive radio telescope originally designed to receive signals from the early communications satellites. This instrument detected a noise of extraterrestrial origin that not only had no dependence on the location of the sun or of the Milky Way in the sky, but came from all directions equally. That is, the noise was isotropic. It did not come from the telescope itself, as Penzias and Wilson verified by carefully cleaning the instrument of various residues such as bird droppings that might have contributed to radio static. Their measurements revealed the noise to be radiation at a wavelength of 7 centimeters, in the microwave region of the radio spectrum. In retrospect, the signal is easily detectable. About 1 percent of the static on a television screen that is not tuned to a local channel is due to this same extraterrestrial microwave radiation.

It very soon became apparent that this microwave radiation originated in the most distant reaches of the universe. If it did not have a very local origin, such as emission from dust within the solar system, its extreme isotropy meant that it had to come from very far away. A group of cosmologists, working at nearby Princeton University, had been simultaneously searching for a microwave radiation background glow that they believed was left over from the big bang. Robert Dicke and his colleagues had argued that most of the helium in the universe, and in the sun and other stars, must have been created by thermonuclear fusion in the very early universe. For this to have happened, the early universe must have been exceedingly hot. In that case, the universe would have been full of energetic photons emitted by the hot electrons and protons. As the universe expanded, the radiation would have cooled, to be observable today in the microwave part of the spectrum.

AT&T Bell Laboratory scientists Arno Penzias and Robert Wilson standing in front of the horn antenna in Holmdel, New Jersey, with which they discovered the cosmic microwave background in 1964.

The Princeton astronomers were unaware that 20 years previously, a similar line of reasoning had been initiated by George Gamow, and had led to similar predictions. Gamow's former students, Ralph Alpher and Robert Herman, estimated in 1949 that the present temperature of the universe, containing the cooled relic radiation field, should be 5 degrees Kelvin. They did not, however, suggest a search for microwave radiation, although in 1963 two Russians, Andrei Doroshkevich and Igor Novikov, went the additional step. They consulted the Bell Laboratories Technical Journals to see whether microwave measurements imposed any interesting limit on the cosmic radiation background. They almost struck gold, when they found a paper from 1961 in which Ed Ohm reported measuring excess static noise at a level of 3 degrees Kelvin when pointing the Bell Laboratories 20-foot-diameter antenna at the sky. Unfortunately, though, Ohm was unable to rule out instrumental noise as the source of the static.

The Princeton researchers did not toil in vain, however, but soon produced a sequel to the Bell Laboratories discovery when they succeeded in measuring the cosmic microwave spectrum at a second wavelength. They concluded that the radiation most likely was blackbody radiation.

George Gamow, pioneer in the 1940s of the big bang theory, also developed primordial nucleosynthesis as the progenitor of the light elements.

Blackbody Radiation

Blackbody radiation is the name given to radiation that is present in an isolated enclosure at a perfectly uniform temperature. Its properties depend only on the temperature, and not on any other property of the enclosure. Blackbody radiation consists of wavelengths spanning a wide range, but its intensity is increased at certain wavelengths. This type of radiation has a unique distribution of intensity with wavelength called the Planck distribution. Although the shape of this distribution is always the same, the distribution peaks at shorter and shorter wavelengths as the temperature increases. The rule for this variation is known as Wien's law, which says that the product of the wavelength of peak emission and the temperature is 0.3 centimeter degrees Kelvin. Hence blackbody radiation whose intensity peaks at a wavelength of 0.1 centimeter has a temperature of only 3 degrees Kelvin, or −270 degrees Celsius.

In concluding that the cosmic background radiation is blackbody radiation, based on measurements at only two wavelengths that were well away from the wavelength of peak intensity, the Princeton researchers took a great leap of faith. A blackbody spectrum is poorly described by measurements at only two wavelengths. One is reminded of a dictum attributed to Enrico Fermi: with four data points, a clever experimentalist can fit an elephant, but with five, he will not only fit the elephant's tail but make it wag. Nevertheless, big bang theory is persuasive: it firmly predicts blackbody radiation. It was to take far more sophisticated technology, and many false alarms, before measurements were successfully performed around the blackbody peak.

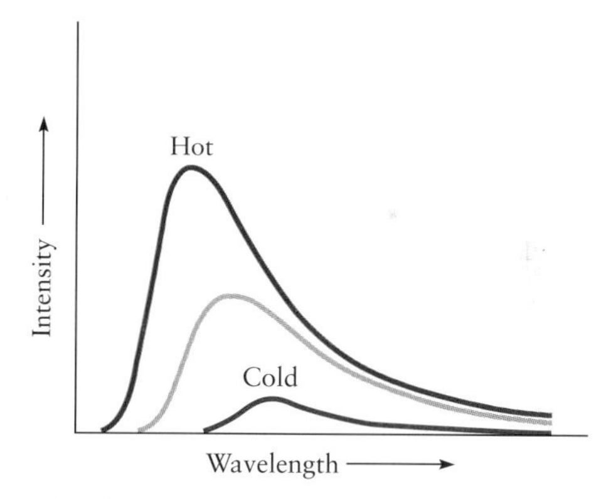

The distribution of light from a blackbody has a universal pattern, regardless of the temperature of the blackbody. The peak of the light intensity is displaced for cooler and cooler blackbodies toward larger and larger wavelengths. The displacement is from blue light toward red light for blackbodies that emit over the visible part of the spectrum.

Indeed, not until 1991 did rocket and satellite measurements show that the cosmic microwave radiation is extremely close in spectrum to a blackbody at a temperature of 3 degrees Kelvin. The dramatic breakthrough came with an instrument designed by John Mather and placed on board the Cosmic Background Explorer (COBE) satellite, launched in November 1989. Mather's instrument was able to measure the temperature of the cos-

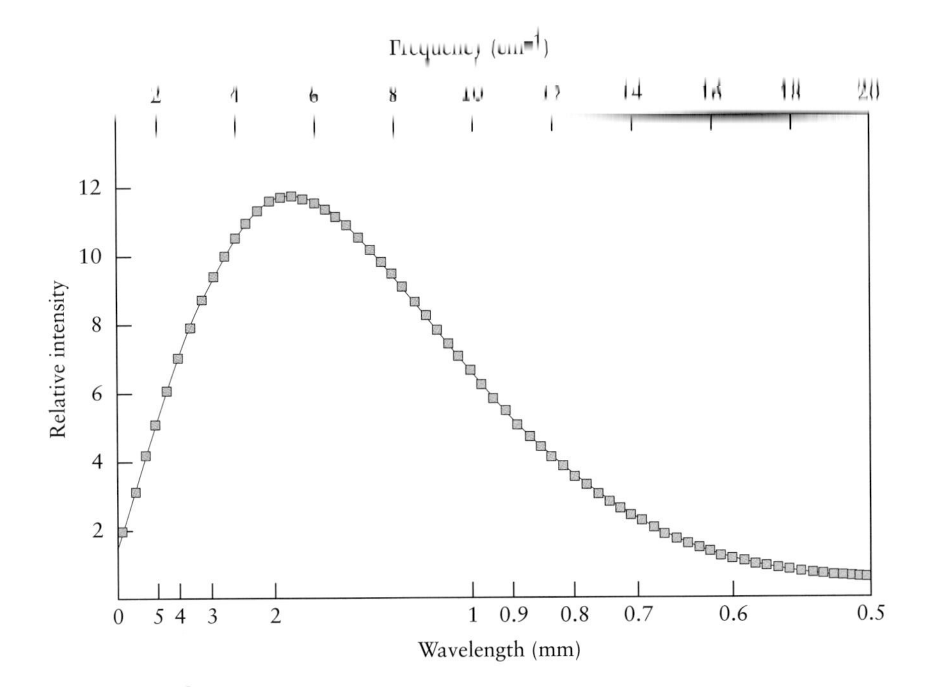

The spectrum of the cosmic microwave background as measured during the first few minutes of observation performed by the FIRAS instrument on board the COBE satellite in late 1989. The plot of intensity against wavelength can be fitted by a blackbody curve with temperature approximately 2.73 degrees Kelvin.

mic microwave background to unprecedented precision. It recorded the precise value of the background radiation temperature as 2.726 degrees Kelvin, with an uncertainty of only 0.005 degree Kelvin. Measurements have also been performed with other telescopes at many different wavelengths, from 30 centimeters (in the radio) to 0.05 centimeter (in the far infrared). Any distortions from a pure blackbody spectrum are exceedingly small: less than a tenth of a percent near the blackbody peak. One cannot measure a more precise blackbody in the laboratory than has been detected in the sky.

The quantum theory relates the energy of radiation to its wavelength or its frequency. Radiation comes in discrete packets, or quanta, of energy, called photons. The energy of a photon is $h\nu$, where ν is the frequency (and $c/\nu = \lambda$ is the wavelength) and h is Planck's constant, equal to 6.626×10^{-27} erg second. Blackbody radiation consists of photons that are as densely packed together as possible. This means that in a box of dimension equal to the wavelength λ, there is about one photon. So the density of blackbody radiation approximately equals $1/\lambda_{\text{peak}}^{3}$ per cubic centimeter. As the universe expands, all length scales increase and the wavelength of any radiation also increases, or is redshifted. Hence the blackbody temperature, by Wien's law, drops proportionately to the factor by which the universe has expanded. At present, there are approximately 400 cosmic microwave blackbody photons per cubic centimeter. These photons of course move at light speed. We may measure the *flux* across a square centimeter to be the number of blackbody photons per cubic centimeter times the speed of light c, or about 1.2×10^{13} photons per square centimeter per second. Early in the history of the universe, this flux would have been much larger.

The Isotropy of the Microwave Background

In addition to its blackbody spectrum, the other remarkable aspect of the cosmic microwave radiation is its extreme isotropy. Recent measurements find precisely the same blackbody temperature, when looking around the sky in different directions, to a precision of better than 1 part in 100,000.

There was expected to be one larger variation, however, and this variation is in effect a Doppler shift caused by the motion of the earth relative to the frame of the microwave background. The earth moves around the sun at 30 kilometers per second, the sun around the galaxy at 220 kilometers per second, the galaxy around the Local Group of galaxies at about 50 kilometers per second, and the Local Group moves in the Virgo supercluster of galaxies at roughly 200 kilometers per second. This all adds up to a net motion of the earth of about 400 kilometers per second, which should result in a blueshift of the radiation in the direction toward which we are moving, and a redshift in the opposite direction. The magnitude of the shift is given roughly by the ratio of 400 kilometers per second to the speed of light (300,000 kilometers per second), or about 0.1 percent of the speed of light. This shift is called a dipole variation, because it smoothly varies over the whole sky with one peak and one minimum in the intensity of the background radiation as viewed over 180 degrees. This smooth variation was the prediction, if the cosmic microwave background were truly of remote origin.

The dipole anisotropy detected by the COBE satellite.

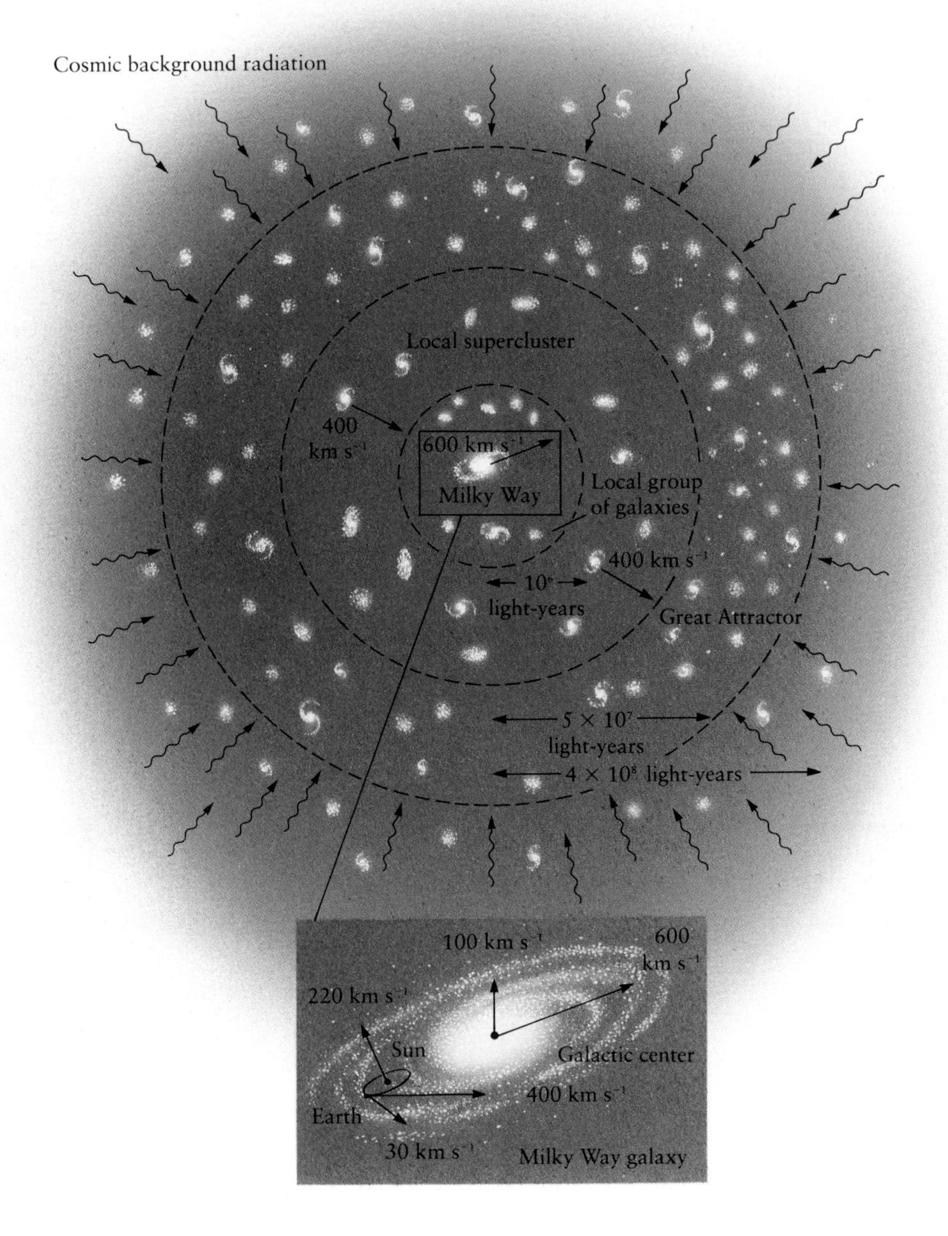

The net velocity of the Milky Way galaxy relative to the cosmological reference frame is about 600 kilometers per second. The actual measured velocity of the earth relative to the background radiation is only about 400 kilometers per second; the higher velocity for our galaxy results when account is taken of the earth's motion around the sun (30 kilometers per second), the solar motion around the galactic center (220 kilometers per second), and the motion of the Milky Way galaxy toward the Andromeda galaxy in the Local Group of galaxies (50 kilometers per second). The entire Virgo supercluster is inferred to be moving at a velocity of about 400 kilometers per second toward an immense concentration of clusters in the southern sky called the Great Attractor.

In 1973, astrophysicists Richard Muller and George Smoot designed a sensitive radio telescope that operated near the peak of the cosmic black-body spectrum. Although terrestrial water vapor would have provided unwanted opacity at this wavelength, Muller and Smoot circumvented the problem by flying the telescope aboard the U-2 spy plane, which would spy on the heavens and discover the motion of our galaxy relative to the cosmic microwave background radiation. The measured dipole anisotropy arising from the earth's motion was, as predicted, a temperature variation on the sky of 1 part in 1000. After correcting for the rotational motion of the sun about our galaxy, the Local Group of galaxies is inferred to be moving at 630 kilometers per second, or about one million miles per hour, toward a point in the direction of Aquarius. This provided the first proof that the microwave background was of cosmological origin and likely to be the relic fireball from the big bang.

The Epoch of Thermalization

The density of ordinary matter in the universe today can be estimated from the observed frequency of galaxies to within a factor of 10. At early epochs this matter was mostly in the form of hydrogen plasma; that is, there were protons and electrons but no hydrogen atoms. By tracing the progress of the expansion backward through time, we can predict the density of this plasma, early in the universe, long before galaxies had formed. From this density, and the measured blackbody temperature, we can deduce what the universe must have been like at different epochs in the past.

Today, radiation travels freely to us from distant galaxies. That is, the universe is highly transparent to electromagnetic radiation. Sufficiently early, however, when the mean density of matter in the universe and the temperature were both sufficiently high, it was as though one were viewing the universe through an exceedingly dense fog. Any radiation could travel barely any distance at all before being absorbed and reemitted. The matter and radiation were in intimate thermal contact, such that their temperatures were identical. This condition is precisely what is needed to create blackbody radiation, and the process by which this is achieved is called *thermalization.*

An example of the physical processes that mediate thermalization of the radiation is *bremsstrahlung*, or "braking radiation." In this process, electrons decelerate in the electric fields of other electrons or protons, thereby either emitting or absorbing photons in order to conserve the total energy of the encounter. Such processes were effective during the first year or so of the big bang but ceased thereafter as the density decreased. Although radiation continued to interact with matter by scattering off free electrons, new photons were no longer created and thermalization no longer occurred. However, the radiation produced in the first year remains as blackbody radiation: no process can destroy it. The expansion of the universe simply results in the gradual cooling of the radiation temperature, but the spectrum retains the form of a blackbody.

The Last Scattering Epoch

At thermalization, the radiation was predominantly in the form of x-rays. As the universe expanded and the radiation cooled, its photons reached lower, ultraviolet wavelengths. Today the background radiation is approximately 3 degrees Kelvin, but about 300,000 years after the big bang, when the universe was about 1000 times smaller than it is today, or more precisely, at an epoch when the expansion factor that denotes the separation of any pair of distant galaxies was 0.001 relative to its present value, the radiation temperature was 3000 degrees Kelvin. This temperature is just high enough that the ultraviolet photons which are present in such hot radiation are capable of ionizing hydrogen, and prior to this moment, matter was still in the plasma state. Hydrogen plasma, if not too diffuse, has the property that radiation cannot pass freely through it. Rather, the radiation is scattered many times. The free electrons receive a tiny impulse as an electromagnetic wave passes by, and consequently the wave is deflected.

About 300,000 years after the big bang, the density of matter dropped and the radiation temperature cooled to the point at which photons were at optical wavelengths. A moment was reached when the radiation could pass freely through the matter. The photon's new freedom coincided with a change in the state of matter from plasma to atomic. As the temperature fell, free electrons and protons were able to recombine to produce hydrogen atoms, and atomic hydrogen became the predominant state of matter. With a negligible number of free electrons remaining, the universe was now perfectly transparent to the radiation. It was as though a veil had lifted. When we look at the cosmic microwave background today, we view the universe as it was long ago, when the blackbody radiation was at 3000 degrees Kelvin. The microwave background offers a glimpse of the universe far earlier than any detectable galaxy or quasar does and, indeed, at a time long before any such objects had formed.

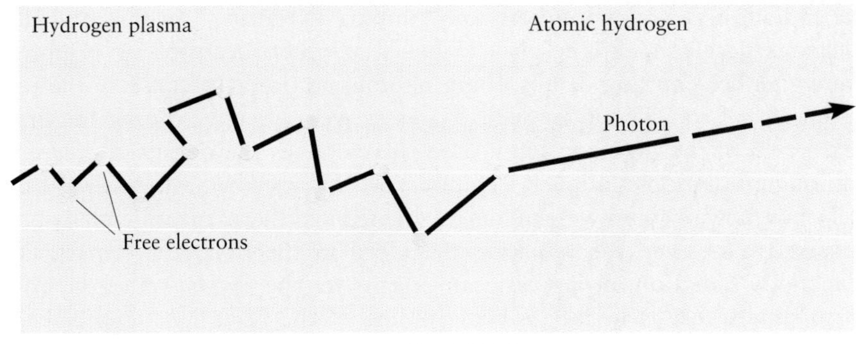

$T \approx 3000$ K

During the first 300,000 years, when the universe contained ionized hydrogen, the photons underwent frequent scatterings off the free electrons. As the temperature cooled below 3000 degrees Kelvin, the hydrogen condensed into atoms, and the electrons were no longer free and able to scatter photons. When we observe the cosmic microwave background photons, we are looking back to when the universe had an age of only a few hundred thousand years.

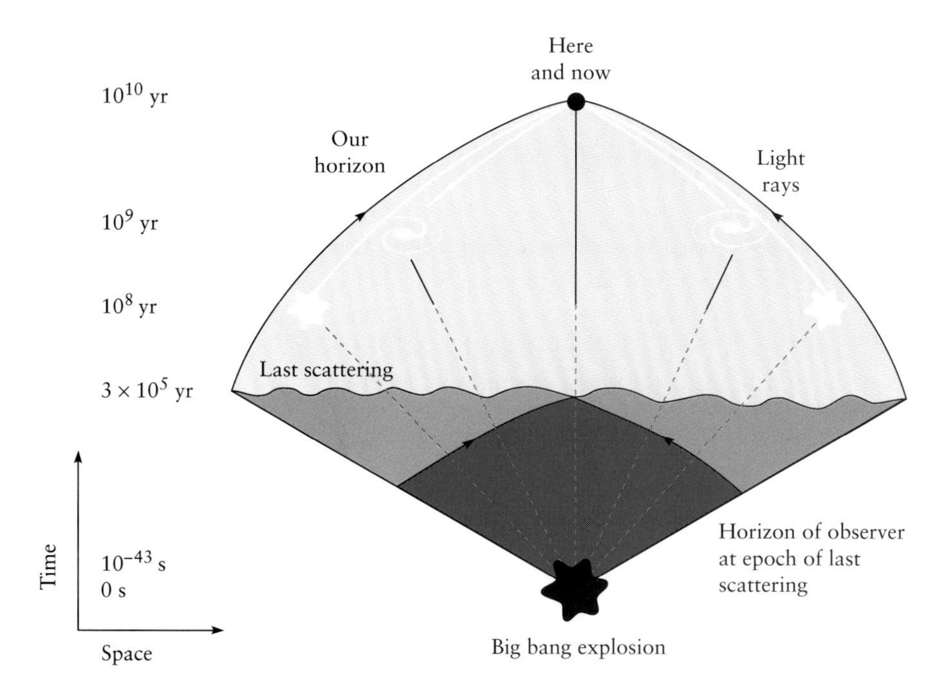

A space-time diagram with time plotted vertically and the three dimensions of space represented by the horizontal axis. Various stages in the evolution of the big bang are shown from conception, to the last scattering epoch when embryo fluctuations are detectable, to the birth of galaxies and their evolution. When we study a nearby galaxy, we see it as it was some time (perhaps millions of years) ago, whereas a more remote galaxy would appear to us as it was billions of years ago. However, the blackbody radiation comes to us directly from the last scattering epoch, only 300,000 years after the big bang.

Fluctuations

All the large-scale structure of the universe today evolved from small primordial fluctuations in density. These fluctuations began as infinitesimal deviations from homogeneity in the very early universe and, by the action of gravitational instability, grew inexorably until by the present epoch they have given rise to galaxies and galaxy clusters. On large scales, such fluctuations, traceable to a primordial origin, may still be discernible in the galaxy distribution. On still larger scales, the cosmic microwave background bears witness to primordial deviations from uniformity.

Since the last scattering, the background radiation is no longer entrained by matter, and travels freely, cooled by cosmic expansion, to now be observable as the cosmic microwave radiation. This radiation is a pale relic of the fireball that dominated the universe for the first 10,000 years, until the end of the epoch of thermalization. The strength of the temperature fluctuations in the microwave background measures the amplitude of the cosmological density fluctuations some 10,000 years after the big bang, when the growth of fluctuations started in earnest. We actually expect to see these fluctuations emerge from the epoch of the last scattering only some 300,000 years after the big bang.

Cosmologists had been expecting to find these fluctuations since shortly after the discovery of the microwave background, but it was not until the spring of 1992 that the detection of the long-sought anisotropies was announced. Fluctuations in the cosmic microwave background of only 30 millionths of a degree Kelvin were recorded by the COBE satellite. The angular scale of fluctuations is so large (from 10 degrees to 90 degrees on the sky) that nothing in standard cosmology, as envisaged by Friedmann and Lemaître, could have generated them.

The instrument on board COBE that detected the fluctuations was the Differential Microwave Radiometer (DMR), designed by George Smoot. What is remarkable about the DMR is that using the same off-the-shelf (in 1975) technology he had used to measure the dipole anisotropy, Smoot was able to make a major breakthrough in cosmology. Theorists had predicted that at a level of 1 part in 100,000, one ought to detect the blemishes in the microwave background that represent the seeds of large-scale structure. To achieve this increase in sensitivity, one hundred times more than the measured dipole anisotropy, required a space-borne platform, and Smoot nurtured the DMR experiment into a reincarnation on board the COBE satellite. Originally destined for the Space Shuttle, the instrument was forced to be redesigned by the *Challenger* disaster and was shrunk to fit into a Delta rocket. COBE was NASA's first mission dedicated to cosmology, and took in data for four years from 1989 to 1993 before NASA turned it off for budgetary reasons.

The DMR measured the radiation intensity at three frequencies, astutely chosen to be both near the maximum intensity (1 cm, 6 mm, and 3 mm) of the cosmic microwave background and near the minimum signal from the galaxy, a source of foreground noise. Cosmological and galac-

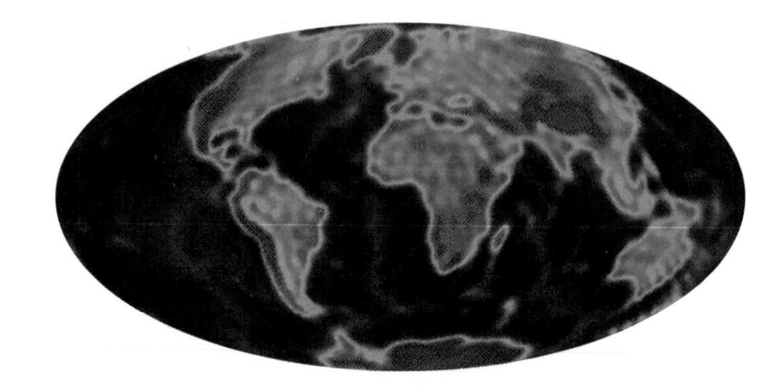

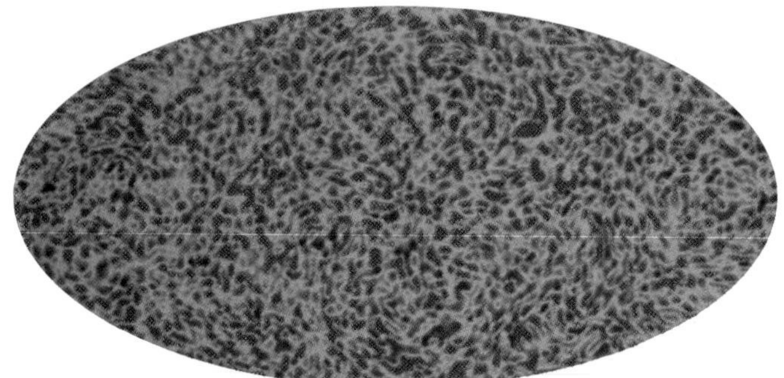

Left: If the earth were viewed with the angular resolution of the COBE DMR experiment, and noise could be eliminated, we would see something like this idealized view. Right: This map gives a more realistic view of how COBE would map the world. The difference is that experimental "noise" has been added, equivalent to two years of data taking. One can see "fluctuations" on various angular scales that correspond to continents and ocean, but most of the definition is lost. This illustrates the difficulties faced by COBE in attempting to reconstruct the underlying patterns of temperature fluctuations in the cosmic microwave background.

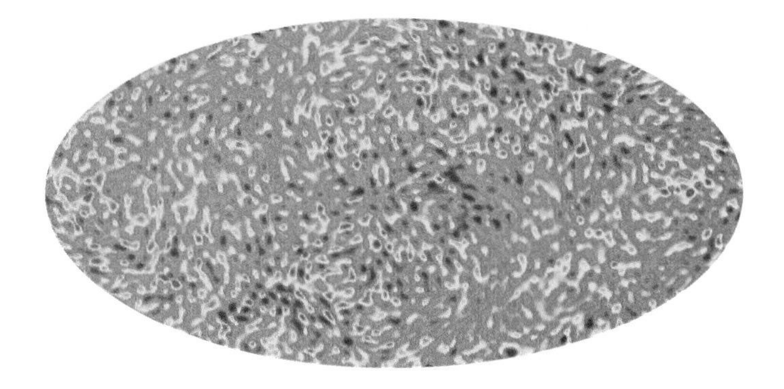

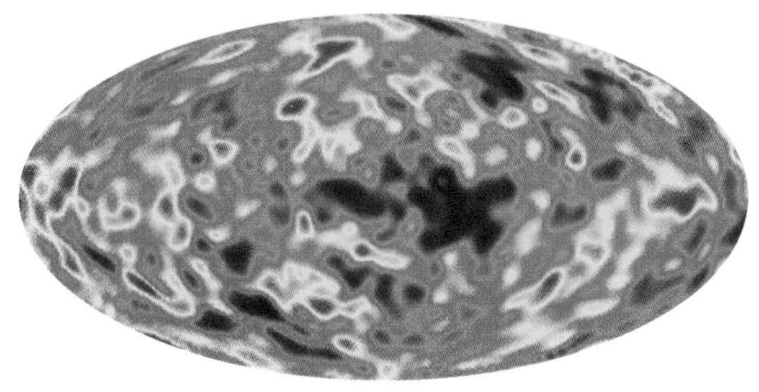

Top left: A raw pixel map of the COBE-DMR data after one year of observation. Top right: When the data is smoothed over 10 degrees and the galaxy contribution to the fluctuations is subtracted, structure can be seen that is mostly instrumental noise. A statistical analysis, however, reveals that there is a true cosmic fluctuation signal.

tic emissions could thus be distinguished by their different spectral forms. The DMR had two receiving horns, each equipped with two separate detectors for all three frequencies, pointing 60 degrees apart on the sky. Noise was reduced to a few parts per million by comparing all the different outputs from the instrument.

The DMR mapped the full sky every six months. After one full year of data was analyzed, the experimenters were able unambiguously to measure anisotropies in the microwave background. Scientists could be sure that these anisotropies were in the cosmic microwave background because the temperature variations possessed the expected blackbody spectrum. The fluctuations on the sky had a characteristic amplitude of 30 (±5) microkelvins measured over angular scales that ranged from 10 to 90 degrees on the sky. There were hot spots and cold spots of various sizes of this typical strength, although the first year of DMR data was sufficiently noisy, the ratio of cosmic signal to noise being only unity, that the fluctuations were only meaningful in a statistical sense. Once foreground contamination from our own galaxy had been corrected for, there was left a cosmological quadrupole (i.e., 90 degree peak-to-minimum) signal of amplitude 17 (±5) microkelvins. This amplitude is equivalent to a level of 6 (±2) parts in a million, just as the theorists had predicted. This level of fluctuation is more than a hundred times smaller than the dipole anisotropy due to our motion relative to the cosmic microwave background. The remaining cosmic fluctuations on scales from 7 to 90 degrees are approximately consistent with the predictions of the simplest models for the formation of large-scale structure.

The link between the theory of structure formation and the observed universe has finally been resolved. In Chapter 10, I will describe how modern cosmology led to the prediction of the fluctuations that were first detected in 1992.

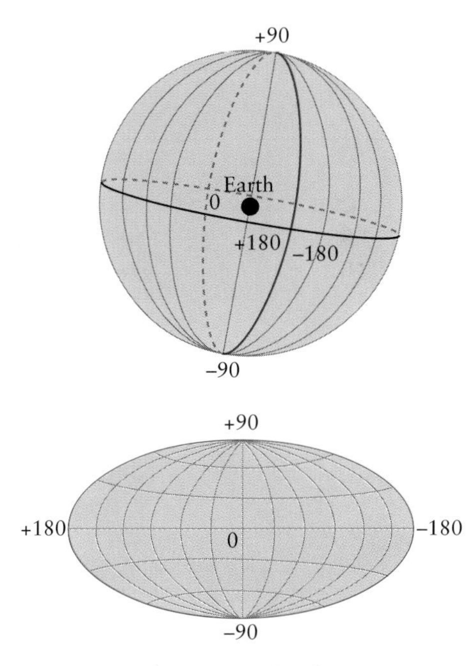

To measure degrees on the sky, imagine a celestial globe with the earth at its center (top) that has been cut along the red seam and flattened (bottom).

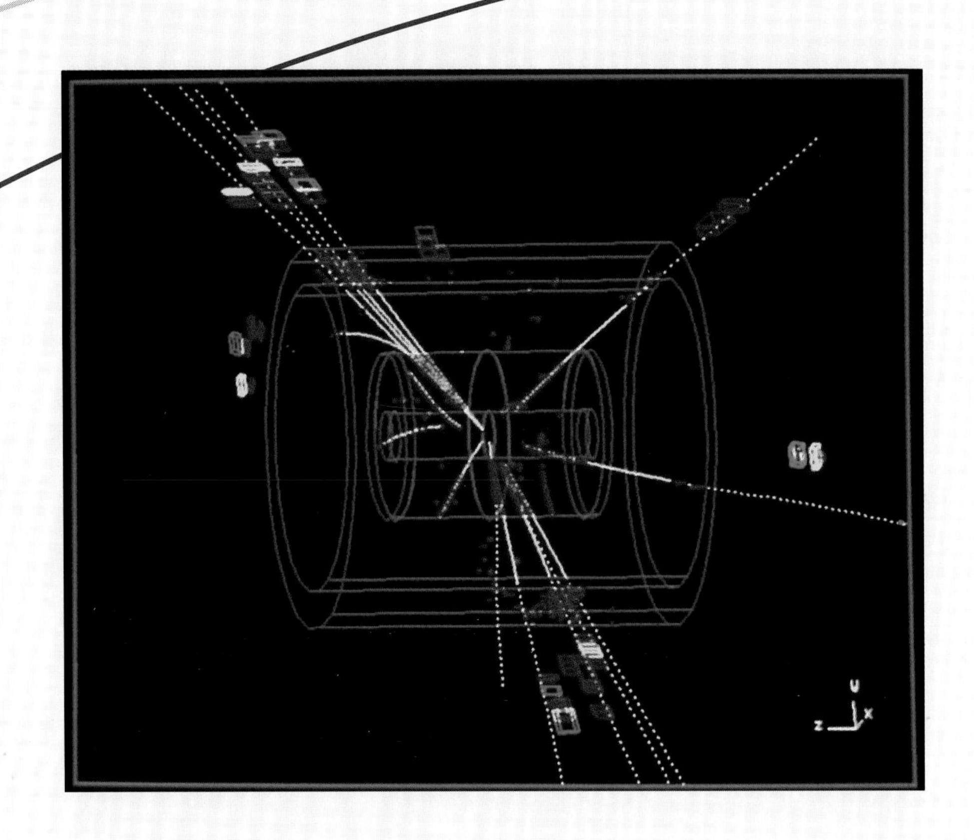

A high-energy proton and a high-energy antiproton collide in the CERN Large Hadron Collider, producing short-lived jets of energetic, decaying particles. Physicists study the detailed composition of the debris, and search for the possible presence of any invisible weakly interacting particles revealed by "missing" events, in an effort to clarify the ultimate nature of particle interactions at ultrahigh energies. Because comparable energies were attained in the first instants of the big bang, attempts to re-create these conditions will help elucidate how matter was created.

4

The First 10^{-33} Second

The cosmic microwave background takes us back to very near the first instants of the universe, but not close enough for physicists. The story of the expanding universe begins earlier still, when matter was in a state of immensely high density and temperature. Understanding the beginning of the universe is a challenge to our knowledge of physics, both the physics of elementary particles and the physics of space and time, yet physicists have been able to elucidate some of the mysterious concepts that shroud the very beginning of time.

Singularities

The universe began at time zero in a state of infinite density. At least the existence of such a singular state is the expectation from extrapolating the present universe back in time. Of course, the phrase "a state of infinite density" is completely unacceptable as a physical description of the universe, infinities being abhorrent to physicists. An infinitely dense universe would be what is called a "singularity," where the laws of physics, and even space and time, break down. The resolution of the paradox simply is that our theory of gravitation has broken down before reaching this extreme state. The first 10^{-43} second is inaccessible to our current theories; the end of that period, called the Planck instant, represents the beginning of time according to those theories. Nevertheless, to the extent that one accepts Einstein's theory of gravity, singularities are predicted to exist in nature.

At a singularity, matter and energy can be created, and come spewing forth, unheralded and intertwined, into the universe. Normally, a singularity is shrouded by a black hole, but there are expected to be relics from the early universe, mini–black holes, that evaporate to probably leave behind a naked singularity. The evaporation of mini–black holes could be observable if they existed. One might, for example, expect to see a burst of cosmic gamma rays. Such bursts are indeed seen, but there are at least a hundred alternative explanations for them!

The existence of a naked singularity would be bad news for physics. Past and future become hopelessly entangled near a singularity, negating the ideas of causality on which physics is based. Wormholes may appear that are gateways to other universes, past, present, and future. The presence of such gateways would lead to bizarre contradictions: one could greet one's mother before one was even conceived. It seems that a physical universe should not contain such entities, at least in our vicinity. Fortunately, no naked singularity has yet been found. It has been conjectured that such objects lie at the centers of the distant objects called quasars, to provide a source of energetic particles for their huge outpourings of energy, although the quasar energy source is much more plausibly interpreted as a supermassive black hole.

Theory suggests that at least one singularity must exist somewhere in the universe. This is the consequence of a powerful theorem proven by Stephen Hawking and Roger Penrose. The existence of an early dense phase of the universe, required to account for the cosmic blackbody radiation, inevitably leads to the prediction of a much earlier and denser phase that itself may have had a singular origin. One must hope that a more complete theory of gravity will come to the rescue and help evade this prediction of a monster lurking somewhere in the sky.

The Quantum Vacuum

The big bang was an act of creation. Was it a singular, unique event, or is the creation of matter a natural occurrence? And what existed before this event? Was the universe created out of nothing? To better understand how to answer these questions, it is necessary to consider what is meant by nothing, or more precisely, by a vacuum.

The vacuum is governed by Werner Heisenberg's uncertainty principle, a fundamental concept in quantum mechanics. The uncertainty principle states that it is impossible to measure simultaneously both the position and velocity of an elementary particle. There is always an uncertainty in one or the other, expressed in terms of Planck's constant, h. If Δp is the ambiguity in momentum (equal to mass times velocity), Δx is the ambiguity in position, and h is Planck's constant, one has the equation $(\Delta x)(\Delta p) \geq h$. This equation tells us that there is an irreducible uncertainty in the product of momentum and position.

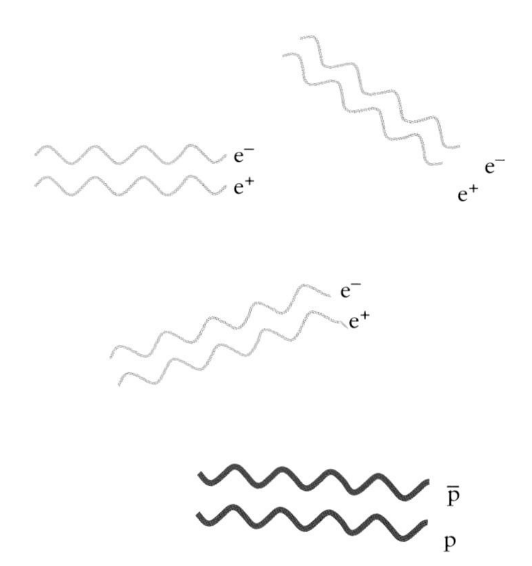

The vacuum is not really empty but, according to the quantum theory, contains short-lived pairs of particles and antiparticles, continuously being created and destroyed.

Suppose, for example, that one were to measure the position of a particle with arbitrarily high precision; then the momentum of the particle could take on any value, however large. When one approaches the quantum limit of measurement where the product of position and momentum uncertainty are nearly h, it becomes possible to improve one measurement only at the expense of the other.

The uncertainty principle has important consequences for the vacuum. It, together with quantum theory, asserts that a vacuum, even the most perfect vacuum devoid of any matter, is not really empty. Rather, the quantum vacuum is a sea of continuously appearing and disappearing particles. However, these particles are "virtual," as opposed to real, particles. Virtual particles are not directly observable. They exist thanks to the uncertainty principle, and the very act of observation would make them real. Energy is "borrowed" from the vacuum to create the particles, and repaid almost instantly. The particles are created in pairs of matter and antimatter particles of opposite charge, to conserve charge, and these pairs annihilate continuously. At any given instant, the vacuum is full of such virtual pairs, which leave their signature behind, by affecting the energy levels of atoms. Their existence has been verified by laboratory experiments that look for such changes in energy levels. For example, hydrogen atoms are jiggled about by virtual particle pairs, and the result is a slight displacement of the atom's lowest energy level. The resulting shift, named after Willis Lamb, has been measured and amounts to a distortion of about one part in a billion.

The influence of virtual particle pairs is also measured in what is known as the Casimir effect. Cool a gas down to a very low temperature, so that all noise and thermal motions are suppressed, and insert two parallel conducting plates. Outside the plates, all possible fluctuations and virtual pairs can exist, but between them, only certain kinds of pairs are present. If quantum motions are represented by waves, then only those pairs are present whose quantum motions expected from the uncertainty principle can allow exactly a whole number of wavelengths to somehow "fit" in the distance between the two plates. There must therefore be fewer waves inside the plates than outside; the result is a net pressure that tends to drive the plates together. At a separation of 10^{-7} centimeter, the quantum pressure is about 10^{-4} of atmospheric pressure, and this tiny effect has been measured in laboratory experiments. The vacuum is modified between the plates, but unmodified outside the plates.

There is a curious consequence of the slightly lower energy density of virtual pairs between plates. Since the pressure in the true vacuum is zero, there must be what in effect amounts to negative energy density in the modified vacuum. The resulting pressure gradient drives the plates together. As the plates come together, the amount of "negative energy" increases. Even fewer wavelengths can now "fit" between the two plates. Correspondingly,

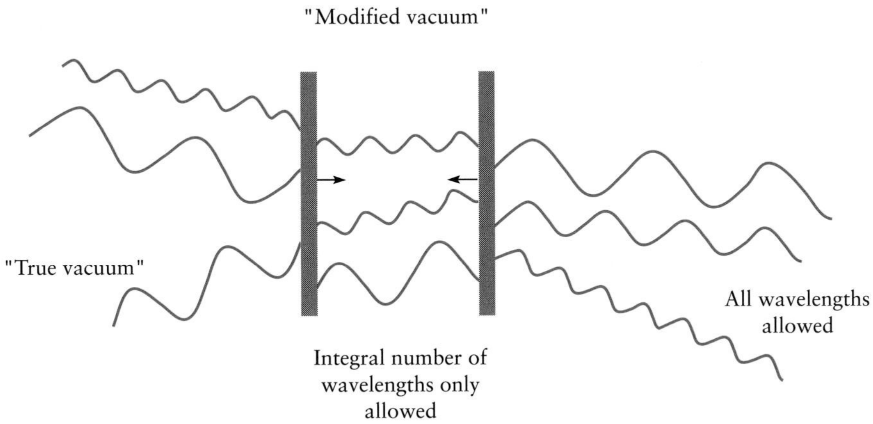

"Modified vacuum"

"True vacuum"

All wavelengths allowed

Integral number of wavelengths only allowed

The Casimir effect demonstrates that the vacuum is not completely empty. Two conducting metal plates approach each other by an infinitesimal amount, because the "modified" vacuum between them exerts slightly less pressure than the surrounding vacuum.

if the plates are moved apart, the amount of "negative energy" decreases. In Einstein's theory of gravity, positive energy density is a source of gravity and exerts an attractive force. Conversely, negative energy density would exert a repulsive force. Hence the Casimir effect illustrates the reality of negative energy density.

These facts tell us that there are regions of negative energy density which can lead to repulsion rather than attraction. This is a quantum phenomenon, and the same effect would be true for gravitational fields if we had a suitable, complete theory of gravity. Despite our lack of such a theory, it seems very likely, as we will soon see, that a similar negative energy density arises naturally for a brief instant early in the universe.

Fundamental Forces

There are four fundamental forces of nature, which account for all known phenomena, ranging from the birth of a galaxy to the birth of a baby. Although the force of gravity is the weakest, it nevertheless controls the motion of the earth around the sun, and of a football thrown in the air. The electromagnetic force is the next strongest force, and it is stronger than the gravitational force by about 40 powers of 10. Consider two electrons separated by distance r. The force of electric repulsion is e^2/r^2, where e is the charge of an electron, and the gravitational attraction is Gm_e^2/r^2, where m_e is the electron mass. The ratio of gravitational to electric forces is therefore Gm_e^2/e^2, or about 10^{-42}.

It would seem that gravity is completely insignificant. However, gravity comes into its own because positive and negative electromagnetic charges usually come in pairs, such as electrons and protons. The positive and negative charges cancel each other, and there is no net electric force on large scales. Moreover, the other two fundamental forces are relatively short-range forces. Hence on sufficiently large scales, gravity inevitably plays a major role. Stars live a delicate balance between the attractive gravitational force and the opposing tendency of pressure exerted by the hot interior, which is a consequence of the electromagnetic force. On much larger scales, including that of galaxies, gravity dominates.

Electromagnetic forces are responsible for all atomic properties. These forces govern the laws of chemistry and the structure of the DNA molecules that determine our genetic characteristics. From muscle power to the explosive force of dynamite, from the fluttering of a butterfly's wings to the charge of a rhinoceros, most familiar natural phenomena lie in the realm of the electromagnetic force. Electromagnetic interactions are controlled by the photon, which is the carrier of the electromagnetic force.

The remaining two fundamental forces are the nuclear forces, which hold the nuclei of atoms together. These are by far the strongest of the fundamental forces. The weak nuclear force manifests itself in the radioactive decays of heavy, unstable nuclei. It is responsible for the emission of the massless particles called neutrinos. It also accounts for the existence and decay of neutrons, particles that, outside nuclei, are unstable. The weak interaction is controlled by a particle that carries the weak force, analogous to the photon. This particle, called the W boson, was discovered in 1985 and is about 80 times the mass of the proton. There are actually two W bosons, carrying either a positive or a negative charge, and a third, related, neutral particle, the Z boson. These particles account for the large difference between the strength of the electromagnetic force and the strength of the weak nuclear force, at low energies.

The strong nuclear force directly controls the binding together of protons inside an atomic nucleus. It acts through the formation and destruction of short-lived particles called mesons. Were it not for mesons, nuclei would fly apart under the electromagnetic repulsion between the protons. These mesons were first discovered in cosmic rays by high-flying balloon-borne experiments, and later produced in particle accelerators.

The proton itself is not a fundamental particle. It has been found to consist of a basic constituent called a quark. A quark is fractionally charged, and a proton consists of three quarks of charges 2/3, 2/3, and −1/3. The quarks are held together inside the proton by massless particles called gluons that are carriers of the strong nuclear force. These particles play a role similar to photons for the electromagnetic force and to W and Z bosons for the weak force. One important difference from photons is that gluons carry a quantum property, analogous to charge, called color. The color force is another way of describing the strong nuclear interaction (quantum color has no relation whatever to the color of light!).

Three of the fundamental forces, the electromagnetic and the nuclear forces, are described by quantum theory, whereas the fourth force, that of gravitation, is described by the theory of general relativity. What is lacking is a quantum theory of gravity that would combine all of the fundamental forces into a single unified description. Such a theory is necessary if we are ever to understand the physics of the beginning of the universe.

Pair Creation

When a sufficiently strong electric field is applied, the virtual pairs in the vacuum can be ripped apart. Real pairs are produced, since once pairs are separated, annihilation is mostly suppressed. To create a pair consisting of

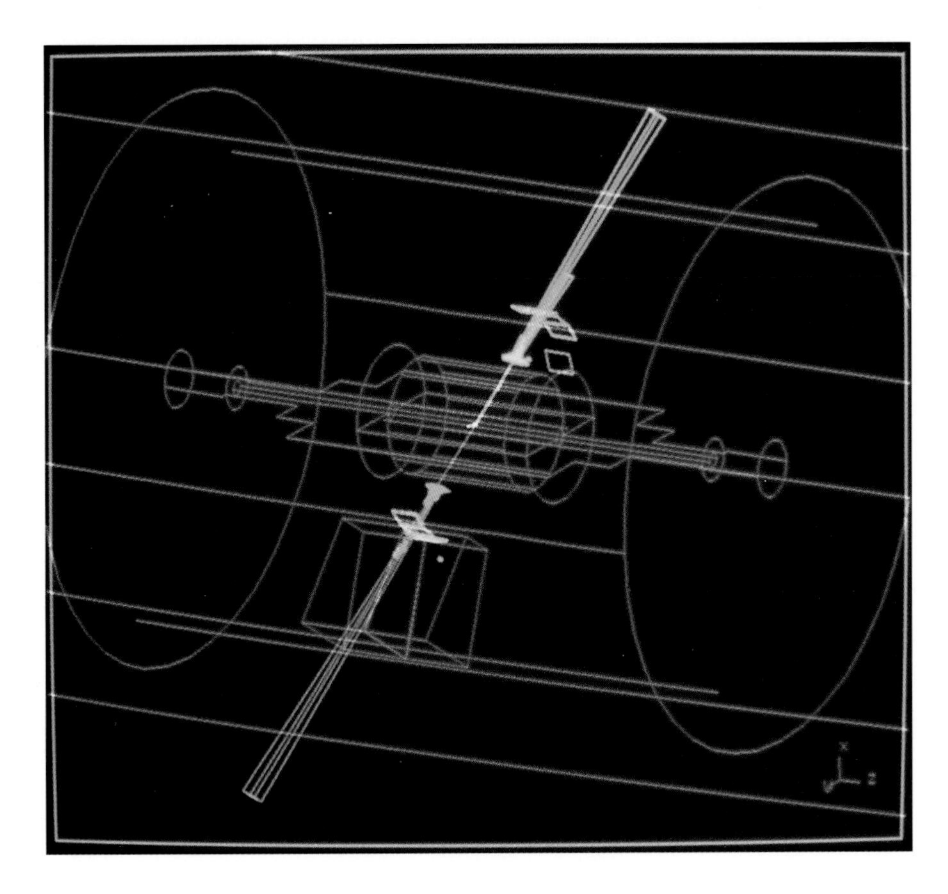

An electron-positron pair is captured within one of the detectors of the Large Electron Positron (LEP) collider at the European Center for Particle Physics (CERN). The pair originated from the decay of a Z boson, which was in turn produced through the annihilation of an electron and positron within the LEP beams.

an electron and its positively charged partner, the positron, requires an electric field or magnetic field whose energy density exceeds the energy required to re-create the mass of the particles. That energy can be calculated using Einstein's famous formula $E = mc^2$; in the case of the electron-positron pair, it is 1 million electron volts. Pair production out of the vacuum has been observed in the laboratory, for example, in particle accelerators.

The early universe is an ideal, although exceedingly remote, laboratory for studying particle pair creation. It is convenient in describing the temperature of the early universe to switch from units of degrees to units of energy, usually electron volts (eV). The reason is that particle masses are customarily expressed in terms of their equivalent energy, according to the $E = mc^2$ formula. For example, an electron has a mass of 0.5 million electron volts (MeV). Both in particle accelerators and in the early universe, electron-positron pairs appear spontaneously at a temperature above 1 MeV, which is equivalent to a temperature of 10 billion degrees Kelvin. This temperature occurs when the universe is one second old. As the universe ages and cools below this threshold temperature, the pairs spontaneously stop forming, those already present annihilating into radiation. The electrons in the universe today were never paired up with positrons, as will be described in the following chapter.

The heavier proton has a mass of 1 billion electron volts (GeV). Proton-antiproton pairs are produced at temperatures above 2 GeV, a mere one-millionth of a second after the big bang. At higher temperatures, more exotic particles appear, particles that ordinarily are very short-lived. The very early universe is teeming with mesons and antimesons, with quarks and antiquarks. In fact, at a temperature of 100 GeV, only a ten-billionth of a second after the beginning, there are more than 100 species of particles and antiparticles existing in thermal equilibrium with the radiation field.

Particles and Waves

Particles could have been created in the early universe from the action of both strong electromagnetic fields, or energetic radiation, and strong gravitational fields. Intense fields of both types are present at the beginning of the universe. To understand how particles are created in an intense gravity field, we first note that elementary particles also act as waves. This is in fact the underlying principle of the electron microscope: instead of using light waves, the instrument makes use of electrons behaving as waves. Particles with large momentum, and corresponding large uncertainty in momentum Δp, are used to measure positions of particles with very small uncertainty in position Δx, compared to what one could measure with ordinary light.

Electrons behave like waves rather than particles on a scale of about one ten-billionth of a centimeter. The wavelength associated with an electron of mass m moving at the speed of light is given by the uncertainty principle, namely $\Delta x = h/mc$. This wavelength is called the Compton wavelength. Waves that are equivalent to particles can be generated in the very high time-varying gravity fields that arise in certain models of the big bang's first moments. In the very early universe, the energies may have been so high that particles were spontaneously created out of the vacuum.

One can even imagine that the entire universe was created out of nothing, if particle creation was effective sufficiently early. Indeed, one could wait an infinite time in a primordial vacuum. Eventually, it has been speculated, a spontaneous quantum fluctuation might give rise to our universe.

Mini–black Holes

The quantum phenomenon of pair production can be stimulated in the present-day universe by the intense gravitational field near a black hole. Long before Einstein, in 1799, Pierre de Laplace noted that if the escape velocity from a material body exceeded the speed of light there would be no means of escape. Such bodies are black holes, a term coined by physicist John Archibald Wheeler. In a black hole, so much mass is present in such a small region of space that the gravity field is powerful enough to trap even light.

Black holes in the present-day universe are a consequence of the deaths of stars more massive than about 50 $M_{\odot}$, and even more massive black holes, containing from 10^6 to 10^9 $M_{\odot}$, form in galactic nuclei. These extremely massive black holes are conjectured to be the energy sources that power quasars. Black holes of mass much below that of the sun are called mini–black holes. Were such black holes to exist, they must have formed near the beginning of the universe, since astrophysical processes that today result in the collapse of gas clouds can only form black holes of mass much larger than the sun.

What is the velocity needed to escape from a body of mass M and radius R? A particle falling through a gravity field, like a rock dropped from a cliff, gains energy; it falls faster and faster. To escape from the gravity field, a particle's initial energy of motion, called its kinetic energy, has to be greater than the energy it would gain by falling through the gravity field, called its gravitational potential energy. To obtain the escape velocity, we set the gravitational potential energy, GM/R, equal to the kinetic energy $v^2/2$ of a particle of unit mass. If the kinetic energy dominates over gravitational potential energy, the particle escapes the gravity field. The escape velocity is then equal to $(2GM/R)^{1/2}$. If the escape velocity equals the speed of light, it follows that the radius of a black hole would be $R_{bh} = 2GM/c^2$. This is the radius within which there is no escape. It is known as the Schwarzschild radius in the case of a spherical black hole. For a body with the mass of the sun, the Schwarzschild radius is equal to about 1 mile, or less than one-hundred-thousandth the radius of the sun.

The powerful gravitational fields of black holes cause the vacuum to effectively self-destruct: a shower of particles is produced in the vicinity of the event horizon, the surface of no return on which light rays are trapped. The vacuum produces pairs, of which one particle escapes while its partner falls into the black hole. This effect drains mass from black holes, leading Stephen Hawking to predict that black holes may not last forever: in essence, black holes can evaporate. The more curved is the boundary of the black hole, the stronger is the effect, so the shower of particles drains mass most effectively from sufficiently small black holes. Since the radius of a black hole is proportional to its mass, the smallest black holes evaporate rapidly. Indeed, the actual evaporation time would be less than a millisecond for a sufficiently small black hole. Much larger black holes should be unaffected by this process.

The Planck Scale

If mini–black holes exist, they may represent the state of matter during, or very soon after, the initial singularity. There is no theory of quantum gravity, however, to describe such extreme conditions, and we can only speculate about these possible relics. However, we can sensibly describe a way of approaching this extreme situation; the approach requires that our theory of gravity be drastically modified to take account of quantum effects. The ultimate state of matter is attained when matter is compressed to where an elementary particle, such as a proton, is on the verge of becoming a black hole. At this stage, the Compton wavelength of the particle equals its Schwarzschild radius. If we could construct such a particle theoretically, we would have the state of matter during the moment of its earliest existence, when the density of the universe was at its greatest value.

The modern concept of particles assumes that they have pointlike interactions, so one can talk of very small scales. To construct the ultimate particle, we choose the smallest length scale from the quantum theory, namely the Compton wavelength (h/mc), and we set it equal to the smallest length scale for a specified mass from gravitation theory, namely the Schwarzschild radius ($2Gm/c^2$). This comparison defines a mass scale given by $m = (hc/2G)^{1/2}$, which is known as the Planck mass (m_{pl}) and is equal to about 10^{-5} gram. The Planck mass is equivalent to an elementary particle of the smallest scale and highest attainable density at which the classical theory of gravity applies. Our cosmology in effect commences at this scale, which amounts to any and all of the following: an energy scale of 10^{19} GeV, a length scale $l_{pl} = 2Gm_{pl}/c^2$ (or $h/m_{pl}c$) of about 10^{-33} centimeter, or a timescale Gm_{pl}/c^3 or $(Gh/c^5)^{1/2}$ of about 10^{-43} second. At that tiny fraction of a second, the density of matter has attained the incredible value of m_{pl}/l_{pl}^3, or 10^{94} grams per cubic centimeter.

A theory of quantum gravity is needed to reach higher densities, shorter scales, or earlier instants of cosmic time. Such theories of quantum gravity are presently beyond our reach, although they are being developed. One version of such a theory is called the superstring. It postulates a space of either 10 or 26 dimensions that represents all possible combinations of elementary particles and their interactions. This space exists only at the Planck scale. Only when the space is compactified, or the extra dimensions collapsed, to produce our present three dimensions of space plus the fourth dimension of time could the present universe be said to have really begun to exist. This instant is the Planck time, when the cosmic clock first started to tick. Prior to this first instant, one can depict the universe as being of indefinite duration, and as consisting of a quantum foam of Planck-mass black holes continuously being created and annihilated. A chance fluctuation, inevitable sooner or later, is believed to have triggered the expansion that led to the big bang.

Grand Unification

The theory of the Planck instant is inevitably pure speculation. There is no connection between quantum gravity and the real world of elementary particles. Superstringers dream of ultimately finding the "Theory of Everything," from which all known properties of particles may be deduced. But they are far from even penetrating the quantum barrier between Planck-scale superstrings and any observable phenomena.

However, as the universe cools below the Planck scale, it enters a regime where the physics is indeed tractable. This is known as the epoch of grand unification, where the fundamental forces, other than gravity, are identical. At very high energy, the electromagnetic force strengthens as the two nuclear forces weaken, and the three forces become indistinguishable. The electromagnetic force increases for the following reason. One can think of the vacuum as a sea of virtual pairs of particles and antiparticles. These become slightly separated by the presence of a charged particle, and hence a proton tends to be surrounded by a cloud of virtual electrons. At high energy, any passing charged particle can penetrate this cloud. It thereby experiences a stronger interaction with the proton than it would were it at lower energy.

In the vacuum, a virtual electron A (top) is surrounded by a cloud of virtual particles of the opposite charge, and a virtual quark B (bottom) is surrounded by virtual quarks and gluons of the same color as B.

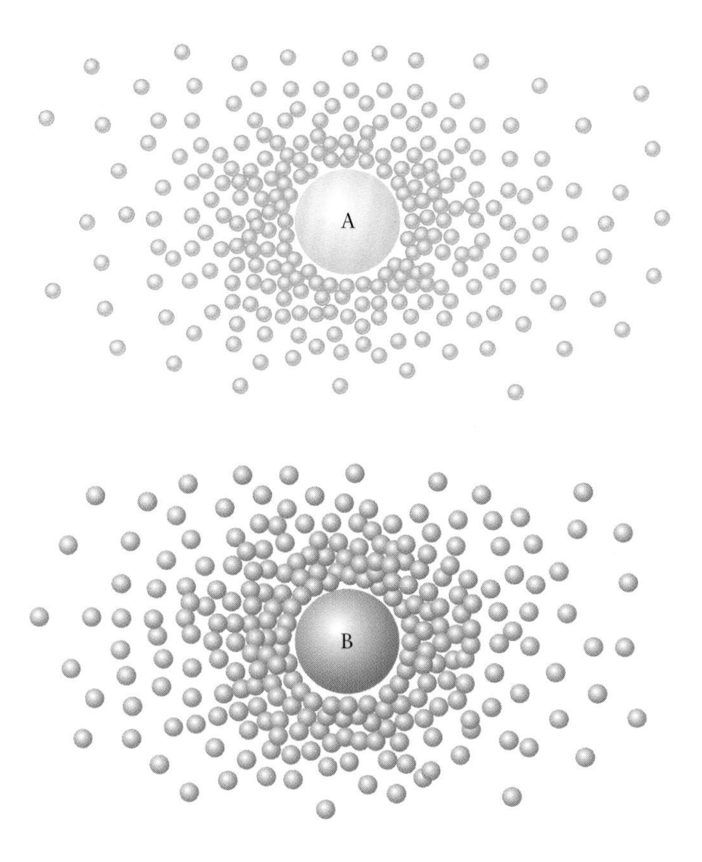

Precisely the opposite occurs with a quark. It tends to be surrounded by virtual quarks and gluons with the same color. So the passing quark at high energy penetrates further, sees less color, and experiences a *weaker* force. This means that one can never expect to find free quarks in nature, since the force between a pair of quarks gets stronger and stronger as they are separated.

The highest energies imaginable are attained in the earliest instants of the big bang. Ultimately, above about 10^{15} GeV, or 10^{15} times the energy that corresponds to the rest mass of a proton, the nuclear and electromagnetic forces attain the same strength. At this energy, we say that grand unification of these three fundamental forces has occurred. Gravity is not unified at this point: it is still the weakest force. One has to go all the way back to the Planck scale, which in energy units means about 10^{19} GeV, before all four fundamental forces are unified. We are still lacking a theory of quantum gravity that describes this ultimate state of unification of the four fundamental forces of matter.

Although the physics of the first instants of the big bang cannot be tested through scientific experiment, cosmologists eventually recognized

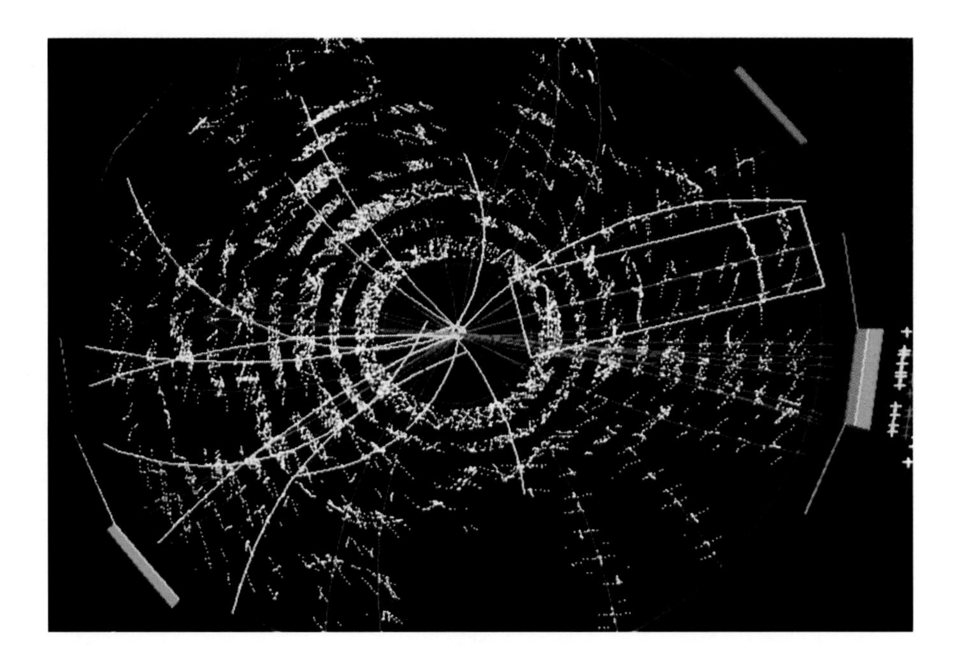

The colliding-beam detectors at Fermilab recorded this proton-antiproton collision. A top quark is thought to have been produced in the collision debris. The decay products include a muon (in the yellow rectangle).

that, very soon afterward, there is a phase that is, however indirectly, experimentally accessible, via elementary particle theory. This realization spawned the first radically new development in cosmology since the 1920s. The inflationary universe was born in 1981.

Symmetry Breaking and Inflation

One consequence of grand unification is that the electromagnetic and nuclear forces are indistinguishable. In the language of particle physics, we say there is symmetry between the forces. As the universe expands and cools down below the critical temperature for unification, this symmetry must be broken: the force balance in the universe changes to one in which one type of force, the strong nuclear force, dominates over the other forces of nature. According to one popular viewpoint, symmetry breaking occurs abruptly and rather violently. Energy is released during the transition in much the way that the energy known as latent heat is liberated when water is frozen. This is the same energy that would have to be supplied in order to melt ice. Indeed, fish survive under the surface of frozen ponds by taking advantage of the latent heat liberated when ice forms, as well as the insulation provided by the ice layer.

The release of energy at symmetry breaking corresponds to a change of the quantum vacuum: the quantum vacuum of the symmetric state is the so-called false vacuum, equivalent to the vacuum of "negative energy" seen

The theory of the fundamental interactions of elementary particles suggests that, at sufficiently high energy, the interactions should become indistinguishable—that is, be unified. The unification of the weak and electromagnetic forces has been measured in particle colliders; experiments with these machines also hint that at much higher energy the strong nuclear force should eventually become unified with the two weaker forces. Quantum gravity theory predicts that ultimately the gravitational force should also be unified at the Planck scale.

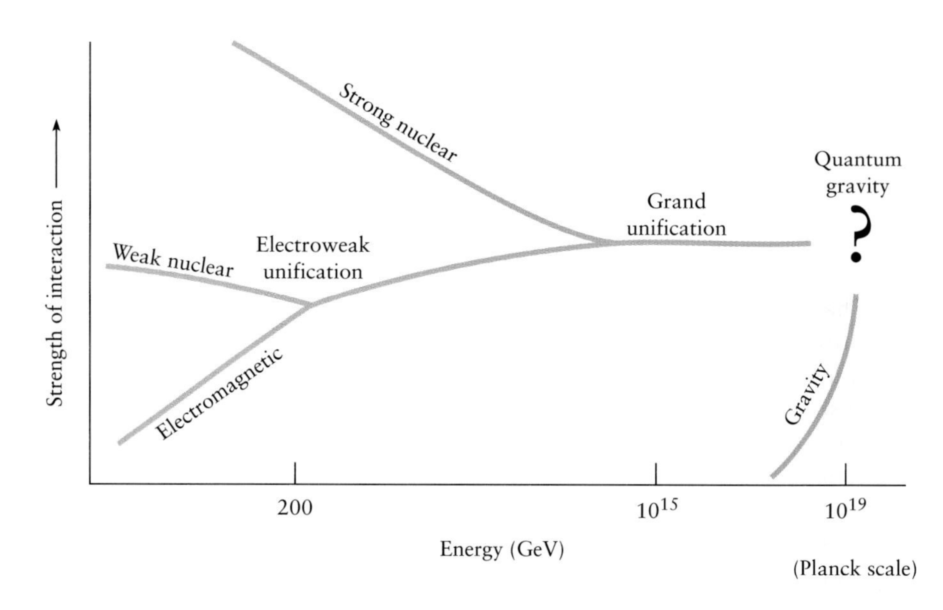

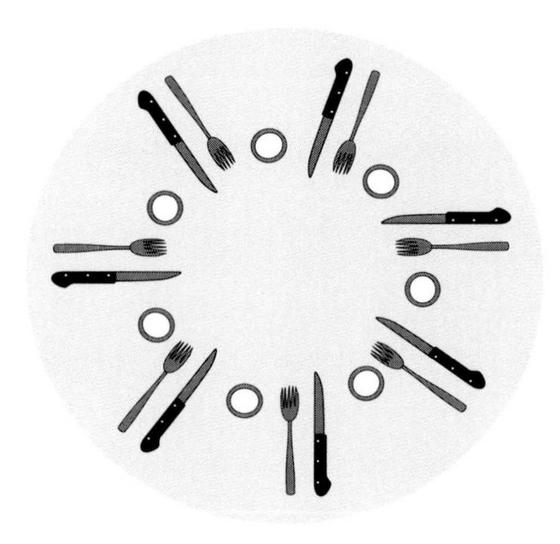

Is your glass on the left or the right? The dinner settings are symmetrical, with left indistinguishable from right, until the first diner chooses a glass and the symmetry is broken.

in the Casimir effect. At the time of symmetry breaking, it changes to the true vacuum of the present epoch, which exists along with the very disparate forces of the universe today. The energy release accelerates the expansion rate of the universe. Indeed, the universe inflates at an exponentially increasing rate until the transition to the true vacuum is completed.

Normally, the expansion rate of the universe is balanced by the gravity associated with the density of matter. As the universe expands, the density drops, and the expansion rate drops in response. However, during the symmetry-breaking transition, the energy input from the false vacuum temporarily holds the expansion rate of the universe steady. This is a subtle effect, and arises as a consequence of the fact that, so Einstein tells us, energy is equivalent to mass, at least from the perspective of gravity. The energy density of the universe is, for a brief moment, constant, and consequently so is the expansion rate. For the universe to be able to expand at a constant rate, and not an ever-decreasing rate as in an explosion, the separation between any two particles must increase exponentially rapidly. This exponential increase of scale is what we mean by inflation.

A physical way to understand inflation is to realize that the false vacuum lacks certain kinds of virtual pairs that are possible in the true vacuum, as was measured in the Casimir effect, and as a consequence, it may be thought of as having constant *negative* pressure. Indeed, the Casimir effect measures a negative pressure that causes the two plates to come together, the vacuum between being depleted of some kinds of virtual pairs relative to the vacuum outside. The true vacuum of course is pressure-free.

At first, the negative pressure of the false vacuum has no effect on the expansion rate; another, stronger and positive, pressure is dominant: the normal thermal pressure of a very hot gas of photons. The early universe is pervaded by energetic photons, and this sea of radiation both has density, indeed giving the dominant contribution to the total density during the first thousand years of the big bang, and exerts pressure on the particles that are present. When the temperature drops to a value such that the radiation pressure is comparable to the negative pressure of the false vacuum, a negative pressure force manifests itself that causes the exponential expansion. Unlike the attractive force exerted by gravity, negative pressure is repulsive. This pressure is not to be confused with the force exerted by differences in pressure, which are irrelevant on a cosmological scale. It is this temporary repulsive boost that is responsible for inflation.

Inflation was initiated only 10^{-35} second after the big bang. An exponential rate of expansion meant that the universe subsequently doubled in scale every 10^{-35} second. Once the transition to the true vacuum was completed, the expansion proceeded normally with ever-decreasing energy density, and decreasing but positive pressure. When inflation ceased, the slow and continually decreasing expansion rate recommenced as the density of matter dropped. The weakened gravity force continues to drive an ever slower expansion rate.

Although inflation began 10^{-35} second after the big bang, it was over by 10^{-33} second. From this point on, the universe resumed its progressive expansion, taking more and more time for each successive doubling of scale. At the instant of inflation, the time required for expansion by a factor of 2 was 10^{-35} second. Today the time required for expansion by a factor of 2 is 3×10^{17} seconds, or 10 billion years.

In the symmetrical state (left), there is a unique point of minimum energy. As the universe cools down, there is no longer a preferred minimum of energy, but rather (right) many possibilities exist. At this time the universe is in an asymmetrical state, when, for example, there may be far more matter than antimatter.

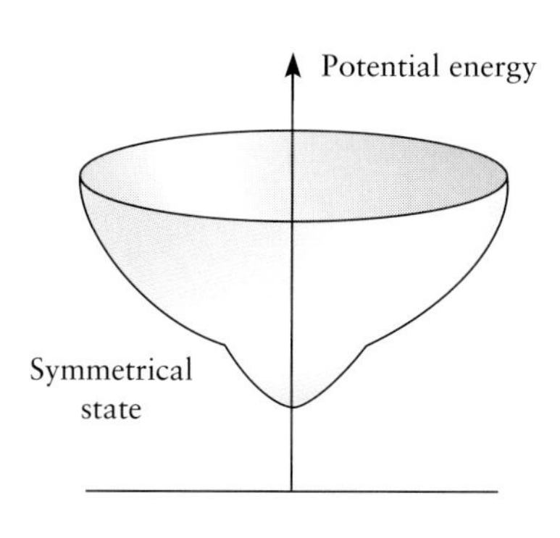

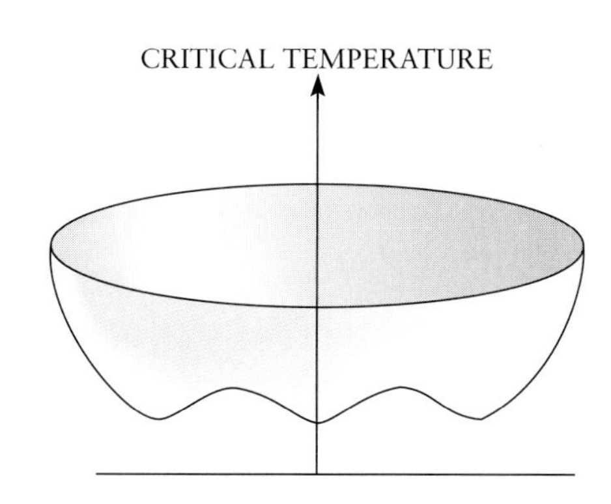

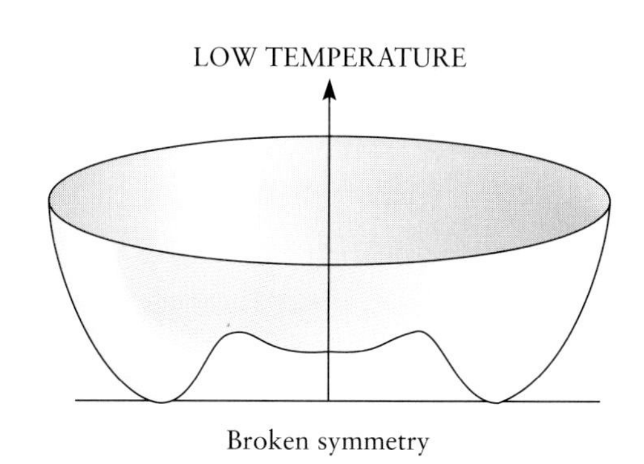

Why Is the Universe So Uniform?

When a young physicist, Alan Guth, conjured up the inflationary cosmology in 1981, one of his primary concerns was to find a rational explanation for why the universe is observed to be so uniform, other than the tautology that it is the way it is because of the way it was. Inflation solves this problem because before inflation has finished, the distance light has traveled increases exponentially to a scale at least as large as our present universe.

There is no recourse to magic in this. Imagine the universe doubling in size every instant τ. If it begins on a scale equal to the distance light has had time to travel, namely $c\tau$, it will soon have a size $2^N c\tau$ after N doublings have occurred during $N\tau$ seconds. At each doubling, the enlarged universe contains matter and radiation throughout its extent. Let N be large, say 100. Then the universe has become vastly larger than $Nc\tau$, which would be the distance traveled by light since the big bang. The doubling in size is like the multiplication of cancer cells. From a negligible beginning in one diseased cell, an entire organ or even a body is soon consumed by virtue of exponential growth.

On reading this description, one may worry that for an object to travel a distance equivalent to the doubling in radius of the universe, it would have to travel faster than light. Yet the size doubling does not involve any material object moving at faster-than-light speed. It is space that is expanding. The apparently superluminal motion is an illusion. It is no more real than the velocity of the pattern traced out on the sky by a rotating searchlight beacon. Photons move along the beam at the speed of light, and there is no physical transverse motion. Analogously, the universe swells exponentially in size because of the local inflation of space. The universe doubles every 10^{-35} second, repeating this doubling 100 times or more.

Prior to inflation, the size of the observable universe was simply equal to the age, 10^{-35} second, multiplied by the speed of light, or 3×10^{-25} centimeter. However, if we take the presently observable universe and follow it backward in time, ignoring inflation, we find that at this same epoch, it would have been immensely compacted, but would still have been about 20 centimeters in extent. Inflation expands the distance spanned by light from 3×10^{-25} centimeter to many meters. It blows up the size of the observable universe to encompass all the matter we can presently see, and even more.

Nevertheless, inflation as originally proposed in 1981 had a fatal flaw. The universe certainly inflated, but it did not stop and in fact continued to inflate forever. Guth's original inflationary theory had no way of resuming the more quiescent expanding phase that we see around us as the big bang. The problem was soon resolved in independent studies by the American

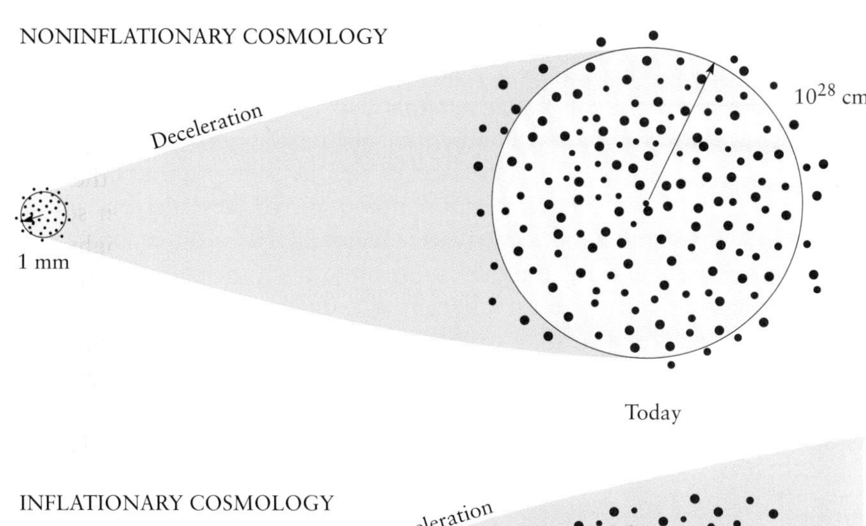

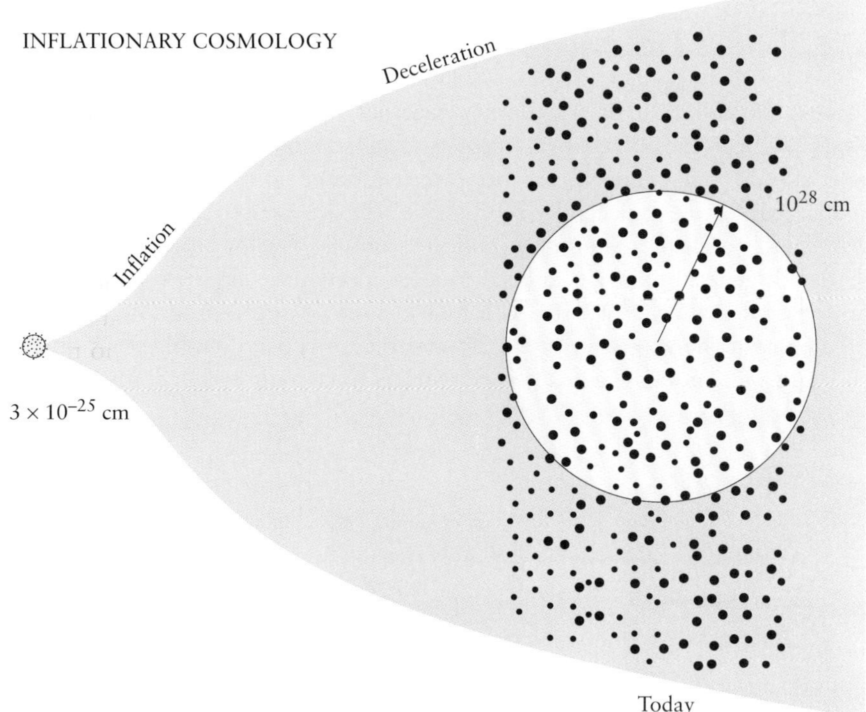

In the old, preinflationary cosmology (top), the observable universe today expanded from a region about 1 mm across at 10^{-35} second after the big bang. Small as this is, it is still vastly larger than the distance traveled by light at that instant, about 3×10^{-25} centimeter. In an inflationary universe (bottom), space expanded during the period of inflation to become larger than the horizon (yellow circle), the distance traveled by light since the big bang. As a consequence, the universe is much larger than we can observe.

physicists Andreas Albrecht and Paul Steinhardt, and by the Russian cosmologist Andrei Linde. They found a way for the universe to gracefully exit from inflation by reheating: the negative energy of the false vacuum was soon swamped by the thermal energy of ordinary matter and radiation, and the inflationary expansion abruptly ceased. The time between the start of inflation at 10^{-35} second after the big bang and its conclusion by 10^{-33} second was enough for a vast increase in size: perhaps 100 e-foldings occurred; that is, the universe expanded by a factor of e^{100} or 3×10^{43} in size. (The symbol e is a mathematical constant approximately equal to 2.71828.) It subsequently increased in size between then and now by another factor of 10^{26}. Had inflation not occurred, the increase in size would only have amounted to this latter factor of 10^{26}, not exactly trivial but not nearly enough to result in uniformity given the tiny scale size of the observable universe prior to inflation. The universe now is far, far larger than the distance to which we can see, some 10 to 20 billion light-years. In fact, the 100 e-foldings mean that its present size is at least equal to 10 million billion light-years!

The Flatness Problem

Just as the inflation of a balloon smoothes out the initial wrinkles, one would expect the inflated universe to appear smooth and uniform. Not only should it be smooth, but because it inflated to a size far larger than is observable, one expects the geometry of the universe to approximate that of a small patch of a very large balloon, namely that of a plane. Einstein's theory of general relativity indeed predicts that space can be curved by the presence of matter. However, to a high degree of approximation, inflation predicts that the geometry of the observable universe resembles that of Euclidean space. Euclidean space is flat space: in two dimensions, it would be an infinite plane.

The inflation of the universe would flatten out space in the same way that the expansion of a sphere flattens its surface.

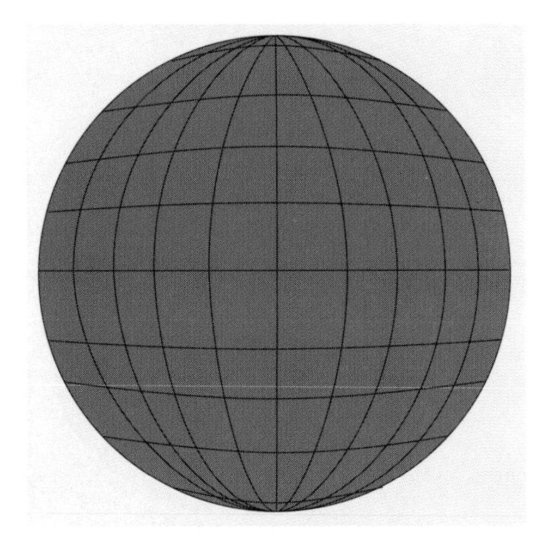
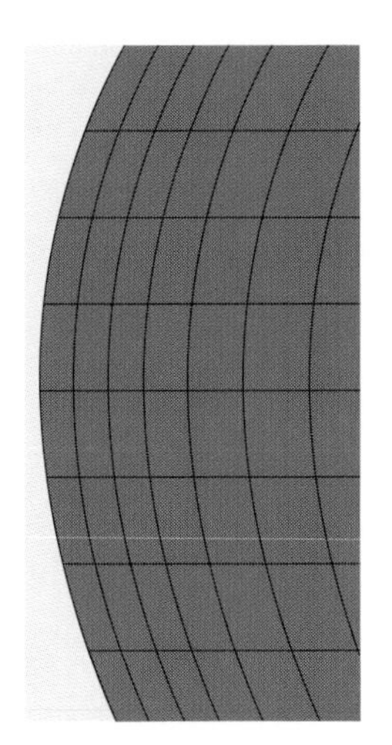
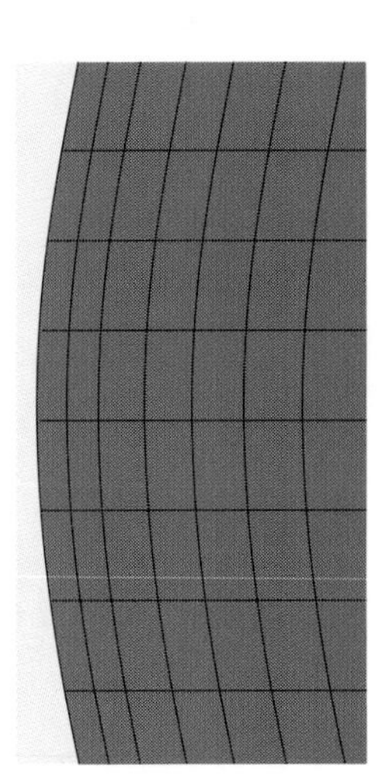
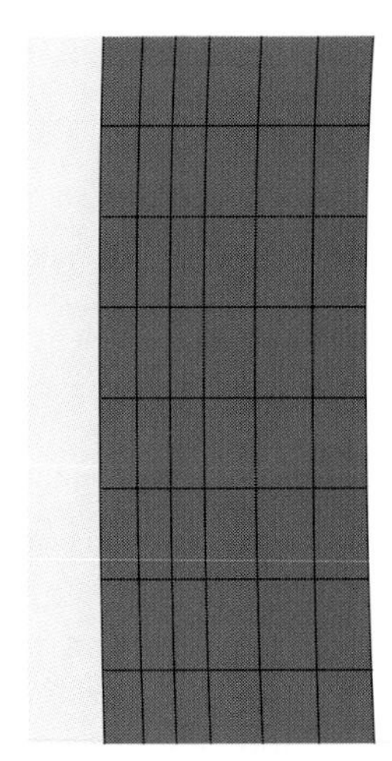

The universe could initially have had a curved geometry. After inflation has done its work, however, the universe, at least over any domain of interest, should be as flat as necessary to explain its present degree of flatness. This is not true for all possible initial curvatures, unfortunately, as there are choices for the initial geometry that fail to inflate sufficiently. Nevertheless, many possible choices for a universe with initial conditions that are inhomogeneous, anisotropic, and curved do end up creating a universe that is uniform, isotropic, vast, and flat. Avoiding the need for precise fine-tuning of the initial conditions of the universe was a major triumph of inflationary cosmology.

The Virgo cluster, at a distance of 15 megaparasecs, contains a thousand galaxies, the brightest of which are clearly visible in this photograph of a region about 1 million parsecs across. Clusters such as the Virgo cluster contain the largest aggregations of ordinary matter (baryons) in the universe, in the form of galaxies and of diffuse, hot, x-ray–emitting gas.

5

The Genesis of Baryons and Helium

Creation is the essence of the big bang theory. We can delegate creation of space and time to the pundits of metaphysics, theology, and even quantum gravity. It is difficult to separate fact from fiction, and faith from fantasy, as any conclusions are far removed from the domain of experimental, testable, or even fully self-consistent physics. A more pressing and relevant issue is the creation of matter. Early-universe cosmology squarely tackles the creation issue by formulating a physical description for the origin of matter, the very stuff that we, the earth, and the stars are made of.

As the universe expands, it inexorably progresses from a state of quantum vacuum to one of plasma soup. Once the temperature drops below 200 GeV, the electromagnetic force becomes distinct from the weak nuclear force. The common particles in the universe at that epoch include photons, neutrinos, electrons, and quarks, the constituents of protons and neutrons. Only after a further phase transition at 300 MeV does the quark soup condense into ordinary composite particles: protons and neutrons.

Our knowledge of the early universe becomes increasingly complete at later and later epochs. As the universe cools below 1 MeV, we finally enter into a regime where there is a direct link with our observations of stars and galaxies. A powerful probe is to observe the abundances of the light elements, for these abundances depend sensitively on conditions during the first seconds of the big bang. Before then, cosmology is entirely speculation. This chapter describes how ordinary matter, from protons and electrons to the light elements, is synthesized in the big bang.

The Matter-Antimatter Puzzle

The fundamental particles of matter are protons, electrons, and neutrons. Protons and neutrons are collectively known as baryons, or ordinary matter.

A CHRONOLOGY OF THE UNIVERSE

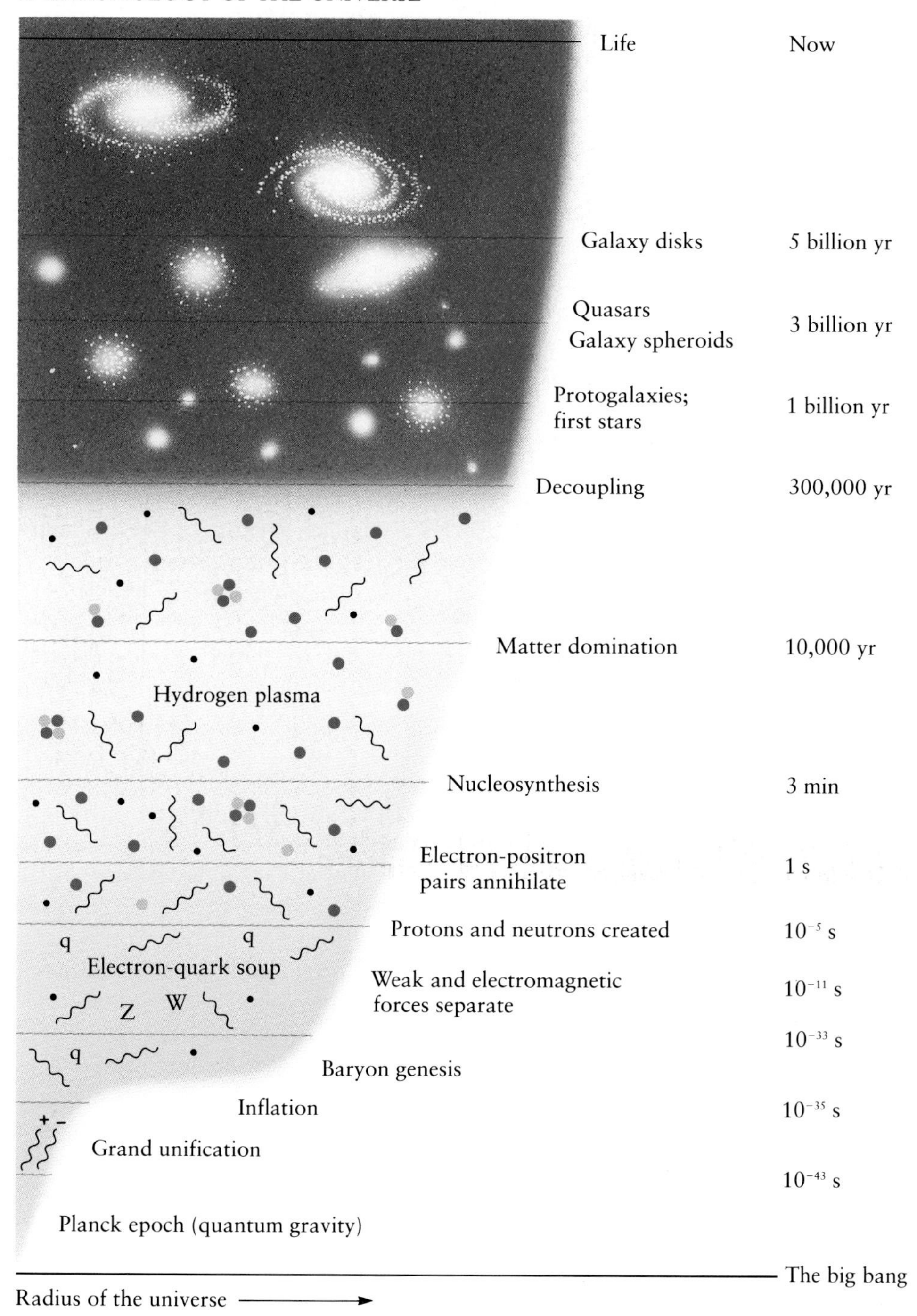

A proton and an electron combine to form a single atom of hydrogen. About 73 percent by mass that we can see in the universe consists of this element. About 25 percent by mass consists of helium, the nuclei of which contain two protons and two neutrons. In our galaxy, about 2 percent by mass consists of heavier elements, notably carbon, oxygen, nitrogen, silicon, and iron. Much of the mass in heavy elements in diffuse interstellar clouds is locked up in tiny, submicrometer-sized grains of mineral-like material, such as olivine and graphite. These grains absorb and scatter starlight, which becomes characteristically reddened and polarized. Because the heavy elements were formed in the hot interiors of stars, primordial gas, out of which the oldest stars formed, was highly depleted in the heavy elements. In addition to the matter, there is diffuse electromagnetic radiation from many distant galaxies; it consists mostly of the cosmic microwave photons and of optical radiation in the form of starlight. Finally, there are neutrinos, both relics of the big bang and those emitted by stars.

Practically all of the matter around us is ordinary matter rather than antimatter. We know this for the following reason. When a proton runs into an antiproton, the two particles annihilate, creating a burst of pure energy. The energy released is twice the rest mass of a proton, or 2 GeV, and comes out in the form of gamma rays. These gamma ray photons are highly penetrating and easily observable. Yet experiments performed above the earth's atmosphere (which fortunately for us absorbs x-rays, gamma rays, and ultraviolet photons) to search for cosmic gamma rays have seen only a few. The conclusion is that at most a millionth of the matter in the universe could consist of antimatter.

Antimatter has been seen directly in space, in the high energy cosmic rays, where about 1 particle in every 10,000 is an antiproton. However, these energetic antiprotons are most probably produced as a consequence of cosmic ray particles smashing into interstellar carbon or oxygen atoms and releasing an occasional antiproton in the shower of debris. Provided the parent cosmic ray has energy in excess of several gigaelectron volts, a proton-antiproton pair will occasionally be produced in such a collision.

There is an intriguing clue that tells us that once, long ago, the universe contained far more antimatter. In fact, there were once nearly equal proportions of matter and antimatter. The photons of the cosmic blackbody radiation today have an energy of only 3×10^{-4} eV, on the average. Once, when the universe was denser, they were much more energetic, but the photons lost energy through the Doppler shift as the universe expanded. For an expansion factor of, say, 10^{10}, their wavelength increases, and their frequency or energy drops, by just this factor. So when the universe was 10^{10} times smaller, or the density of matter was 10^{30} times larger than it is today, their energy was 3 MeV. But 3 MeV exceeds the rest mass of an electron-positron pair, and so such energetic photons can create particle-antiparticle pairs. These are real particles, as opposed to the transient virtual particle pairs created in the quantum vacuum. Go back a little earlier and even proton-antiproton pairs are created. These pairs also annihilate, and there is a balance between creation and destruction, which we refer to as *thermal equilibrium,* that results in nearly equal amounts of radiation, of particles, and of antiparticles. Then the universe expands, the photon energy drops below the threshold for creation, and essentially all the particle pairs annihilate. Today there remains only 3×10^{-10} proton for every photon in the cosmic blackbody radiation.

This leads to a puzzle: we are here today, in a matter-dominated universe. If the early universe had been precisely symmetric between matter and antimatter, we would not be here since everything, or almost everything, would have been annihilated. Calculations show that in a symmetric big bang, only about 10^{-20} proton would survive for every photon in the cosmic blackbody radiation. This would be a disaster! There is a way to resolve the puzzle, however. There is a huge asymmetry today between matter and antimatter; in the past this asymmetry, though much smaller, was nevertheless finite. How big was the initial asymmetry? For every pair of photons in the blackbody radiation today, a particle-antiparticle pair would have been produced in the early hot phase of the universe. So let us count the number of photons at 3 degrees Kelvin, on the average, for every proton we see in the universe. We find that there are about a billion photons for every proton, so that the initial asymmetry was only about one part in a billion. The particle pairs annihilated, leaving behind photons plus the tiny excess of matter over antimatter that survives as pure matter after all else annihilates. It is this tiny excess that accounts for our existence. Where indeed do we come from?

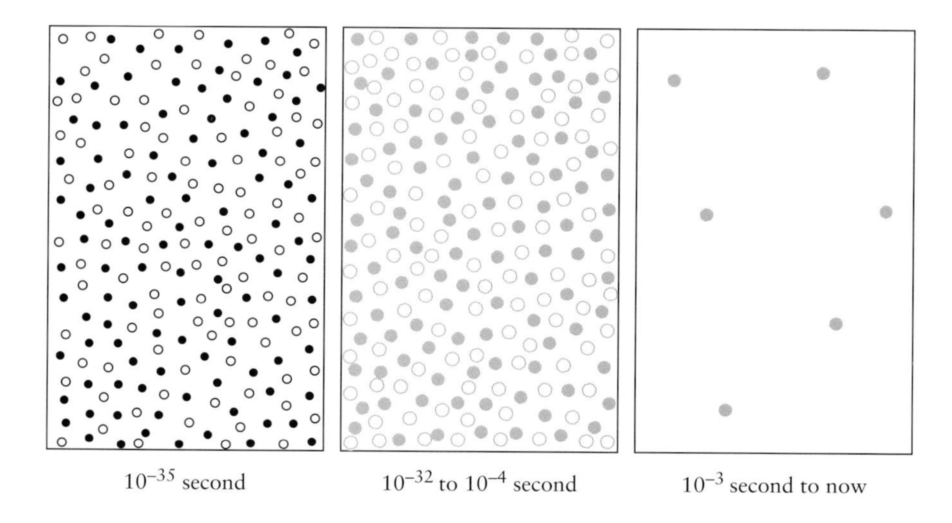

At 10^{-35} second after the big bang (left), there would have been equal quantities of particles (dots) and their antiparticles (open dots). By 10^{-4} second (center), almost all the particles have decayed, but a slight imbalance of protons over antiprotons remains. Pairs of particles and antiparticles annihilate each other, leaving only the excess protons behind (right).

Neutrino

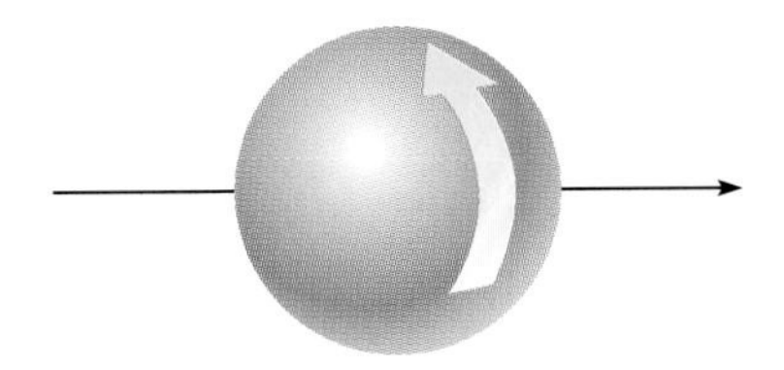

Spins to the left.

Antineutrino

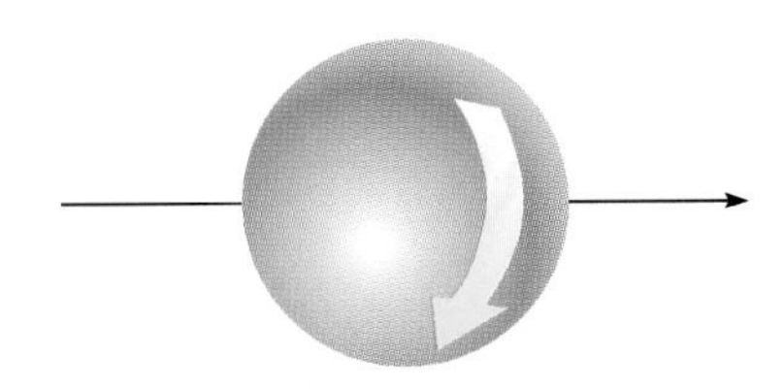

Spins to the right.

All elementary particles have spin. For example, neutrinos spin to the left in the direction of motion, while antineutrinos spin to the right.

Generation of the Baryon Number

The ratio of the number of baryons, meaning particles such as protons, to the number of photons in the cosmic blackbody radiation is called the baryon number of the universe. As long as photons or baryons are not created or destroyed, as is approximately true over the recent universe, the baryon number is constant. Of course, photons are created in stars. However, the photons of starlight amount to only a small fraction, less than 1 percent, of those in the cosmic blackbody radiation.

Physicists do not like the assumption that a number as small as 3×10^{-10} for the baryon number of the universe must be an initial condition for the very early universe. There are three natural numbers: 0, 1, and infinity. Anything else, so the physicist argues, needs an explanation. In any case, such an assumption is not necessary: we saw earlier that the grand unification state is likely to be one of perfect symmetry, and in that situation, the baryon number of the universe should have been produced after the symmetry-breaking transition. There is indeed a simple theory to explain how.

At the epoch of the grand unification, heavy particles (such as protons) were treated on an equal footing with light particles (such as electrons or photons). This meant that at the grand unification epoch, some 10^{-35} second after the big bang, baryon number was not necessarily a conserved quantity. Protons could decay, and be created, without necessarily being accompanied by the decay or creation of a matching antiproton. Every process need not have been matched by the reverse process. Exotic particles are present at the end of the grand unification epoch that could decay into protons without matching antiprotons. When grand unification symmetry is broken, irreversible decays become possible. A few of the exotic particles are left behind: they decay and generate a slight excess of protons (or of quarks, to be more precise) over antiprotons (or antiquarks). Calculating exactly how many quarks remain is difficult, but the theory has been shown to provide a plausible explanation of the baryon number of the universe.

The late Russian physicist and dissident Andrei Sakharov pointed out that three fundamental conditions are necessary for baryon genesis. First, the universe must be out of equilibrium. This is a natural consequence of the expansion of the universe. Second, the conservation of baryon number must be violated. This, too, is inevitable during the epoch of the grand unification of the fundamental forces above an energy of 10^{15} GeV. The third and crucial condition is more subtle: conservation of two properties called "charge conjugation" and "parity" must be violated.

This piece of particle physics jargon is not as intractable as it seems. Charge conjugation is the property that antiparticles undergo identical interactions to their counterpart particles of opposite charge. Thus in terms of their interactions, matter and antimatter would be indistinguishable. A

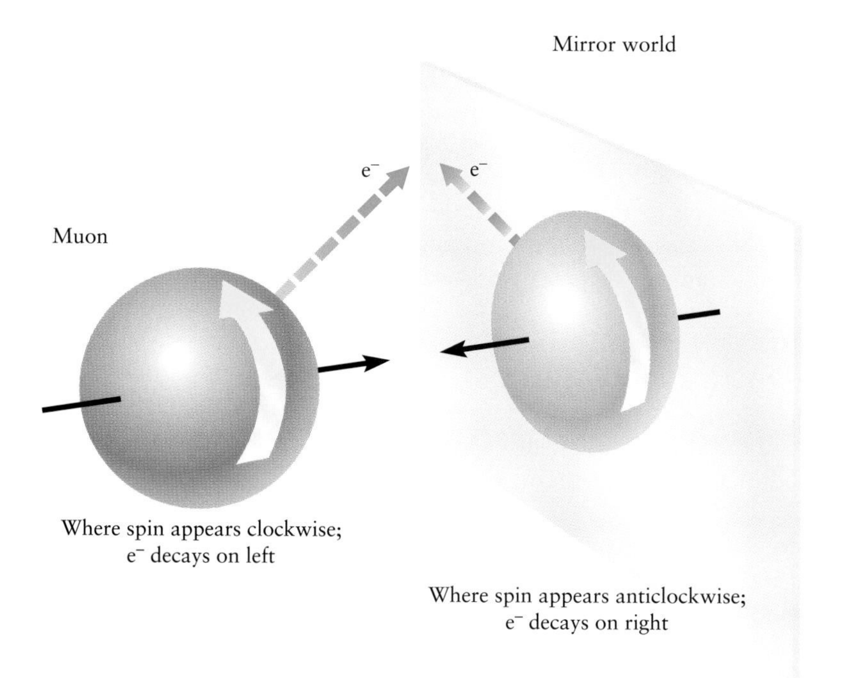

The electron released by a decaying muon violates left-right symmetry (or parity): a real decaying muon releases an electron to the left in the direction of spin, and a true mirror image would emit electrons to the right of the direction of spin. Instead it too emits electrons to the left.

star made of matter would appear identical to a star of antimatter. It is, however, well known from experiments probing weak interactions that particles and antiparticles need not undergo identical interactions: in other words, charge conjugation is violated.

Parity is a measure of the amount of left-handedness or right-handedness; when this quantity is conserved, the mirror reflections of particles are symmetrical. Elementary particles have spin, which, like the turning of a corkscrew, can be to the left or to the right, in the direction of forward motion. The mirror reflections of particles are found to be nonsymmetrical in spin in some elementary particle interactions. Thus parity is also not an invariant. However, particle physicists had fondly hoped that the combination of charge conjugation (C) and parity (P) would provide a quantity, referred to as CP, that is invariant in interactions of elementary particles. In the mid-1960s, physicists James Cronin and Val Fitch found to their surprise, while studying the decays of a short-lived meson called the kaon, that CP violation occurred in about one decay in a thousand. The inference was that CP violation was a natural property of at least some elementary particle interactions.

The corollary for the universe is that during its early evolution, CP was violated. In the complex broth of particle decays, annihilations, and creations, there was a net increase in the number of baryons. It only required

CP violation to occur during one particle decay in a million in order for baryons inevitably to be generated. Precisely when this phenomenon occurred is a contentious issue among theorists. Baryon creation certainly occurred in the sequel of asymmetry-breaking transition that rearranged the matter content of the very early universe. This may have happened 10^{-35} second after the big bang, when grand unification broke down, or much later, at 10^{-10} second after the big bang, when the electromagnetic and weak interactions went their separate ways.

Proton Decay Experiments

There are interesting consequences of the theory of baryon genesis for physics today. One is that slight, exceedingly weak asymmetries (CP violations) should be observed in certain nuclear decays. This effect has been detected for kaons, which are rare particles. The hope is that a new generation of particle accelerators will find evidence for CP violation in other, more common high-energy particle interactions. Greatly magnified, this same effect accounts for the exotic particle decay at grand unification that may have led to baryon genesis. Another implication is that protons should decay today, but very, very slowly. It turns out that in today's low-energy universe, the lifetime of a proton in the simplest grand unification scheme is about 10^{30} years.

There is no immediate danger to us, since the universe is only 10^{10} years old. It is interesting that in the human body, which contains about 100 kilograms or about 10^{29} protons, there could be approximately one "proton" decaying per decade. The resulting gamma ray flux is not enough to produce any risk of cancer. Indeed, one can perform the following thought experiment. Suppose that the proton lifetime were as short as 10^{14} years: we would all be dropping dead of cancer.

Although slow, the rate of proton decay predicted by grand unification is measurable. Take 100 tons of water containing about 10^{32} protons, and, on the average, 100 protons should decay per year. Since the decay of a proton releases a lot of energy (in the form of energetic muons and gamma rays), this effect can be detected, at least in principle. However, one has to shield the experiment from cosmic rays that induce muon showers in the earth's atmosphere.

Several experiments to detect proton decays are under way around the world. Each makes use of thousands of tons of water stored in vats deep underground, surrounded by scintillation counters that search for light flashes from the rare decay events. To date, however, no protons have been observed to decay. From the absence of proton decay, it is inferred that the lifetime of the proton exceeds 10^{32} years. Diamonds could be forever, if this is any consolation. For elementary particle theory, the conclusion, for the moment, is that the simplest models of grand unification are wrong. This presents no problem for particle physicists, however. Physicists are ingenious: they simply concoct slightly more complicated schemes that lead to lifetimes of 10^{32} years or longer.

The Weak Interaction Freeze-out

We can now trace the complete progress of baryon generation, from the epoch of grand unification through the final symmetry breaking. At high temperature, the matter in the big bang consisted only of its most elementary constituents, particles like quarks, the building blocks of protons, and electrons. Above a temperature of 10^{15} GeV, there was no distinction between strongly and weakly interacting particles, such as quarks and neutrinos. The symmetry between the strong, weak, and electromagnetic forces was first broken as the temperature cooled below 10^{15} GeV. As the temperature continued to cool below this energy, the weak nuclear and electromagnetic forces remained indistinguishable until a temperature of about

The W boson was discovered during the October–December 1982 run of the LEP proton-antiproton collider at CERN. Although the particle was not seen directly, its decay products, an electron and a neutrino, were. The collision produced an electron of high transverse energy, indicated by the pink arrow. Some of the expected energy is "missing," a sign that an invisible neutrino was also emitted.

200 GeV was reached. Above this energy, the W and Z bosons exist in great numbers. Recall that these particles unite the weak and electromagnetic interactions much as the photon unifies electricity and magnetism into a single force. As the energy drops below their rest mass, these particles decay and it is energetically unfavorable to re-create them. Consequently, below 200 GeV, the weak nuclear force is quite distinct from the force of electromagnetism, as it indeed is in our low-energy universe.

Immediately after the baryon genesis epoch, there are still none of the recognizable heavy particles like protons and neutrons that make up atomic nuclei. Called hadrons, these are composite particles that can first form only when the temperature drops below about 300 MeV. Before then, one had a sort of quark soup, and afterward, some 10^{-5} second after the big bang, there was a phase transition, much like water freezing to ice, that led to ordinary nucleons (or hadrons) being the preferred state of matter. These are necessarily protons and neutrons since no heavier nuclei would survive the high temperatures.

In addition to hadrons, there are the light particles called leptons, such as electrons, neutrinos, and photons. Together, hadrons and leptons constitute the particles of ordinary matter. Neutrons and neutrinos interact with electrons and protons by means of the weak nuclear interaction. When the temperature of the universe drops below about 1 MeV (or 10 billion degrees Kelvin), the weak interaction rate becomes slower than the rate of expansion of the universe, and the weak interactions are effectively halted. At this stage, about one second of cosmic time has elapsed since the big bang. Once the weak interactions have ceased, the residual number of neutrons (and neutrinos) is fixed. There is approximately one neutron remaining for every 10 protons.

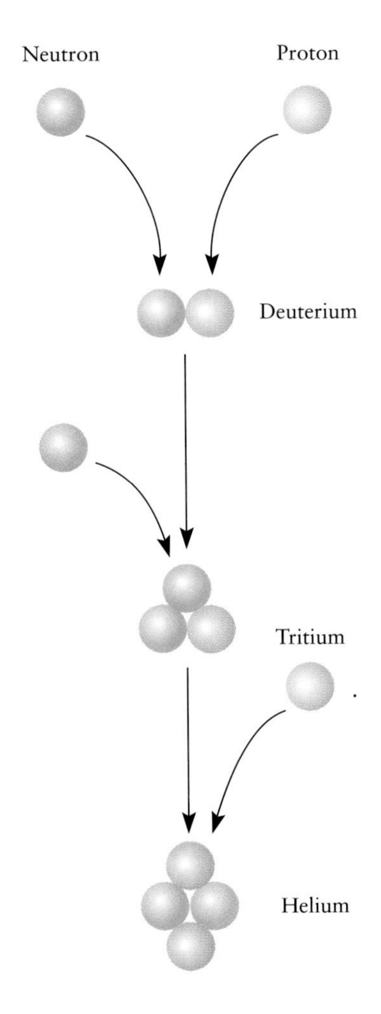

At high temperatures, like those during the first few minutes of the big bang, neutrons are captured by protons and form deuterium, which in turn captures a neutron to form tritium, which finally reacts with a proton to form helium nuclei.

Primordial Nucleosynthesis

The lifetime of a free neutron until it decays is about ten and a half minutes. Most neutrons do not have time to decay, however, since after only about three minutes have elapsed, something else happens to them. They interact with protons to form nuclei of deuterium, or heavy hydrogen. The deuterium soon gains another neutron to form tritium, which in turn rapidly absorbs a proton to form a helium nucleus of mass 4, consisting of two protons and two neutrons. There is no stable element of mass 5, or of mass 8, so additional nucleosynthesis by a helium nucleus combining with a proton or by a helium nucleus combining with another helium nucleus is generally not possible. Trace amounts of one or two heavier elements, most notably lithium (of mass 7), do form.

One finds that practically every neutron ends up in a helium nucleus. The big bang theory therefore predicts that there should be one helium nucleus for every 10 protons, created in the first three minutes of the expansion. We expect the big bang legacy of helium to be slightly augmented by the helium produced in the interior of stars. And indeed, approximately 27 percent by mass of the matter in the young stars is now in the form of helium nuclei; the rest consists of protons. The abundance of helium turns out to be a robust prediction of the big bang theory. It depends only on the fact that the very early universe passed through a high-temperature, high-density phase, much like the center of a star.

Helium Synthesis

Helium is synthesized inside stars by thermonuclear fusion. However, most stars, like the sun, are still burning hydrogen and so have made little helium, and certainly have dispersed none of it. What helium has been synthesized is still deep inside the stellar interior. Yet the universe is indeed observed to contain one helium atom for every 10 or 11 atoms of hydrogen. By mass, it is about 27 percent helium.

The sun, for example, is known to contain this percentage of helium. Because the sun's surface is not hot enough to excite the helium atoms, helium absorption lines are not observed in its spectrum. Nevertheless, helium was discovered in the sun before it was identified on the earth. Bright emission lines from this element were observed in the spectrum of the solar chromosphere, a layer of hot gas that surrounds the solar surface that we normally view from earth. Only when the main body of the sun is blocked by the moon during a solar eclipse does the weakly glowing chromosphere reveal itself.

Direct evidence for the abundance of helium in the sun is also obtained both from solar cosmic rays and from the solar wind. Solar cosmic rays are high-energy particles that are accelerated in the corona of the sun, an even hotter, very extensive envelope of rarefied plasma around that body. Lower-energy particles from the corona itself stream outward from the sun to form the solar wind. For every 10 protons in solar cosmic rays and the solar wind, there is one nucleus of helium. These measurements provide the best estimate of the solar helium abundance. The helium abundance is also about 27 percent for interstellar gas in ionized hydrogen (HII) regions, where emission lines of helium are produced during the recombination of a helium ion with an electron, and for hot stars, where the helium becomes sufficiently excited to produce absorption lines. To within a few percent, the helium abundance is everywhere the same. Even the differences in abundance are attributed to local enrichment by galactic stars.

Proof of a relatively constant helium abundance comes from comparing older, metal-poor stars with younger, metal-rich stars. One finds a helium abundance in metal-poor stars that is only slightly lower, by 2 or 3 percent, than in the sun. There are metal-deficient galaxies, with 10 or even 3 percent of the heavy element abundance of the sun, that contain, again to within 2 or 3 percent, the same helium abundance as the sun. This confirms that most of the helium has not been synthesized in stars along with the heavier elements, such as the metals, but was made before the first stars formed. The coincidence between the predicted helium abundance and the abundance actually observed provides one of the major pieces of evidence for the big bang theory. It justifies our extrapolation from the cosmic epoch of today, 10 billion years after the big bang, back to a mere second after the big bang.

The solar chromosphere appears above the edge of the moon during a total eclipse of the sun. Its red color is created by its emission of light at the wavelength of the hydrogen alpha line.

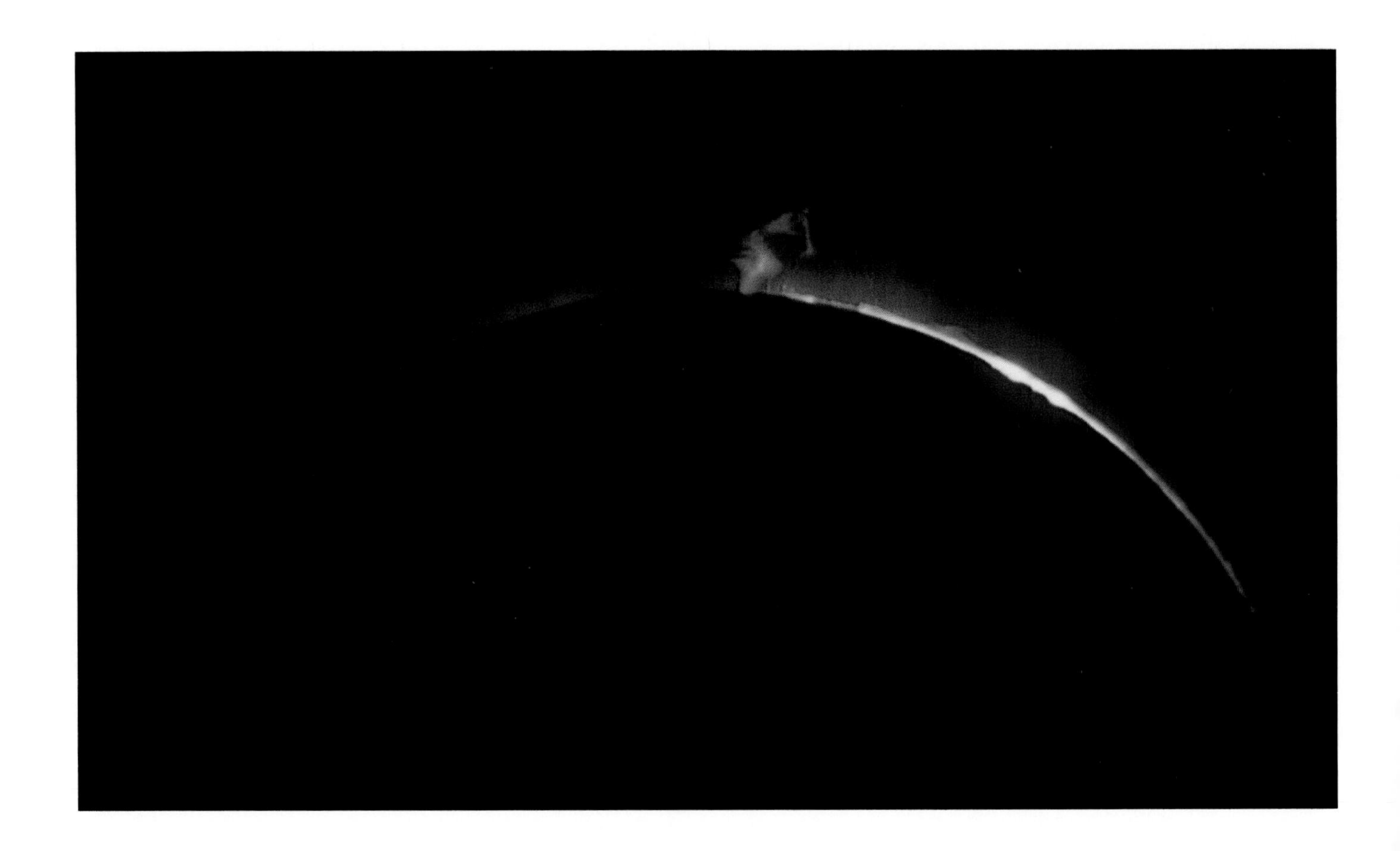

Other important predictions include small amounts of deuterium and lithium. The final abundances of these elements, deuterium especially, depend on the precise value of the present-day density of the baryons that compose ordinary matter. If the density of ordinary matter is high, early nucleosynthesis is efficient. The early universe then makes essentially no deuterium. If the baryon density is low, however, the early universe makes an amount of deuterium that is comparable to what is observed by the astronomers. Precisely what is meant by "high" and "low" densities is central to the future evolution of the universe, as described in the following chapter.

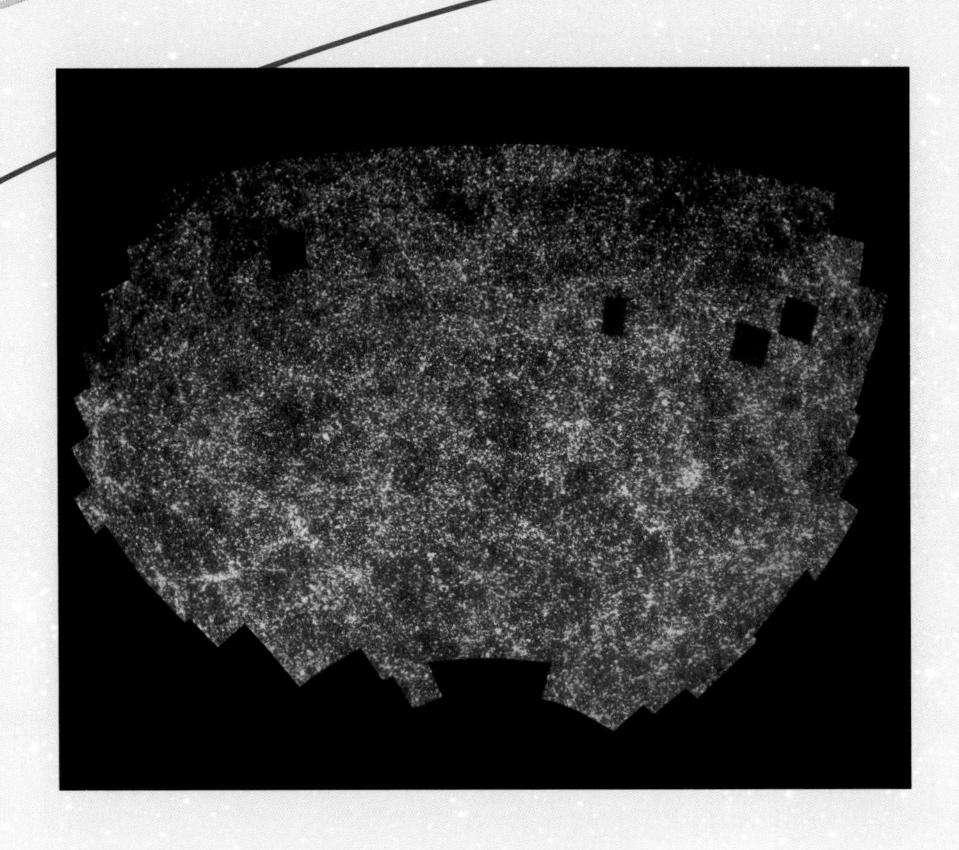

Some two million galaxies are shown in this survey of about 4000 square degrees in the southern sky.

6

Models of the Expanding Universe

The universe is emerging from a hot and dense fireball and rushing toward a bleaker and colder future. Will it continue to expand forever, or will its headlong expansion falter and turn into a phase of collapse that re-creates the fiery past in the distant future? The answers to these questions are hidden in an equation, named after its discoverers, Alexander Friedmann and Georges Lemaître, that describes the theory of the expanding universe. Once we understand what the equation says about the possible futures awaiting the universe, we are able to develop a battery of cosmological tests to see what fate actually awaits us.

The Density of the Universe

One of the key components of the Friedmann-Lemaître equation is the mean density of the universe. It is density that measures the mass of the universe and helps to determine whether the universe is destined to expand forever or some day to collapse. But how can we tell what the density of the universe is?

The average density of the universe is evaluated by taking a sufficiently large volume and measuring the observed mass. To obtain this mass, one counts the total number of luminous galaxies and multiplies by the mass of an average galaxy. The mass of an individual galaxy is assumed to match the average for its given type, such as an elliptical galaxy or a spiral galaxy. That average mass is arrived at by studying nearby galaxies in detail. By one method, for example, the 21-centimeter hydrogen line from orbiting gas clouds is measured, and the rotation rate deduced from the line width, for various distances from the galaxy center. One thereby infers a mass from the requirement that centripetal and gravitational forces must balance.

How large a volume must be surveyed is still a slightly contentious issue. Certainly one needs a volume that is at least 100 megaparsecs across in order to obtain a fair sample of the universe. One finds that, on the average, there is about one luminous galaxy per 100 cubic megaparsecs. Our nearest luminous neighbor, which is the Andromeda galaxy at about only 0.75 megaparsec distance, is nearer to us than the average distance between galaxies because it is in the Local Group of galaxies. On the average, each galaxy has a mass of 100 billion $M_\odot$ and radiates 10 billion $L_\odot$. This translates to an average mass density of 1 billion $M_\odot$ per cubic megaparsec and an average luminosity density of 100 million $L_\odot$ per cubic megaparsec. Equivalently, the mean density of luminous matter, ρ, is found to be about 10^{-31} gram per cubic centimeter, or 10^{-7} atom per cubic centimeter.

Armed with the mean density of luminous matter, ρ, one can now set up the basic theory for the big bang. First, the theory must describe how the density of matter decreases as space expands, by accounting for the growing separation between two distant mass points as the universe increases in size.

In an expanding universe, the distance between any pair of points continuously grows larger. We need a way to describe the instantaneous distance d between the two points, which is the distance at any instant. To help us do so, we make use of the coordinate distance R and use it as a benchmark against which to measure the changes in d. We refer to R as the comoving distance, because it is always the same for any two specific points. To measure the expansion at any instant, we simply look at the ratio of the instantaneous distance d between any two points in an expanding universe to their separation R in some nonexpanding system. That ratio is the scale factor $a(t)$. Since the scale factor $a(t)$ is the ratio d/R, then $d = Ra(t)$ is the actual distance between the two points. The (t) indicates that the scale factor will have a different value at each time t.

The universe, of course, is three-dimensional in space. An expanding three-dimensional volume can be described by the volume d^3. It can also be compared to a comoving, nonexpanding volume R^3, to obtain the expression $d^3 = R^3a(t)^3$.

For the mass within a specified region to remain constant as the universe expands, the density must decrease as the volume increases. Conservation of mass, the fact that the matter within a specified volume is being neither created nor destroyed, tells us that the density ρ must satisfy the condition $\rho \times$ (volume) = constant. Since the volume of the region is d^3, we infer that ρd^3 is constant as the universe expands. In fact, ρd^3 just equals the mass within a volume of coordinate dimension R. Because R is con-

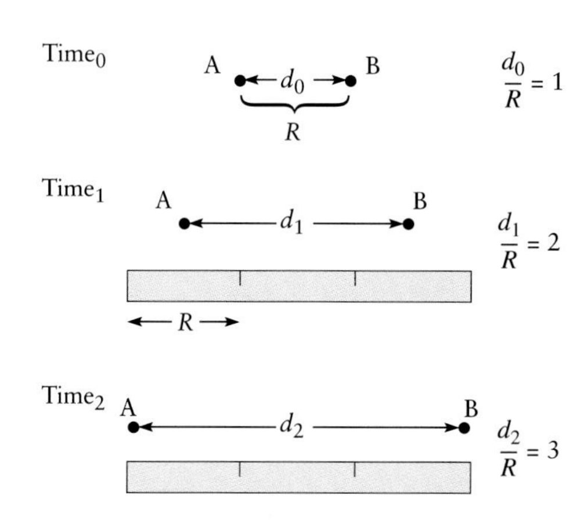

The instantaneous distance d *between two points continuously changes as the universe expands, whereas the comoving distance* R *remains unchanged. The ratio of* d *to* R *is the scale factor* a(t).

A cube containing a specified mass grows in volume as the universe expands. The density decreases by an amount proportional to the reciprocal of the scale factor. Thus as the universe expands by a factor of 2, the density decreases by a factor of 8.

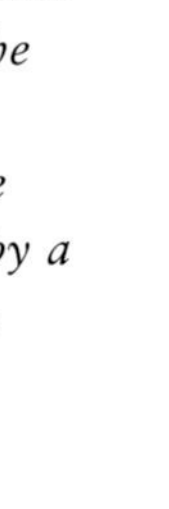

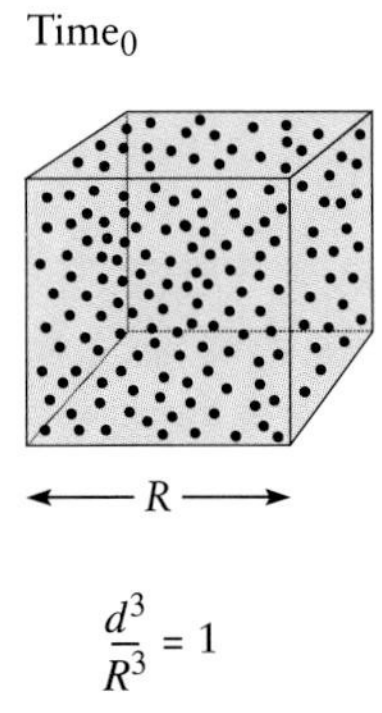

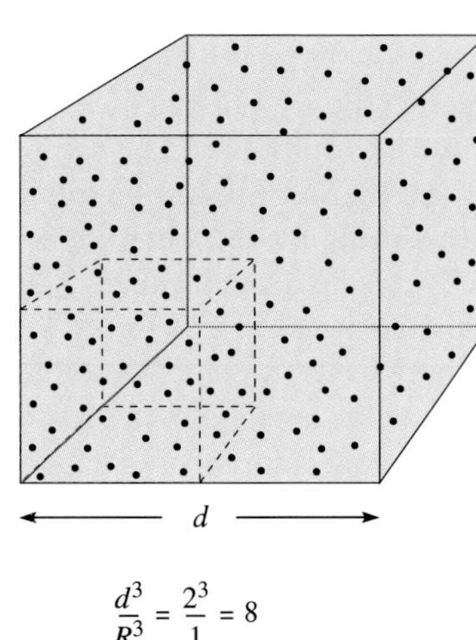

stant, we infer that the density is proportional to $a(t)^{-3}$. This is computed by substituting ρd^3 in the expression $d^3 = R^3 a(t)^3$. The result tells us that when the universe expands by a factor of 2, the density decreases by a factor of 8.

Next, the cosmological principle requires that the velocity of expansion between any two points satisfies $v = Hd$ as the universe expands, where, in general, H is a function of time (but not distance). At any specified time, this expression reduces to Hubble's law for the expansion of the universe. In particular, if the specified time is *today,* we write H as equal to Hubble's "constant," H_0. Once, a time $1/H_0$ ago, if there was no deceleration, the universe was infinitely dense if one takes the Hubble law at face value. Of course, we previously saw that any such singular beginning can be averted, if only by our ignorance of the appropriate physics of quantum gravity. Hubble's law tells us that for points that are farther and farther apart, the velocity of recession is progressively larger.

Now take any arbitrary volume of the universe. Divide it into concentric shells. There are two types of energy associated with each shell: the energy of motion, called kinetic energy, and the energy that can be acquired by falling through the gravitational field felt by the shell, called gravitational potential energy. The motion within any shell must satisfy the principal of energy conservation:

$$\text{kinetic energy} + \text{gravitational potential energy} = \text{total energy of shell.}$$

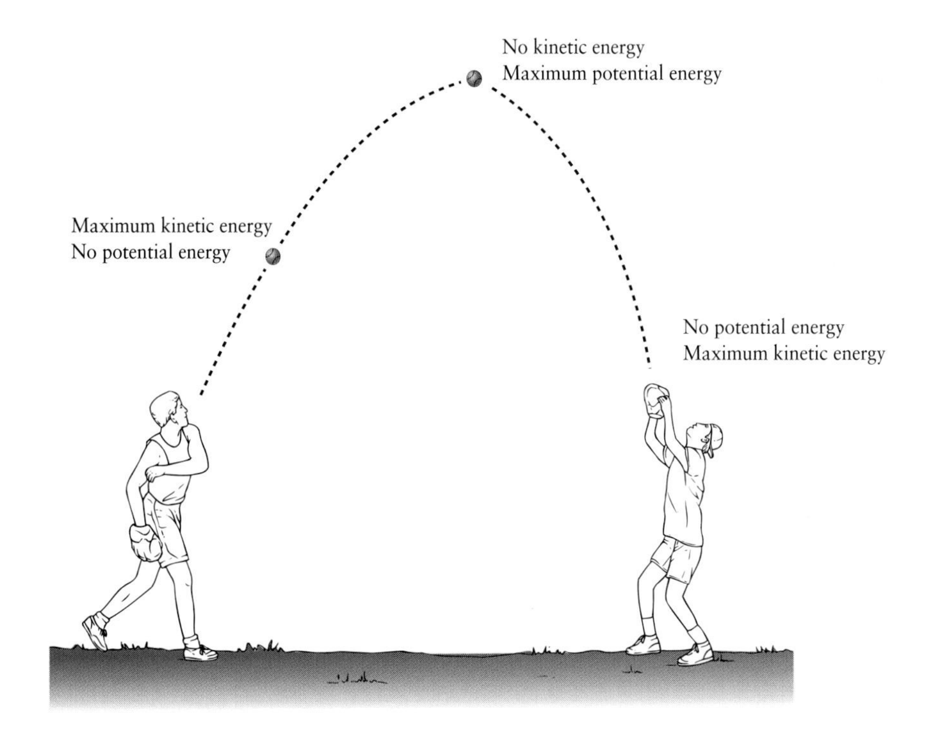

A ball thrown upward into the air illustrates the principle of energy conservation. As the ball travels upward, its kinetic energy is transformed into gravitational potential energy. At the top of the path, the energy of the ball is entirely potential energy; as gravity acts on the falling ball, its energy changes back to kinetic. Throughout the ball's journey, the sum of the two forms of energy remains constant.

Here the kinetic energy of a shell is $m_s v^2/2$, where m_s is the mass of the shell. The gravitational potential energy of the shell is $-Gmm_s/d$, where all the mass filling the spherical volume bounded by the shell is density $\rho \times$ volume of shell ($\frac{4}{3}\pi d^3$), or $m = \frac{4}{3}\pi\rho d^3$. We conclude that the total energy of the shell equals

$$m_s v^2/2 - Gmm_s/d,$$

or equivalently,

$$m_s[H^2 d^2/2 - [G(\tfrac{4}{3}\pi\ \rho d^3)/d].$$

The total energy of the shell is of course constant, and so is the mass of the shell. This means that we can divide out the constants m_s and R, or $d/a(t)$, to obtain the equation

$$H^2 - (8\pi G\rho/3) = \text{total energy of the shell} \times 2/m_s d^2.$$

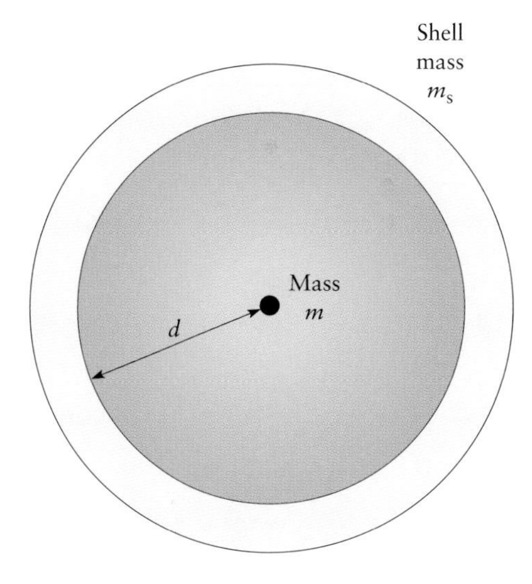

The universe can be depicted as a series of concentric expanding shells centered on the observer.

We set the constant k to denote twice the total energy of a shell, divided by its mass and by R^2, to obtain

$$H^2 - (8\pi G\rho/3) = k/a^2.$$

This equation is commonly known as the Friedmann equation, after its discoverer, the Russian cosmologist Alexander Friedmann, who published his result in 1922. Because the Belgian cosmologist Georges Lemaître independently made a similar discovery in 1927, this equation is also referred to as the Friedmann-Lemaître equation.

The evolution of the universe is governed by a delicate balance between kinetic and gravitational potential energy. If kinetic energy dominates, the universe expands forever; if potential energy dominates, it is destined to recollapse. It all rests on whether the density is less than a critical value, in which case kinetic energy wins out, or vice versa. A useful analogy is the escape velocity from the earth. Launch a rocket with a speed more than 11 kilometers per second, and it escapes the earth's gravity, to continue forever. Launch it below this velocity, and it returns to earth. We know the earth's mass and are free to adjust the rocket velocity. In the case of the universe, we measure the velocity of expansion, but do not know the mass of the universe. Indeed, it is the average density of matter that plays a role in the universe analogous to that of the mass of the earth. If we knew the average density of matter in the universe, we would know whether or not the universe had the necessary "escape velocity" to expand forever.

Suppose the kinetic energy just balances potential energy; in this case, the total energy left over, k, vanishes. The Friedmann-Lemaître equation now reduces to $\rho = 3H^2/8\pi G$. This expression defines a critical density, at any given epoch. In general, the term H is equal to 1/(time since the big bang). Measured at the present epoch, the present value of H, which we refer to as H_0, is approximately 1/(15 billion years), but early in the universe, H was much larger because the time elapsed since the big bang was much smaller. Hence the density was very much larger near the big bang. We therefore see that the density was proportional to H^2, or inversely proportional to the age of the universe ($\rho \propto H^2 \propto t^{-2}$), to a very good approximation, since the term in ρ (proportional to a^{-3}) becomes much larger than the ka^{-2} term, whether or not $k = 0$.

That the density of the universe was larger near the big bang is no surprise, but the interpretation becomes more interesting when we look at the energy constant k. This energy constant can be positive or negative. If positive, then at late times (large a), H^2 is proportional to k/a^2, so there is always a well-defined value of H. We infer that the universe expands forever. But if k is negative, there is no longer a solution with a real value for the

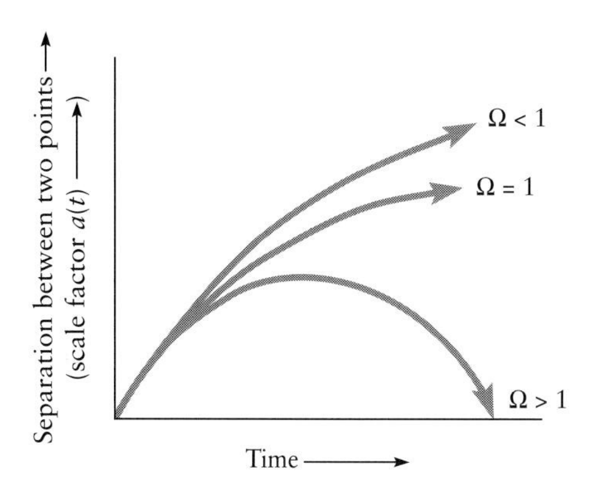

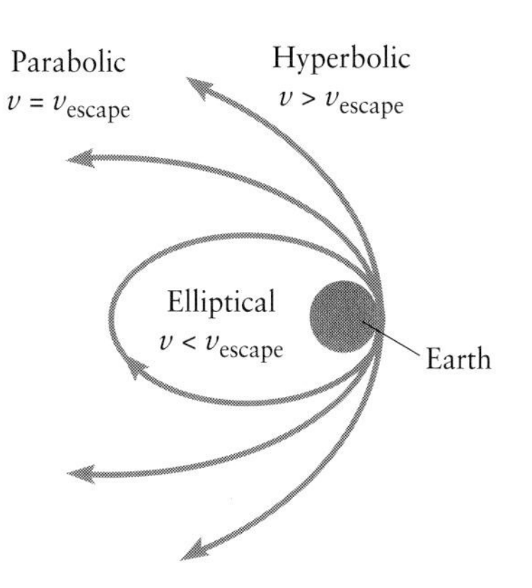

The future of the universe depends on the ratio Ω *of the actual density to the critical density in the same way that the fate of a rocket fired from the earth depends on the rocket's velocity.*

expansion rate H at late times. We infer that the universe expands to a maximum radius and then starts to contract.

The fate of the universe, determined by the value of its energy constant k, depends on the value of the actual density of the universe relative to the critical density. The scale factor a reflects an arbitrary choice of measuring units, and most simply its value today can be set equal to 1. Then the Friedmann-Lemaître equation becomes a simple equation that relates quantities observable at the current epoch, namely

$$H_0^2 - 8\pi G\rho_0/3 = k.$$

Here the subscript zero refers to present-day values of H and ρ. Clearly there is a critical density today, defined by

$$\rho_{\text{crit}} = 3H_0^2/8\pi G.$$

We usually define the cosmological density parameter Ω to be equal to the actual density ρ, whatever that may be, divided by the critical density ρ_{crit}. Thus $\Omega = \rho/\rho_{\text{crit}}$. If the density today is less than the critical density, or $\Omega_0 < 1$, k is positive, and the universe expands forever. If the density today is more than the critical density, or $\Omega_0 > 1$, k is negative, and the universe is destined to eventually collapse. For a Hubble constant $H_0 =$ 100 kilometers per second per megaparsec, the critical density at present is about 10^{-29} gram per cubic centimeter.

Because we can easily detect luminous matter, we can observe its density and calculate that Ω for luminous matter is about 0.01, far less than is needed for the universe to collapse. However, we cannot conclude therefore that the universe will continue expanding indefinitely. There is much reason to believe that a large proportion of the mass of the universe is in the form of invisible "dark" matter. Since the contribution of dark matter could be much larger than that of luminous matter, we cannot say for sure whether $\Omega < 1$ or even whether $\Omega > 1$.

If $\Omega < 1$, the universe will expand forever. Galaxies will exhaust the gas supply for forming new stars, and all the stars will eventually fade away. The universe will be a very bleak, cold place in a few tens of billions of years. By contrast, if $\Omega > 1$, the expansion of the universe will slow, and in a few billion years from now, the collapse phase will begin. All galaxies and all stars will be obliterated as the density of matter increases and the universe becomes hotter and hotter. The universe will end in a fiery catastrophe. There will be a final "big crunch" that mirrors in reverse the original big bang.

Will the universe end in fire or ice? The future of the big bang may be either the big crunch or the big chill. It all depends on whether the average density in the universe exceeds the critical value for ensuring a recollapsing universe. Density is notoriously difficult to measure, since we only see the luminous component of matter. However, there is another possible way to predict the future universe: by observing its geometry.

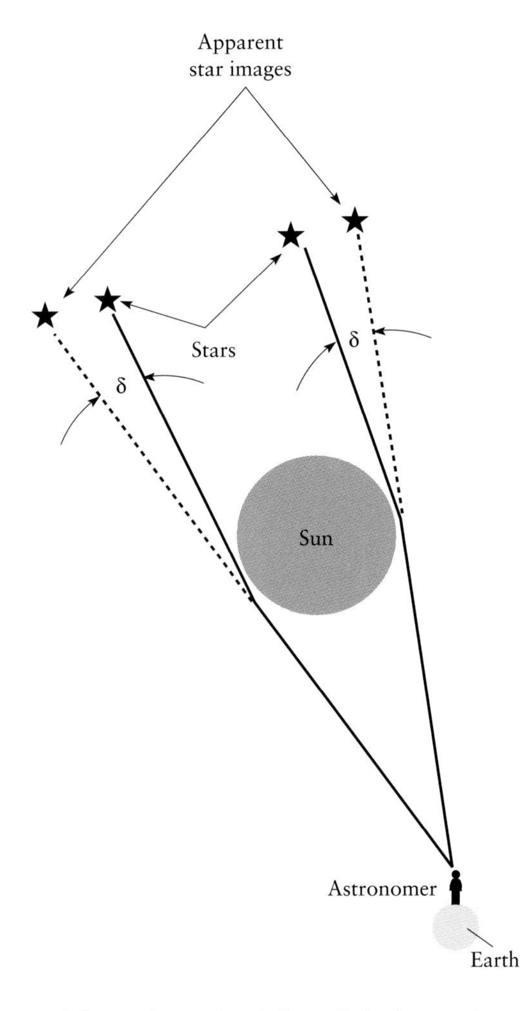

The sun's gravity deflects light from other stars, altering their apparent positions in the sky. The deflection, δ, is about 0.0005 degree.

Curved Space and General Relativity

Einstein's theory of gravitation reinterprets the energy constant k in terms of the curvature of space. One of the important concepts in Einstein's theory of general relativity is that space is distorted by gravity. Space and gravity in fact have a mutual influence: gravity curves space, and space causes gravity. It is the curvature of space that causes matter to move under the influence of gravity, according to Einstein.

Because gravity curves space, light no longer travels in a straight line. This effect was first measured in 1919, three years after Einstein's theory was published, by studying how the light from background stars was deflected near the sun during a total solar eclipse. It was found that light does not travel in a straight line when passing close to the sun. In fact, since light has energy, even the Newtonian theory of gravity predicts a deflection of light. However, in Einstein's theory, space is curved, and consequently the light deflection is double the Newtonian value.

In the presence of matter, there are no more "straight lines": the shortest distance between two points must be a curve. The familiar theorems of Euclidean geometry no longer apply: parallel lines can meet, and the angles of a triangle no longer sum to 180 degrees. Near the sun, the effect is

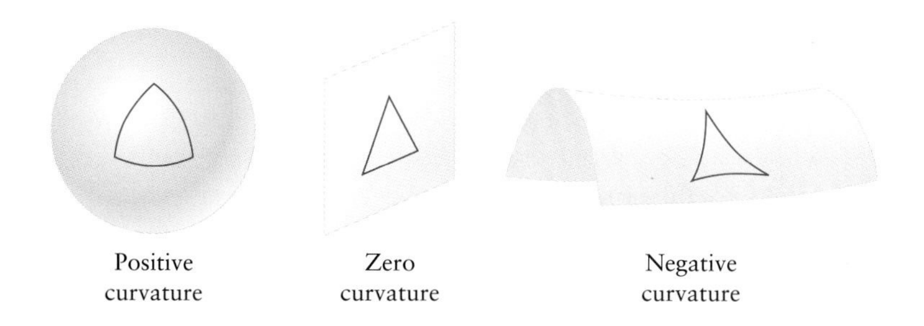

The three-dimensional geometry of space may be spherical, flat, or saddle-shaped, analogous to these two-dimensional surfaces. The curvature constant k *is equivalent to what in Newton's language would be energy* −k.

small, since the sum of three angles of a triangle drawn around the sun exceeds 180 degrees by about one part in a million.

The implication from this revolution in our thinking about the nature of space is that "gravity curves space, and space moves matter," thereby providing a new paradigm for reinterpreting the geometry of the universe. The curvature of space plays an important role in the big bang, since it reflects the strength of gravity. First, we need to reinterpret the constant k. It is no longer energy, but now in Einstein's language a measure of the curvature of space. The theory of general relativity assigns three possible values to k, namely $+1$, 0, or -1, that are identical to the more classical concepts of negative, zero, and positive energies for the universe. These values correspond to spaces that are positively curved (like the surface of a sphere), flat (like a plane), or negatively curved (like a saddle-shaped or hyperboloidal surface). In terms of total energy, they are equivalent, respectively, to negative total energy (the gravitational potential energy dominates), zero total energy (the potential and kinetic energies exactly balance each other), and positive total energy (the kinetic or expansion energy dominates). A positive energy for the universe corresponds to a negatively curved space geometry. The universe of negative total energy, which only expands to a maximum size, has positive curvature. It is a closed space, and finite, while the positive energy universe, which expands forever, is an open space, and infinite. In the former case, the three angles of a triangle add up to more than 180 degrees, while in the latter case they add up to less than 180 degrees.

An open universe is infinite today and always was infinite. It is infinite simply because in an open geometry an edge would violate homogeneity or, in other words, the cosmological principle, for which there is mounting observational evidence. This does not mean that the galaxy distribution within an infinite universe may not have an edge: such an edge would simply signify that galaxies had formed at some time in the past, before which there were no galaxies.

An infinite universe can always expand further, because there is no bound to infinity. Consider a hotel with an infinite number of rooms, all occupied. An infinitely larger number of visitors can always be accommodated; just ask each guest to move along to the next room, freeing up the first room. Because the number of rooms is infinite, there is always a next room to which each guest can move. An open universe contains an infinite mass, because of the infinite volume for a specified average density of matter.

If the universe is open now, it was also open at the beginning of time, $t = 0$, an instant that is unreachable by way of the general relativity theory, but one that we can speculate about. At that first instant, an open universe was infinite in volume, but the density of matter was infinite too. We say an open universe began as a singularity of infinite density, where known physics breaks down. More realistically, one can begin the history of the open universe at the Planck time, 10^{-43} second, when its density was already finite, but it still occupied an infinite volume. In contrast, a closed universe begins as a singular point, also of infinite density, but in this case of zero volume. More realistically, one would also begin the history of the closed universe at the Planck time, when the density was finite. The volume of a closed universe is and always was finite.

The density of the universe today is remarkably close to the critical density, and its geometry is remarkably close to being flat. For this to be true, the geometrical analogy of space curvature tells us that there must be a specific balance between the gravitational potential energy and the kinetic energy of matter in the expanding universe. The existence of such a finely tuned balance presents a major quandary. The imbalance would have to be tuned to a high degree of precision when the universe was very young, since the earlier the epoch, the closer are the two types of energy. Today they may differ by a factor of 10, but at the Planck time, they must have been nearly equal to within 1 part in 10^{60}. Why should the two forms of energy have been so nearly balanced?

The quandary is resolved with the help of inflation. Inflation, at least in its most optimistic incarnation, commences with an arbitrary mismatch between the kinetic and the potential forms of energy. Without inflation, there would be an imbalance between the potential and kinetic energies of the universe. With inflation, however, any excess of one form of energy over the other is reduced to a very small value. To see why, consider that in Einstein's language, such an imbalance is equivalent to a curvature of space. The enormous increase in the size of the universe during inflation flattens out this curvature, much as inflating a balloon removes any wrinkles, and much as any curved surface is approximately flat over small regions to a high degree of precision. A universe with a flat geometry is one in which there is indeed an almost precise balance between the kinetic and the potential forms of energy.

Cosmological Tests

A big crunch universe is characterized by spherical space, and a big chill universe is associated with hyperboloidal space. The geometry is quite distinct between these two options. We can probe the geometry of the universe by looking far away at remote galaxies, in order to test the isotropy and homogeneity of the universe. It is of course the deviations from an idealized uniform and isotropic cosmological model that make the universe interesting, and these blemishes from perfection can be used to explore its geometry. Unfortunately, no firm conclusions can yet be drawn, but we have learned much about the universe in the process of exploration.

Homogeneity from Galaxy Counts

The starting point of modern cosmology is the cosmological principle, which assumes that the universe is uniform, when viewed with sufficiently coarse resolution. It is as if the lumps and bumps that are the galaxies and galaxy clusters all blur together as though seen out of focus. A test of the uniformity, or homogeneity, of the universe is to count galaxies to fainter and fainter limiting magnitudes. If the density of the universe is uniform and constant with time, as expected in a nearly flat universe, the number of galaxies counted should increase proportionately to the volume sampled. Such a proportionate increase is indeed observed out to about 1000 megaparsecs, and demonstrates the large-scale uniformity of the universe up to that scale.

At larger distances, however, the density should not remain constant. We can define a local volume of the universe to encompass, say, a specified number of galaxies. At larger distances, such a local volume shrinks as one looks back in time to when the universe was more compact. The same number of galaxies is counted, but in a smaller volume, since it is only the fabric of space, not matter itself, that is expanding when we loosely talk of the expanding universe. Since a closed universe is decelerating, distant elements of volume would shrink more rapidly with increasing redshift than would be the case for an open universe.

Many deep images have been taken of distant galaxy fields, and the number of galaxies counted. The results turn out to be inconclusive, because the most distant galaxies seemingly have different properties from their nearby counterparts. They are younger, and not surprisingly, their youthful attributes such as their brightness must affect the counting, which always is done to a fixed apparent brightness.

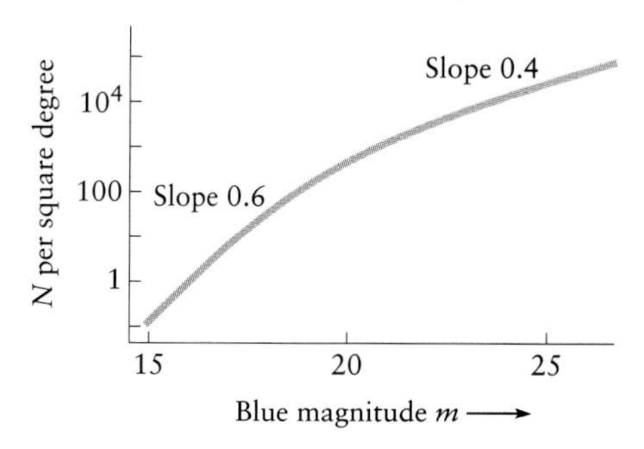

The number of galaxies brighter than a certain magnitude m *increases as the magnitude is decreased, but eventually begins to level off.*

A very long exposure (10 hours) of a distant galaxy field taken with the 4-meter NTT telescope of the European Southern Observatory in Chile. The density of galaxies in the sky is about 100,000 per square degree; the field shown is 5 arcminutes across.

Isotropy of the Microwave Background

Another assertion of the cosmological principle is that there should be no preferred direction in the universe, another way of saying that the universe is isotropic. If this were true for any arbitrary observer, then the uniformity of the universe would be an inevitable consequence. The isotropy of the universe can be tested from galaxy counts by dividing the sky into small segments of equal area, and repeating the counts for each segment. To within perhaps a factor of 2, identical counts are measured.

The most stringent test of isotropy comes from examining the microwave background. Precisely the same temperature, 2.73 degrees Kelvin, is measured in different directions. No variations are seen to within less than a few thousandths of a percent, apart from the dipole anisotropy due to the motion of our galaxy. Since the microwave background originated from very distant regions of the universe, at a redshift of perhaps 1000, or about 300,000 years after the big bang, one concludes that the universe was homogeneous and expanded at a uniform rate to this degree of precision. This is the most rigorous justification for the cosmological principle to date.

Recently, there have been reports of temperature fluctuations at a very low level, about 1 or 2 parts in 100,000. These measurements, to be described in a later chapter, tell us about the strength of the weak primordial inhomogeneities from which the large-scale structure of the universe developed.

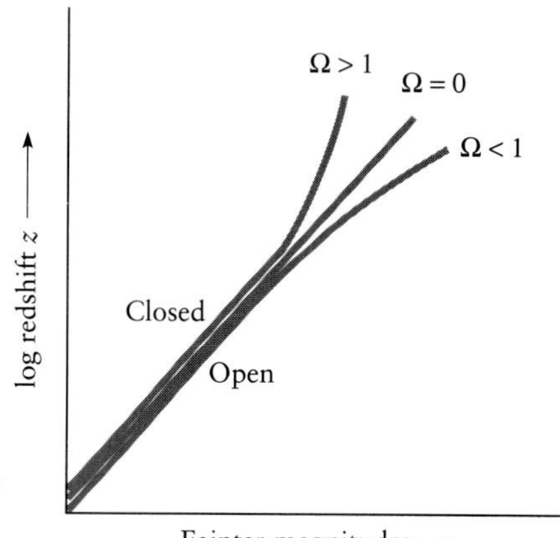

The magnitude of a galaxy decreases (the galaxy becomes fainter) as the redshift increases. In a closed universe, galaxies are slightly closer, and brighter at a given redshift, and in an open universe they are more distant, and dimmer.

Measuring the Curvature of the Universe

We may be able to measure the degree of curvature of the universe from studies of distant objects. An important test is to measure the redshifts and magnitudes of distant galaxies. Suppose one knew the absolute luminosities of these galaxies, and suppose also that we were living in a closed universe: such a universe would have significantly decelerated during the passage of light from distant galaxies. In this case, one expects these galaxies to appear brighter than in an open universe, where the deceleration is less and hence the effective distances are greater. Redshift measures distance, so at a given redshift one expects galaxies to be fainter in an open universe than in a closed universe.

Another test using distant objects exploits the fact that light travels on a curved path through the universe, because the presence of matter inevitably curves space. Consider the case of the surface of a sphere as a two-dimensional analog to three-dimensional curved space. Imagine two neighboring points as viewed by an observer on the sphere, who sees the points by means of light rays that have traveled from the points to his eyes. These light rays only propagate on the sphere's surface. As the observer moves away from the two fixed points, their apparent separation, or angular size, decreases with increasing distance. There soon comes a moment, about halfway around the sphere, when the effect of the curved surface is important, however, and it causes the light rays to diverge. The result is that the angular size reaches a minimum value and then increases. If one knew the intrinsic sizes of galaxies, or the intrinsic mean separation of galaxies in a cluster of galaxies, one could apply this test to very distant galaxies to measure the curvature of space. An open universe curves light less than a closed universe.

Finally, the volume of space could in principle be measured directly by counting galaxies and measuring their redshifts over a finite but deep volume. This should be an ideal way to measure the geometry of the universe. However, while all of these tests seem attractive, their application is made very difficult by the effects of galactic evolution.

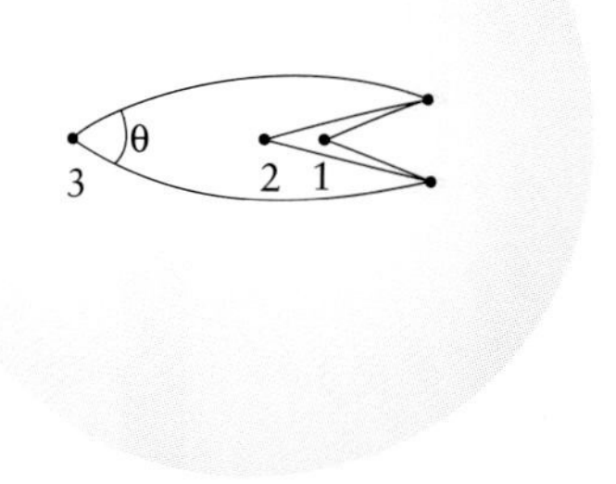

When two fixed points on the sphere are viewed from greater and greater distances (at positions 1 through 3), their apparent separation first decreases, and then increases, as a consequence of corresponding changes in the angle ø between the converging light rays. That angle has been affected by the bending of the light rays on the sphere's curved surface.

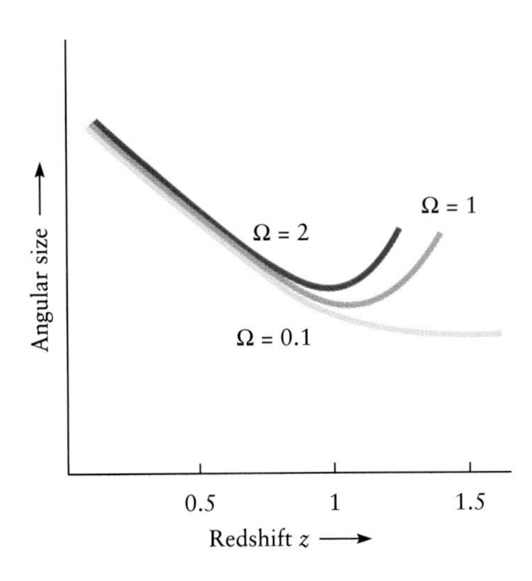

The angular size of an object decreases with increasing redshift, but eventually increases. It will increase more strongly in a universe of density greater than the critical value, because such a universe is more curved.

Impediments to the Tests

Light from a galaxy at a redshift of one has taken about seven billion years to reach us. We view this galaxy as it was before our solar system formed, only 4.6 billion years ago, less than a third of the age of the universe. We have no guarantee that a galaxy's luminosity or size has remained constant over this time span. For example, in their distant youth, galaxies may have been much more luminous, containing many more massive stars than galaxies do today. Galaxies would appear larger or smaller, depending on whether the first stars formed out in the halos or in the central bulges. Galaxies may have undergone mergers in the past, and in this case, distant galaxies would have been smaller and less massive than nearby galaxies.

There is an even more dramatic effect as we view the remote past, which further confounds attempts to use galaxies as measuring tools. Most of the galaxies we see in the remote depths of space, and therefore in time, may have no detectable nearby counterparts. One arrives at this conclusion by going an additional step beyond just counting galaxies. From the spectrum of a galaxy, one obtains its redshift. While one can count galaxies down to the 27th magnitude, about the light received from a candle on the moon, galaxy redshifts can be measured to about the 24th magnitude, sufficient to probe the universe to a redshift of approximately one-half, or to look back in time some five billion years. One learns that there are more galaxies out in the distant depths of the universe than are seen locally. Even after allowing for the effects of the expansion of space, there would be three or even five times more galaxies at redshift unity, some seven billion years ago, than are seen nearby. The excess galaxies are not especially luminous, but they are blue, the signature of massive stars and of youth. A large population of old galaxies is not seen in the early universe. This implies that galaxy formation is most likely a recent phenomenon, and that much of the action occurred over the past five billion years. At least this is likely to be the case for spiral galaxies, which dominate the galaxy counts.

We cannot correct our attempts to measure the curvature of space for these various effects, because we cannot precisely compute their amplitude. Because of this uncertainty, any direct measurement of the geometry of the universe seems impractical in the near future.

With large enough telescopes, one might hope to eventually circumvent these difficulties. For example, with adequate spectroscopic information about the types of stars in a distant galaxy, one could compute how much more luminous it is as compared to a nearby, unevolved galaxy. The most promising technique presently appears to be to study a deep sample of galaxies, and use many galaxy redshifts to directly measure the shrinkage of an element of volume in a distant segment of the universe. The change in volume with distance depends on the curvature of space. The main problem is that very many redshifts of faint galaxies are needed. Collecting these redshifts is a time-consuming task that is awaiting the development of dedicated telescopes and specialized instruments able to obtain hundreds of galaxy redshifts in a single exposure.

While astronomers are frustrated at looking far afield in order to ascertain the fate of the universe, an alternative strategy is beginning to pay dividends. One can try to measure the density of the universe directly by performing relatively local measurements. Since matter determines the curvature of space, one can evaluate the possible forms of matter that may constitute the density of the universe. The goal, in other words, is the direct detection of dark matter.

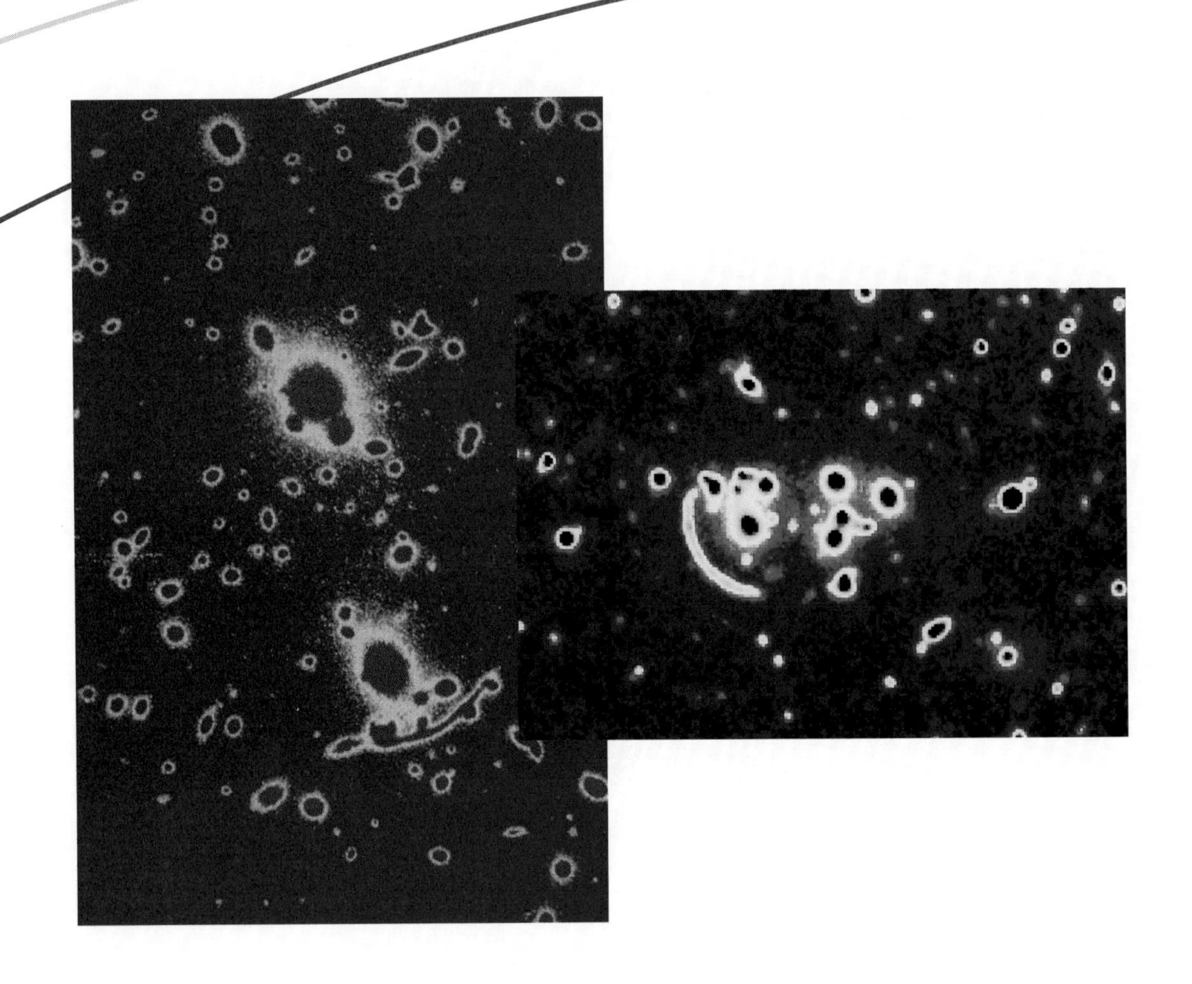

Left: *The cluster of galaxies Abell 370 is at a redshift of 0.374, about 6 billion light-years away. The cluster acts as a gravitational lens to produce the arc seen in this false-color image; that arc is the distorted image of a distant background galaxy at redshift 0.724, about 10 billion light-years away. The curvature of the arc measures the cluster mass, revealing how much of that mass is invisible dark matter.* ***Right:*** *Here, too, the giant arc is the magnified and distorted image of a distant background galaxy, which has been gravitationally lensed by the cluster of galaxies Cl 2244-02.*

7

Dark Matter

Nearly 50 years ago, Fritz Zwicky realized that clusters of galaxies consisted predominantly of matter in some nonluminous form. The search for this dark matter has dominated cosmology for half a century. Precise measurements were first obtained over 20 years ago, when dark matter was mapped in galaxy halos. Only recently has the existence of dark matter over much larger scales than even galaxy clusters been confirmed.

The astrophysicist Fritz Zwicky, who pioneered such ideas as dark matter and neutron stars, was also the preeminent cataloguer of supernovae and bright galaxies.

It is not possible to understand how dark matter was investigated without knowing something of how galaxies are structured. Our galaxy consists of a disk of stars, of radius 10 kiloparsecs (1 kiloparsec = 1000 parsecs) and thickness 500 parsecs. The stars in the disk, called Population I, are the younger stars in the galaxy. They may be found, moving in approximately circular orbits about the galaxy center, in star-forming regions, in loose groupings of young stars termed stellar associations, and in *open clusters,* more numerous groupings of older and younger stars. Distributed throughout the disk is almost all of the galaxy's interstellar gas and dust, which has a mass equal to about 5 percent of the mass of the galaxy's visible stellar component. In fact, the molecular gas and the youngest stars occupy a disk that is only about 100 parsecs thick.

The oldest stars in the galaxy, called Population II, are found in a central volume of nearly spherical shape. This galactic *spheroid* of stars occupies the central kiloparsec, but trails off gradually so that some spheroid stars are found 50 kiloparsecs or more from the center of the galaxy. Population II stars include densely crowded star clusters, called *globular clusters,* known to be old from their Hertzsprung-Russell diagrams, as well as stars scattered throughout the halo of the galaxy. Population II stars are chemically distinct from Population I stars, in being metal-poor: their spectra reveal abundances of the heavier elements that are a factor of 10 or more below those found for the sun. This is to be expected if Population

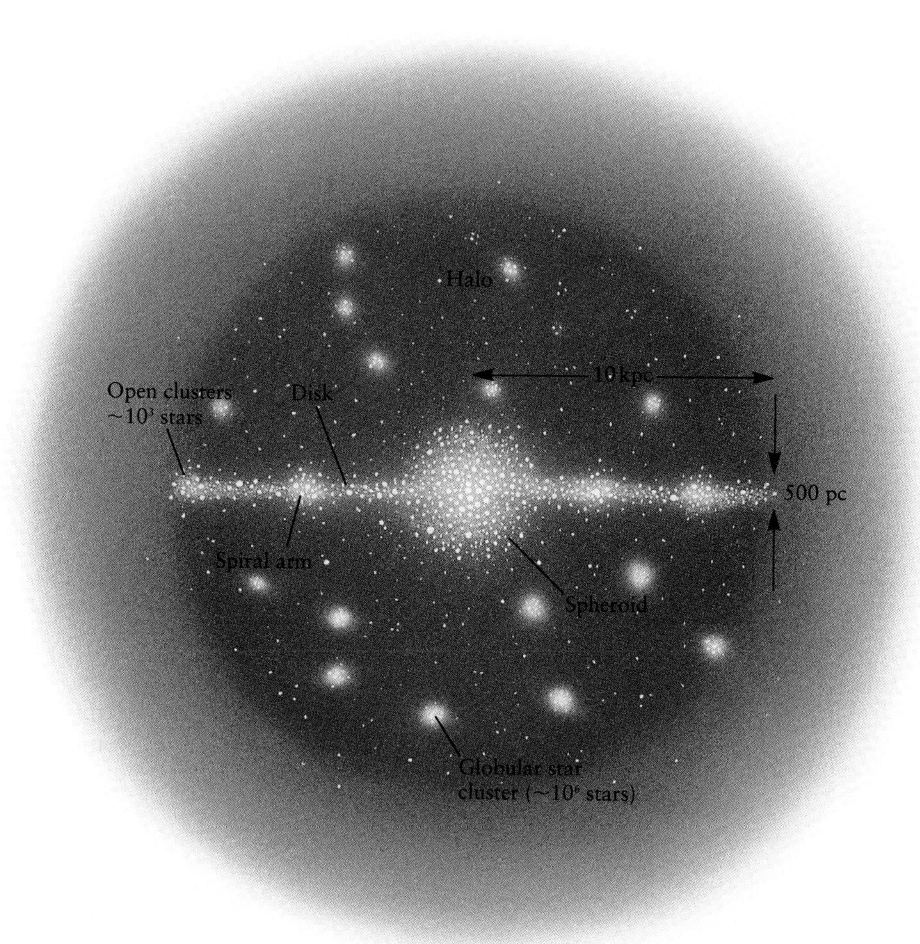

The structure of a spiral galaxy, viewed edge on.

II stars are older, since early in the history of the galaxy the abundance of metals was lower than at the present epoch. The *halo* of the galaxy is the outermost region, which extends out to about 100 kiloparsecs in all directions from the center of the galaxy. It contains the outer spheroid stars, but is otherwise nearly invisible. Spheroid stars have elliptical orbits and are *not* confined to the plane of the galaxy. Consequently, in the solar neighborhood, Population I stars have low velocities, since they are nearly in the same orbit as the sun, while the Population II "interlopers" are high-velocity stars, on very different orbits.

Local Indicators of Dark Matter

The density of matter in the solar neighborhood is measured by sampling a uniform population of luminous stars that extends well above the disk

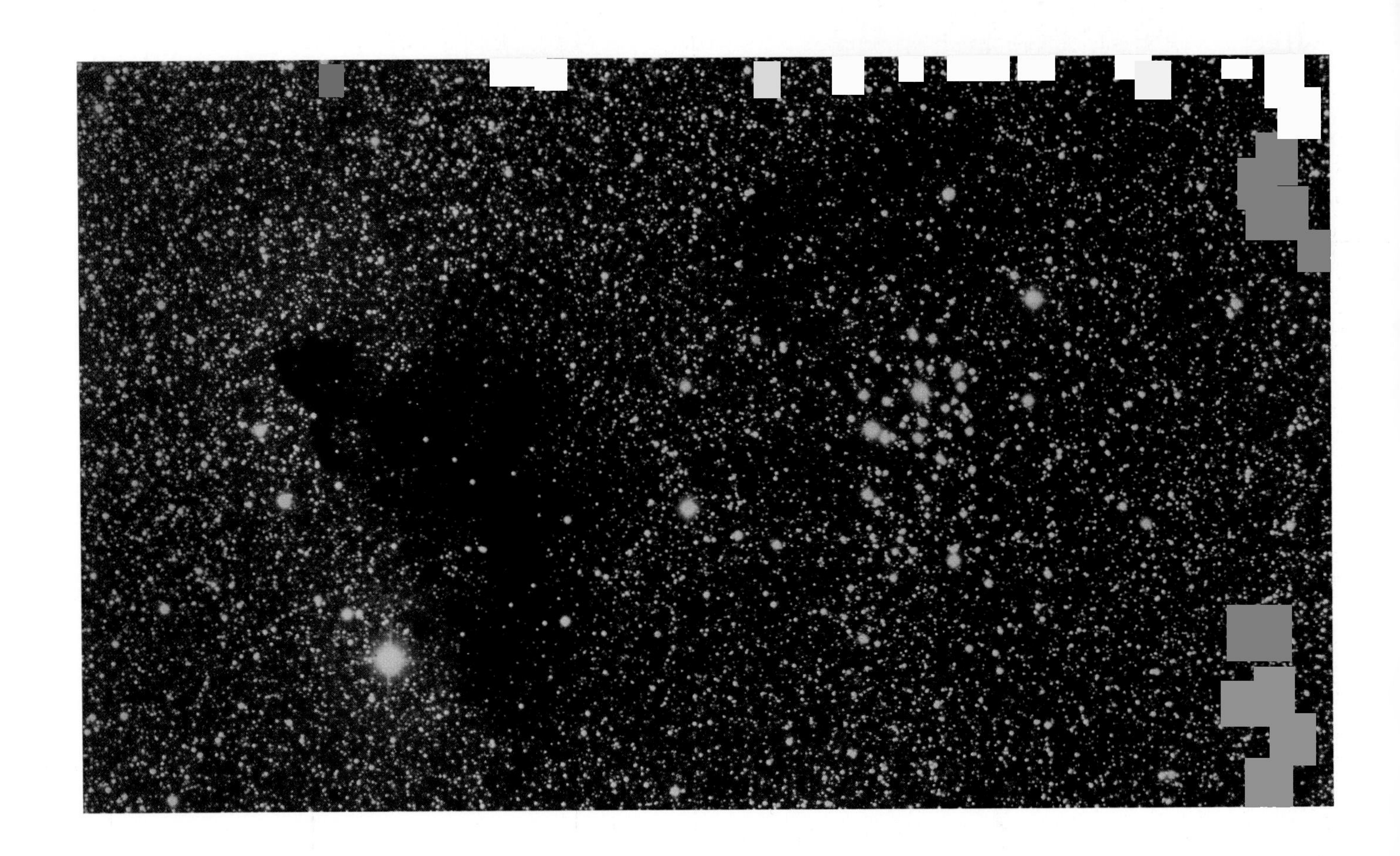

An open star cluster, NGC 650, and the dark cloud Barnard 86, seen against the Milky Way.

of the galaxy. The average velocities of the stars and the vertical distances they traverse about the disk provide a measure of the gravitational restoring force that keeps these stars in the disk. From the strength of this force, one can deduce the total density of matter that exerts this gravitational pull. Comparing this density with actual counts of stars, one finds that the number of observed stars falls short, by perhaps as much as a factor of 2, of the number needed to account for this density. This is the first hint of any dark matter, and it is present in the vicinity of the sun.

It should be noted that the amount of such a shortfall in the disk matter is controversial. There is at most an amount of dark matter in the disk equal to the amount of luminous matter. A more conservative estimate might place the amount of dark matter at about 25 percent of the amount of luminous matter. In fact, this additional component of matter need be nothing very exotic.

What could the dark matter be? The dark matter in the disk most likely consists of very dim stars, such as white and even black dwarfs. White dwarfs are the destiny of stars like the sun, attained when the nuclear fuel supply is exhausted. A typical white dwarf has a mass of about 0.6 $M_\odot$ but a size smaller than that of the earth. It formed as the hot core of a planetary nebula, the final luminous phase of stellar evolution when the envelope of a red giant is ejected as the core burns the last remnants of nuclear fuel. A white dwarf fades slowly to oblivion as it cools down to become a black dwarf.

A useful measure of mass is obtained by taking the ratio of the mass of all stars to the luminosity emitted by all stars, in a volume of a few hundred parsecs around the sun. Luminosity is approximately proportional to a high power of mass (M^4), so stars less massive than the sun are much less luminous and more massive stars are much brighter. If the typical star near the sun were equal in mass to the sun, the ratio of total mass to total light would be unity, but that ratio would be larger than unity if the typical star were less massive and smaller than unity if it were more massive. Since the resulting ratio is found to be 2 for nearby stars, in solar units of $M_\odot/L_\odot$, we conclude that the average star near the sun is slightly less massive than the sun. In the solar vicinity, there is little necessity for any dark matter other than ordinary stars and their dark dwarf remnants.

Luminous Regions of Galaxies

There is little evidence for any dark matter over the larger scales that correspond to the luminous regions of galaxies such as the Milky Way. The luminous region of a galaxy extends over a radius of about 10 kiloparsecs. The sun, for example, is on the outskirts of the Milky Way galaxy, at about 8 kiloparsecs from its center. The rotation of the galaxy enables us to calculate the region's mass.

Both clouds and stars participate in the circular rotation of the galaxy. Interstellar HI clouds can be detected throughout the entire Milky Way by observing their absorption lines at a wavelength of 21 centimeters. The clouds are found to be moving in nearly circular orbits in the disk. From the 21-centimeter line width, one calculates the rotation velocity of the galaxy at various distances from the galactic center: that velocity is found to be approximately constant in the disk between 3 and 8 kiloparsecs out from the center. This means that the inner parts of the disk have shorter rotation periods.

To calculate the mass of the galaxy's luminous region, we first consider the rotation velocity near the location of the sun. The rotation velocity of the sun is about 250 kilometers per second. Hence the sun takes 200 million years to make one orbit around the galaxy, and has orbited the galaxy about 25 times since its birth. One measures the total mass interior to the orbit of the sun from the sun's rotation velocity around the galaxy and its galactocentric distance. This gives the centripetal force, which must be balanced by the gravitational force due to all the mass interior to the sun's orbit. If we consider the rotation velocity in the outer part of the region of the disk occupied by stars (the disk extends out to about 15 kiloparsecs), we can infer that the total mass of the galaxy increases with radius to a value of about 10^{11} $M_{\odot}$. In contrast, the cumulative luminosity of all the stars in the Milky Way is about 10^{10} $L_{\odot}$. The ratio of mass to luminosity is therefore equal to 10. The disparity does not have to suggest the presence of dark matter, however. Instead, one concludes that the average star is about half the mass of the sun (and thus considerably less luminous). This is not a great surprise. After all, the solar neighborhood contains younger, relatively more massive and luminous stars, as do other spiral arms.

Galaxy Halos

The first real surprise emerges in the outermost parts of galaxies, known as galaxy halos. Here, there is negligible luminosity, yet there are occasional orbiting gas clouds, both atomic and ionized, which allow one to measure rotation velocities and distances. The rotation velocity is found not to decrease with increasing distance from the galactic center. If the mass had the same distribution as the light, then the gravity field would be weaker far from the luminous region of the galaxy, and less velocity would be necessary to maintain a circular orbit at this distance. This constancy of velocity implies that the galaxy's distribution of mass cannot be as concentrated as its distribution of light. The cumulative mass must continue to increase with the radial distance from the center of the galaxy, even though the cumulative amount of light levels off.

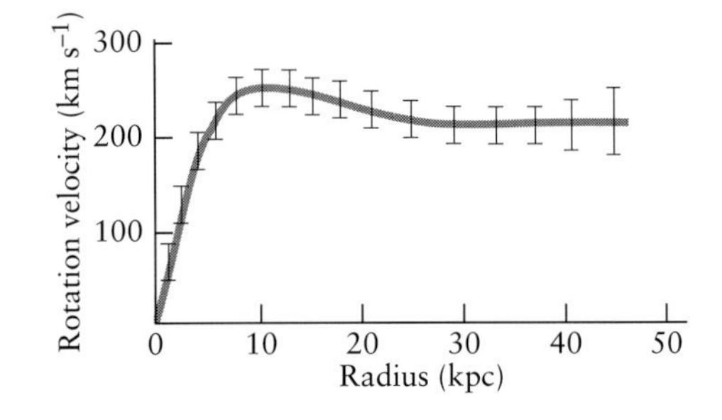

The rotation curve of the dwarf spiral galaxy DDO 154 is a plot of its rotation velocity against the distance from the center. The rotation velocity is constant at large distances from the center.

Just how much additional mass is in the halo? Since the rotation velocity satisfies the equation $v^2 = GM/r$, where M is the mass within radius r, we infer that M increases proportionally to r. This rise appears to stop at about 50 kiloparsecs, where, at least for the Milky Way, the halo seems to be truncated. We infer that the mass-to-luminosity ratio of the galaxy, including its dark halo, is about five times larger than estimated for the luminous inner region, or equal to about 50. This is the first solid, incontrovertible evidence for dark matter. The rotation velocities throughout many spiral galaxies have been measured, and all reveal that dark matter is dominant.

The rotation velocity of a spiral galaxy, pictured here face on, is constant at large distances from the center, indicating that the galaxy is embedded in a halo of dark matter.

Galaxy Groups, Clusters, and Superclusters

Moving further afield, the mass-to-light ratio can also be evaluated by studying pairs, groups, and clusters of galaxies. In each case, one measures velocities and length scales, from which one determines the total mass required to stop the system from flying apart. The inferred ratio of mass to luminosity is about 100 $M_{\odot}/L_{\odot}$ for galaxy pairs, which typically have separations of about 100 kiloparsecs. The mass-to-luminosity ratio increases to 300 for groups and clusters of galaxies over a length scale of about 1 megaparsec. Over this scale, 95 percent of the measured mass is dark.

The largest scale on which the mass density has been measured with any precision is that of superclusters. A supercluster is an aggregate of several clusters of galaxies, extending over about 20 megaparsecs. Our local supercluster is an extended distribution of galaxies centered on the Virgo cluster, some 10 to 20 megaparsecs distant. The Milky Way galaxy together with the Andromeda galaxy forms a small group (the Local Group) that is an outlying member of the Virgo supercluster. The mass between us and Virgo tends to decelerate the recession of our galaxy relative to Virgo, as expected according to Hubble's law, by about 10 percent. This deviation from the uniform Hubble expansion can be mapped out for the galaxies throughout this region, and provides a measure of the mean density within the Virgo supercluster. Over the extent of our local supercluster, about 20 megaparsecs, one again finds a ratio of mass to luminosity equal to approximately 300.

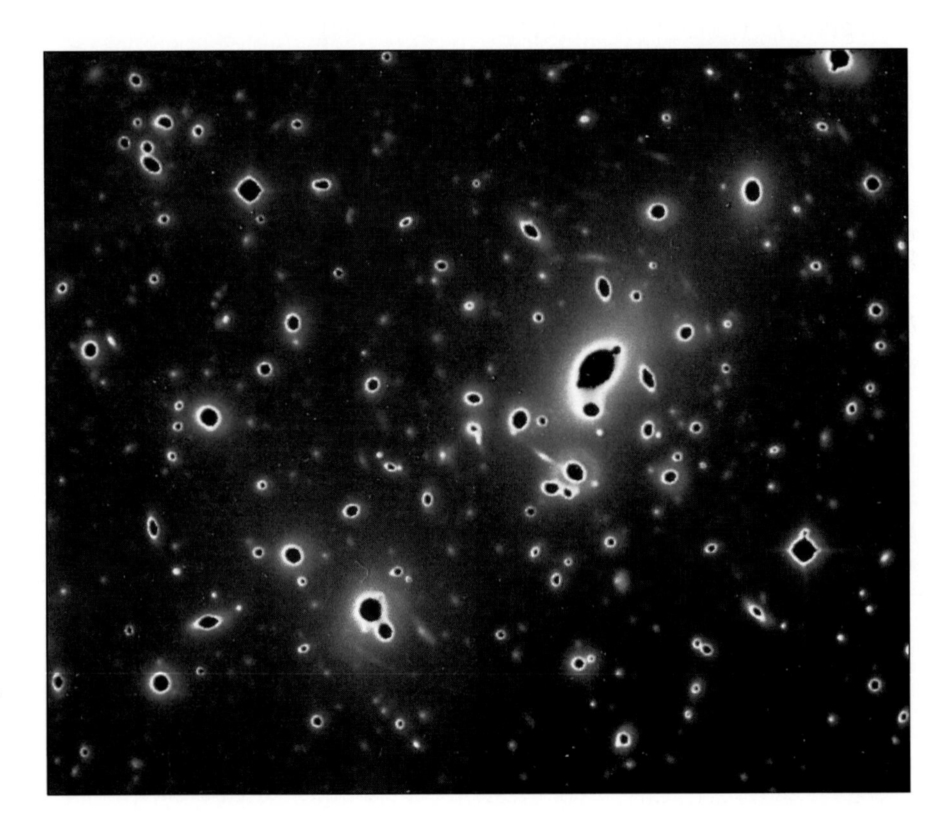

The galaxy cluster Abell 2218 contains arcs that are the gravitationally lensed images of distant, background galaxies, created as the cluster's mass bends light coming from those galaxies. The distribution of the arcs provides a map of the dark matter distribution in the cluster.

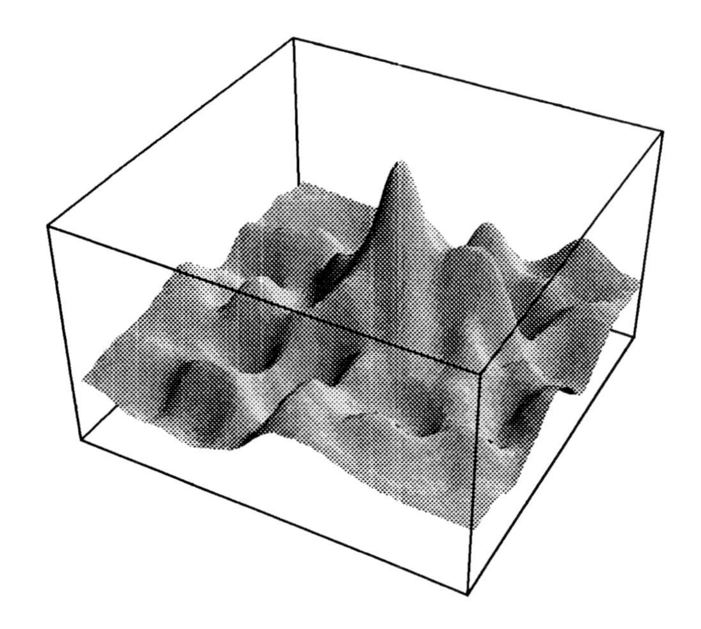
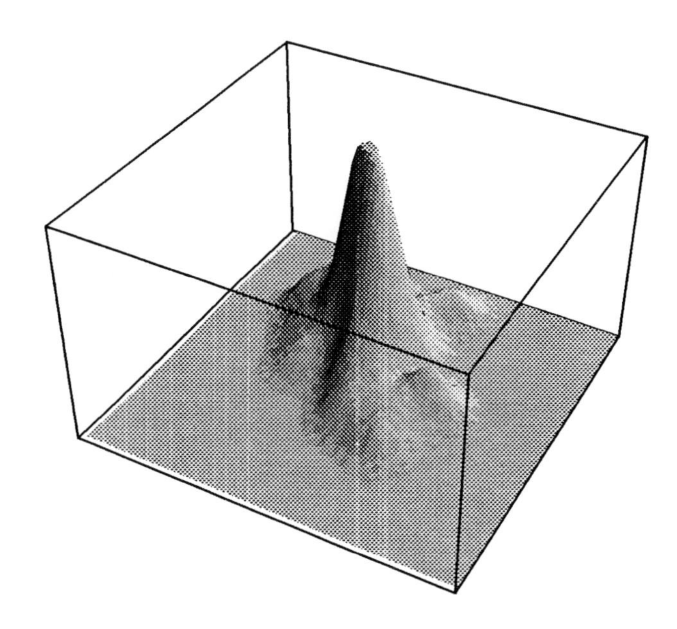

The dark matter distribution in a distant galaxy cluster (left), inferred from the distortions of faint background galaxy images by gravitational lensing, is possibly more extensive than the distribution of luminous matter (right). The horizontal scale of the box is approximately one quarter of a degree.

Large-Scale Flows and Dark Matter

On the very largest scales, there are no longer any gravitationally bound objects. Yet the galaxies are not distributed perfectly uniformly: there remain small density fluctuations that have persisted since the very earliest epochs of the universe. The dark matter that accounts for the critical density may well be uniform over galaxy cluster and supercluster scales. However, it should, at least in the case of some kinds of dark matter, participate in the density fluctuations on larger scales. Simply measuring the luminosity density, by counting galaxies, will not capture the dark matter fraction. There is a technique, however, that does measure the dark matter content of the universe, over scales of 10 to 100 megaparsecs.

Where these tiny fluctuations show an excess in density, they exert a slight gravitational attraction on their surroundings, and where they show a slight deficit in density, they exert a slight repulsion. This effect manifests itself on nearby galaxies as slight deviations from the uniform Hubble expansion around us. If one measures these "peculiar" velocities of galaxies relative to the Hubble flow, they will trace the fluctuating component of the dark matter. The Tully-Fisher correlation is especially useful in this regard. This relation between a galaxy's luminosity and its rotation velocity, $L \propto V_{\mathrm{rot}}^4$, yields a measure of the galaxy's distance. The redshift also gives a distance, by way of Hubble's law. However, the distance calculated from the redshift is distorted to be larger or smaller than the true distance, depending on whether the galaxy's peculiar velocity adds to, or subtracts from,

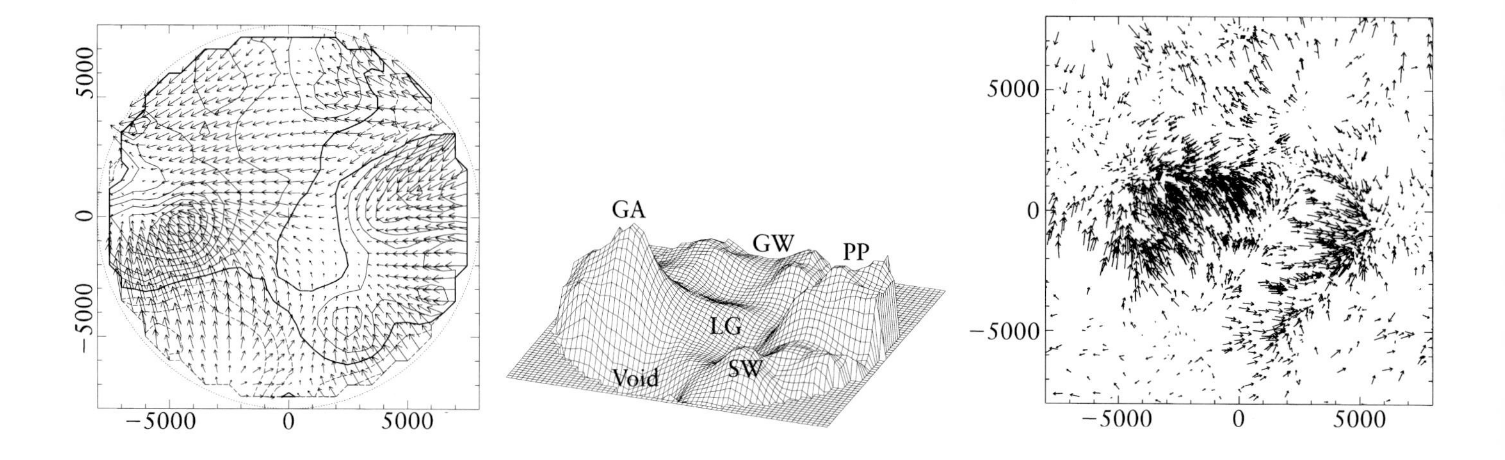

Upper left: This reconstructed large-scale velocity field, compiled from measurements of optically visible galaxies, reveals a bulk flow toward the Great Attractor on the left. The length scale is measured in kilometers per second: each 75 kilometers per second is equal to 1 megaparsec. Upper center: The large-scale density field has been reconstructed for the same group of galaxies. If you compare the two views, you can see that galaxies have peculiar velocities toward the areas of high density. Upper right: This large-scale velocity field has been inferred from measurements of a sample of galaxies emitting at infrared wavelengths. The projection is onto the midplane of the Virgo supercluster; the Milky Way is at (0, 0).

its Hubble velocity. Comparing the two distances for thousands of galaxies, one can map the peculiar velocity pattern out to 100 megaparsecs.

The resulting maps reveal large-scale bulk flows that amount to about 10 percent of the Hubble expansion and are coherent over 30 megaparsecs or more. The flows are induced by the gravitational attraction from all matter present, and therefore probe the total amount of clumped matter, dark as well as light. Preliminary indications are that an amount of dark matter about equal to the critical density must be present in order to account for the amplitude of the observed flows. One can even pinpoint the sources of the flows, since there are vast concentrations of matter that must be responsible. The nearest one has been dubbed the Great Attractor; it is located at a distance of 40 megaparsecs from us. If real, it must consist of more galaxies than would be found in a dozen rich galaxy clusters. Our

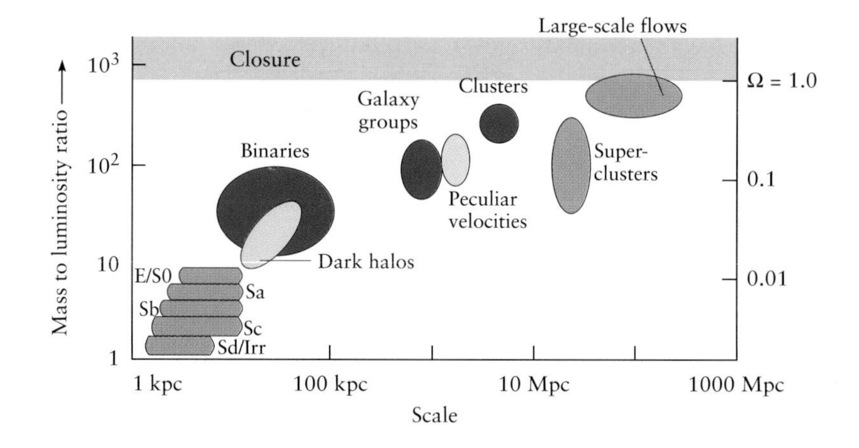

The ratio of mass to luminosity is an indication of the amount of dark matter in a system. Here, the ratio of mass to luminosity is plotted for various systems on scales from a few kiloparsecs, as measured for galaxies, to about 200 megaparsecs (Mpc). The larger the scale, the more dark matter seems to be present.

galactic plane obscures a large part of the Great Attractor, so one cannot count the number of galaxies directly. There may well be other similar complexes of galaxies that help generate the bulk flows.

A Universe at Critical Density?

The theory of inflation predicts that we live in a flat universe, where the density parameter Ω is equal to unity. That is, the density of matter in the universe should just equal the critical density at which the universe is closed. Any deviations from the critical density at the epoch of inflation can be visualized as a slight curvature of space, which would be smoothed out as the universe expands by a large factor during the inflationary epoch. Can we tell from the amount of dark matter observed whether the universe is indeed at critical density, as the theory of inflation predicts?

One can translate the Ω parameter, which measures mass density in terms of the critical value $3H_0^2/8\pi G$, into a ratio of mass to light. One does this by taking the ratio of the critical density to the observed luminosity density of an average, and large, volume of the universe. The result is that the mass-to-light ratio equals 1500 Ω. In other words, if $\Omega = 1$, one needs a mass-to-light ratio of 1500 to close the universe. This amount of mass is far greater than is observed directly. Alternatively, if we adopt the mass-to-light ratio of 300 measured on large scales as being a universal value, we would conclude that $\Omega = 0.2$, far less than the value predicted by inflation. One can reconcile inflation theory with observation only if the bulk of the dark matter is uniformly distributed over scales up to 10 megaparsecs. In this case, the dark matter would not have shown up, since on these scales only the clumped component of the matter has been measured. Indeed, a critical density would be compatible with the density measured by the bulk flows, which only sample scales larger than 10 megaparsecs or so.

The nature of the dark matter predicted by inflation is a profound and unresolved puzzle. We have two choices. Either the dark matter consists of ordinary, baryonic matter, or else it consists of some more exotic form of matter. The history of the universe during the first few minutes provides an interesting measure of the total amount of baryonic matter in the universe that may help resolve the puzzle.

Deuterium and the Baryon Density

For a significant clue to the composition of the dark matter, we look to the abundance of the heavier isotope of hydrogen created during the big bang, weighing twice the mass and called deuterium. Unlike helium, deuterium

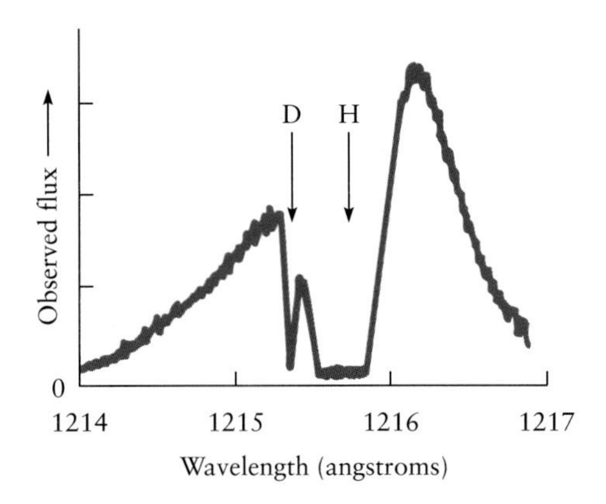

A spectrum of the nearby star Capella taken with the Hubble Space Telescope reveals the presence of cold interstellar hydrogen (broad dip) and a small amount of associated deuterium (narrow dip). The deuterium atoms, being twice as heavy as hydrogen, produce Lyman alpha absorption that is slightly shifted, by 80 kilometers per second, to the blue. The shift enables deuterium to be distinguished from hydrogen.

is a very fragile element. It burns at a temperature of only 1 million degrees Kelvin, well below the temperature in the solar core. By now, a considerable fraction of any primordial deuterium present at the birth of the galaxy would have been destroyed inside stars. This is confirmed by observation: interstellar clouds contain deuterium, as do gravitationally powered stars that have not yet developed nuclear burning cores; on the other hand, evolved stars have no deuterium.

To estimate how much deuterium was created in the big bang, one has to factor in all the deuterium that has since been destroyed. The percentage of the isotope destroyed since the big bang can be calculated if one knows its rate of destruction, which can be found by comparing the abundance of deuterated molecules in the atmosphere of Jupiter with the abundance of deuterium in interstellar clouds. A deuterated molecule is one in which a hydrogen atom is replaced by a deuterium atom: for example, deuterated, or heavy, water is HDO. The deuterium detected in Jupiter samples the interstellar gas as it was at the formation of the solar system, some 4.6 billion years ago. The abundance of deuterium in Jupiter, about 2 parts in 100,000, is approximately twice its abundance in interstellar clouds, which contain gas that has been processed through stars that have formed and died over the past lifetime of the galaxy. Observations with the Hubble Space Telescope, and by earlier satellite experiments, find a somewhat lower abundance of atomic deuterium in interstellar clouds than is detected in Jupiter, or in other words, than was present in our galaxy some 6.6 billion years ago.

The net trend in deuterium is unmistakable: it decreases with time. This is expected, since stars burn deuterium and do not generate a fresh supply. But because not all of the interstellar gas has passed through the hot cores of stars, some primordial deuterium survives. Allowing for the destruction of deuterium by stars, one infers a pregalactic deuterium abundance of 0.01 percent relative to hydrogen.

How does this figure compare with the deuterium abundance predicted by the big bang theory? Unfortunately, we cannot answer that question without knowing the value of the present-day density of baryonic matter, which cannot be stated within a factor of 10. Rather, we turn the problem around, and try to estimate the density of baryons in the universe from the measured deuterium abundance.

About 25 percent by mass of the hydrogen is synthesized into helium within the first three minutes, and deuterium is formed as an intermediate product in the reaction chain. A small fraction of this deuterium survives. If the density of ordinary matter is high, then the early universe synthesizes helium so efficiently that it makes essentially no deuterium. The density of baryons cannot exceed about a tenth of the critical density for closure of the universe, or too little primordial deuterium would have been synthesized. Conversely, the density of baryons cannot be too low, below 2 or 3

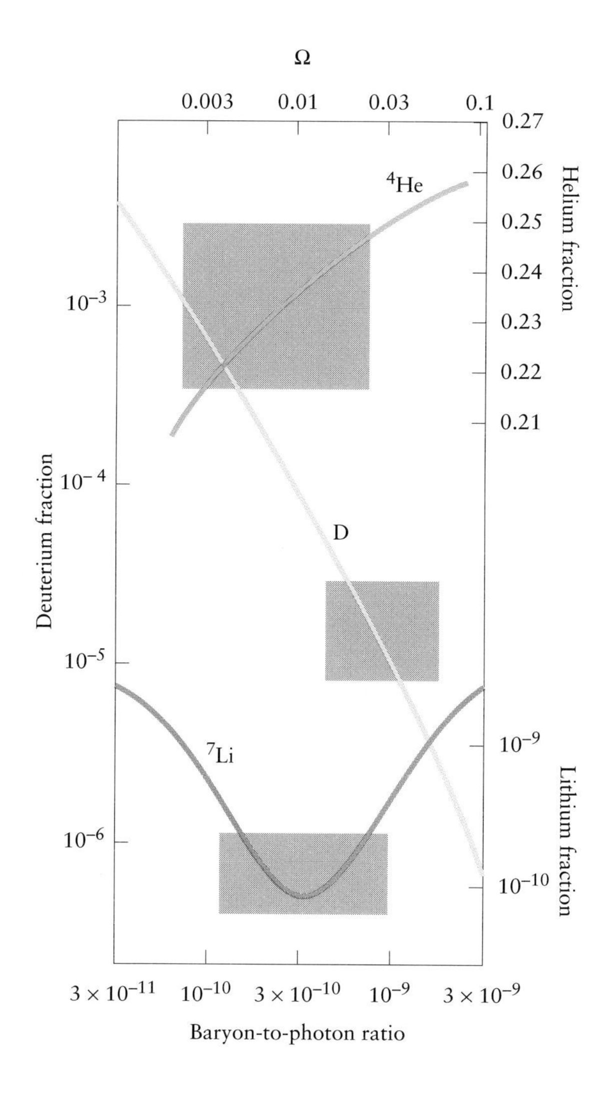

The abundances of helium, deuterium, and lithium-7 today depend heavily on the density of the universe at the time of their creation, measured here by the baryon-to-photon ratio (bottom axis) and the density parameter Ω (top axis). The conversion of deuterium to helium is more efficient in a high-density universe, so there is much less deuterium created and somewhat more helium. Shaded areas indicate uncertainties in observed abundances.

percent of the critical density, or else the early universe would overproduce deuterium. If the universe is at critical density, 90 percent of the matter in the universe must be nonbaryonic. That is to say, most of the dark matter consists of weakly interacting neutral particles that did not participate in the nuclear reactions that led to deuterium production.

The Lithium Puzzle

Helium and deuterium are not the only light elements produced in the big bang. Lithium is much rarer, and like deuterium, it too is destroyed by stars.

One actually measures lithium in T-Tauri stars. These stars, named after a prototype star in the constellation of Taurus, are very young, gravitationally powered stars that are often embedded in dense interstellar clouds of gas. The turbulent, convective atmospheres of such stars contain a high abundance of lithium. This latter element, only destroyed by stars, is a sure signature of stellar youth. As stars age, the lithium is destroyed. Convection churns up the atmospheric gases during the early phase of evolution and carries the pre-existing lithium down to hotter regions where it systematically is burnt. A middle-aged star like the sun has no detectable lithium in its atmosphere.

Lithium is synthesized both by the big bang and by cosmic rays in interstellar clouds. Cosmic rays are energetic particles that induce nuclear reactions when they occasionally collide with interstellar atoms of carbon, nitrogen, or oxygen. These heavy atoms are broken up and lithium nuclei are spewed out. A telltale signature of this process, called cosmic ray spallation, is that it produces nuclei of two isotopes of lithium, one of mass 6 as well as the common isotope of mass 7. The oldest stars, in Population II, show a residual lithium abundance of about 1 part in 10 billion relative to hydrogen; that abundance seems to be independent of the abundances of other elements such as iron. Moreover, most of the lithium is the

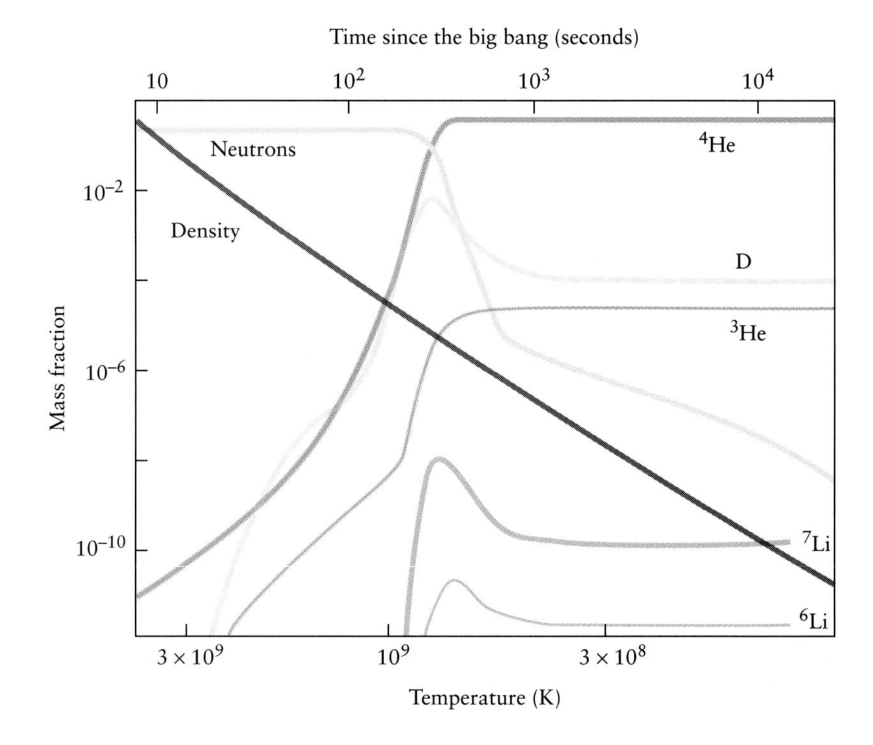

The history of element synthesis during the first three minutes. For example, while helium abundance built up gradually to roughly 25 percent by mass, deuterium attained a peak abundance of about 1 percent by mass, then was reduced to a very low value as helium was synthesized.

isotope of mass 7; any mass-6 lithium is too rare to have been produced by spallation. By contrast, in young Population I stars, about 10 times more lithium is measured. The lithium of these younger stars is believed to be of spallation origin, produced in interstellar clouds from which the stars were subsequently formed. A self-consistent picture emerges of lithium formation and destruction. We can be fairly confident that in the old spheroid stars, we are seeing lithium that was produced in the big bang.

While we infer that there most likely is a predominance of nonbaryonic dark matter, the abundances of lithium, deuterium, and helium require a minimal amount of baryonic matter. This amount is more than is directly measured in galaxies, and we conclude that there must also be a few percent of the critical density in the form of baryonic dark matter.

WIMPs: Exotic Particle Dark Matter

If, in a universe at critical density, most dark matter could not be baryonic, what other forms could it take? Likely relics of the early universe are species of stable, weakly interacting particles. One example is the neutrino, if it possesses a small mass. Normally, the neutrino is assumed to be practically massless, but a finite mass is not implausible. There are so many neutrinos left over from the big bang that a neutrino mass of even 50 eV, or one ten-thousandth the mass of an electron, would suffice to close the universe. Laboratory experiments are under way in several countries to determine a definitive mass for the neutrino, but at present these experiments are inconclusive. The current upper limit on the electron neutrino mass, which is obtained from tritium decay experiments, is about 10 eV. Other species of neutrinos could have higher masses.

A special name has been coined for a weakly interacting particle that is so massive it exceeds the mass of, say, a proton: the WIMP, for weakly interacting, massive particle. Exotic WIMPs such as the photino have been postulated to exist in sufficient quantity to close the universe. The problem is that there is no guarantee that these particles do exist. Disregarding this uncertainty, the big bang theory predicts their density today, if they do exist and are stable over the age of the universe.

The existence of the photino is predicted in a theory called supersymmetry. This theory doubles the number of known particles by postulating the existence of partner "-ino" particles. These particles are almost all short-lived, and exist in large numbers only in the very early universe, when the temperature was high enough to exceed the energy scale characteristic of supersymmetry, affectionately abbreviated to SUSY. As the universe cools, supersymmetry is broken. The relevant energy scale is not known from theory, but it must exceed 100 GeV to avoid conflict with particle experiments.

In our low-energy universe today, the lightest supersymmetric particle should still survive. It is expected to be the partner, in the sense of having a complementary spin, of the photon, and is therefore known as a photino. Its mass is expected to be 10 to 100 times that of the proton. The photino is uncharged and interacts very weakly with matter.

There is strong evidence of SUSY from experiments at CERN that measure the strengths of the nuclear interactions. At high enough energies the strengths of the weak and strong nuclear forces are expected to converge to the same energy, although there is no guarantee that they will do so. That they do converge at very high energy is the thesis of grand unification of the fundamental forces, whose breakdown in the very early universe gave rise to inflation. While this energy, some 10^{15} GeV, is very much higher than is directly accessible by experiment, the trend toward convergence is already apparent. Only if SUSY describes the high energy world do these two fundamental forces become indistinguishable at a unique energy. Only therefore with SUSY could one construct a strong case for the inevitability of grand unification.

Despite the weakness of the photino's interactions, several experiments are being designed to search for this particle. These experiments are of four types. One uses the atom-smashing machines, called particle accelerators, to verify the particle's existence. The high-energy collisions in these machines normally cause jets of energetic hadrons, including particles and antiparticles, to be ejected. So that momentum is conserved, the hadronic jets go off in two opposite directions, transverse to the collision direction. Although the weakly interacting photino would be invisible, it carries off momentum that must be balanced on the other side by a detectable jet. Hence a one-sided jet would be evidence for a supersymmetric particle. In another type of experiment, sensitive laboratory detectors search directly for photinos in the galaxy's halo that have been intercepted by the earth or the sun as the sun orbits the galaxy. Photinos that are trapped by the sun actually annihilate in its core. The heat they produce can slightly, but perhaps significantly, affect the sun's evolution. The annihilations generate as a by-product some energetic neutrinos that are quite distinct from those produced by thermonuclear fusion in the solar core. These high-energy neutrinos, as well as neutrinos produced by photino interactions in the earth, may be detectable in some of the underground detectors that are searching for solar and supernova neutrinos.

Radically different methods are used to search for the debris of photino interactions in the halo. Space or balloon-borne telescopes hunt above the earth's absorbing atmosphere for particles such as cosmic ray antiprotons and positrons produced in the halo by photino annihilations. However, cosmic ray protons interacting with heavy interstellar atoms also generate relatively low energy antiprotons and positrons. A way is needed to disentangle the two signals. Of course, the detection of a single heavier antinucleus, even antihelium, would be a phenomenal discovery and would require the existence of antistars and even antigalaxies. No such particles have been detected, needless to say. A similar strategy is to search for another relic of photino interactions; these are photons, specifically gamma rays, also produced when photinos annihilate in the halo.

If the dark matter is abundant enough to be the dominant contributor to the mean density of the universe, it will have important gravitational effects. Massive particles, whether light neutrinos or heavy photinos, inevitably play significant roles in the early evolution of structure in the universe. But before turning to the gravitational effects of exotic dark matter, we look at the more ordinary, and far more massive, forms it can take.

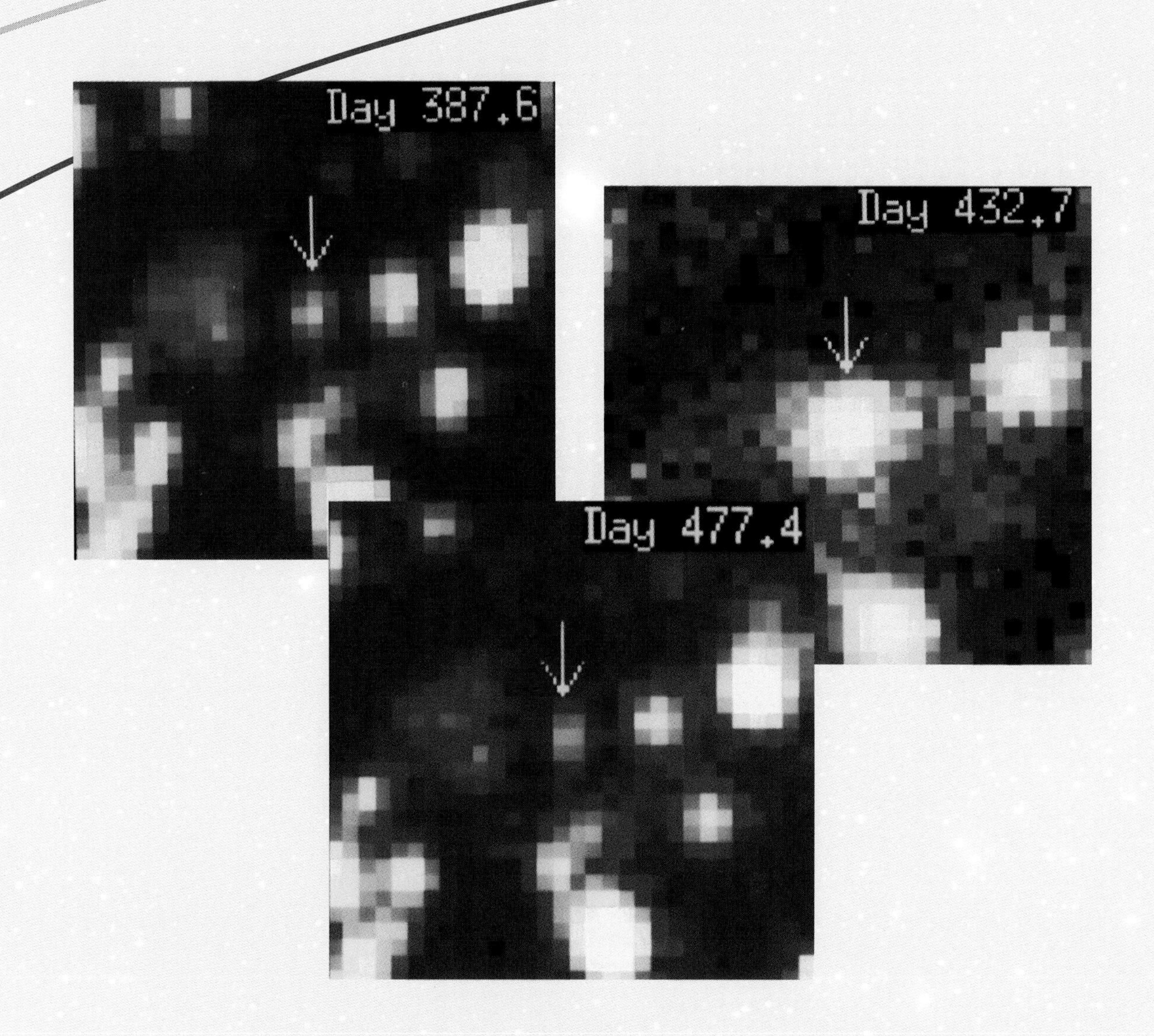

A MACHO event is evidence of dark matter: a star in the Large Magellanic Cloud is amplified when an invisible massive compact halo object (or MACHO) in our galaxy halo passes across the star's line of sight.

8

Baryonic Dark Matter

The most natural form for dark matter is matter that we know exists, namely baryons. The big bang explanation of the light element abundances requires the existence of some baryonic dark matter. Although these same abundances imply that most dark matter is *non*baryonic, the amount of dark baryonic matter is still most likely several times that we can see in luminous matter, or about 3 percent of the critical density for closing the universe. But where do we look for the baryonic dark matter?

One's first expectation might be that baryonic dark matter consists of burnt-out stars in the galactic halo, yet other forms, such as planets and black holes, are also possible. Baryonic dark matter does exist: it is far more uncertain whether there exists enough to account for all the dark matter in galaxy halos, galaxy clusters, and galaxy superclusters, or to close the universe. However, baryonic dark matter is a serious candidate for dark matter at least in galaxy halos. To account for the closure density, on the other hand, one must appeal to WIMPs, or some other weakly interacting particle. In acknowledgment of the rivalry between these two forms of dark matter, the favored candidates for baryonic dark matter have been dubbed MACHOs, for massive compact halo objects.

Life after the Main Sequence: White Dwarfs

A star like the sun is destined to die in an incendiary display of stellar brilliance. Once the core of the star is depleted of hydrogen, the star's source of nuclear energy is temporarily exhausted. Nuclear reactions cease in the core itself, although hydrogen burning continues in the surrounding shell. Meanwhile, the hydrogen-burning shell itself becomes hotter, so that helium is produced ever more vigorously. The injection of extra heat into the

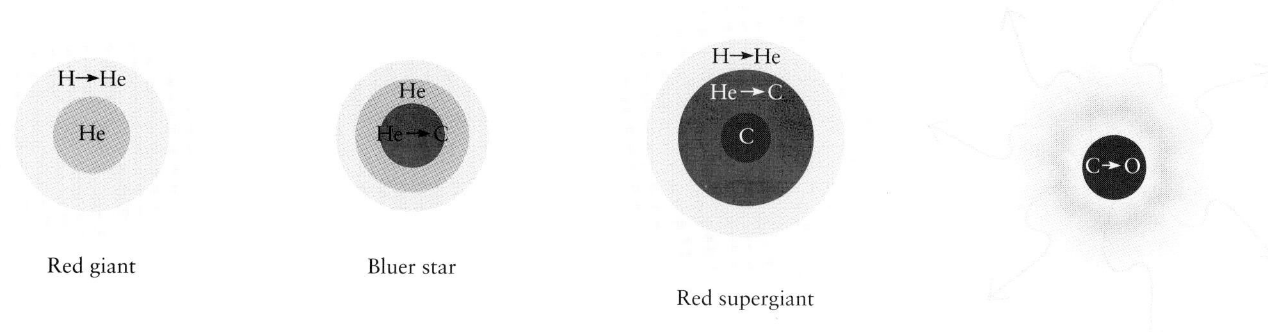

Late stages in the evolution of a star of low mass.

shell causes the envelope of the star to expand. The star's radius increases by a factor of about 100, and the star becomes a red giant.

As the envelope expands, it cools. This cooling of the star's outer layers of gases is explained by a property of stellar luminosity. The luminosity of a star is proportional to the fourth power of its surface temperature, T, and to the square of its radius. As the stellar radius increases, the star's energy supply, and therefore its luminosity, remains the same, so its effective temperature has to decrease.

While the envelope is cooling, the helium core, containing about 10 percent of the star's mass, contracts and heats up. Once it attains a 10-fold increase in temperature, to about 100 million degrees Kelvin, the helium ignites. Three helium nuclei combine to form a carbon nucleus in the triple alpha process, which results in the release of fusion energy. During the helium-burning phase, which lasts for about 100 million years, the star moves onto the horizontal branch in the Hertzsprung-Russell diagram.

As the hydrogen-burning shell eventually exhausts its fuel and weakens, the star shrinks and becomes bluer. Its time spent on the horizontal branch ends when the helium in the core is exhausted. The core is now composed of carbon, but the carbon core is not hot enough for further fusion to take place and it contracts. Meanwhile, helium outside the carbon core becomes hot enough to ignite. This helium shell burns vigorously, and its heat ignites the fusion of hydrogen in an additional surrounding shell. The heat from both burning shells causes the envelope to swell and form an even larger red giant, called a red supergiant star, as luminous as a thousand suns. Now the carbon core heats up sufficiently to produce energy by carbon fusion. Carbon, of atomic mass 12, combines with helium, of mass 4, to form oxygen, of mass 16. Energy is produced so vigorously that the star becomes unstable, and the outer layers are ejected in a stellar wind. Eventually, a core of carbon is left surrounded by an expanding shell of ionized gas, containing about 10 percent of the star's original mass. The

The evolution of a star of 5 $M_\odot$ is traced on the Hertzsprung-Russell diagram, from its leaving the main sequence to its final dimming as a white dwarf.

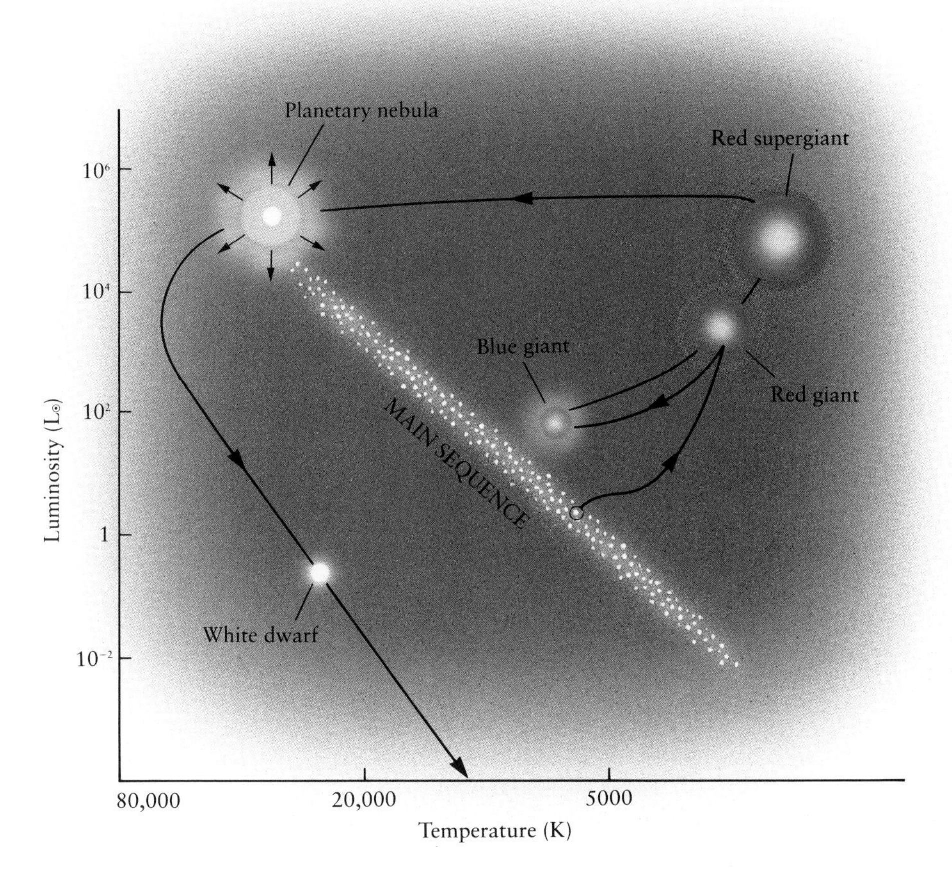

former supergiant has become a planetary nebula. The shell of gas is chemically enriched by the debris from the nuclear reactions that occurred during the life of the star. In the center of the planetary nebula is the hot, rapidly cooling stellar relic.

The central star in a planetary nebula is a white dwarf. Its temperature is not hot enough to ignite the remaining fuel, which is mostly carbon and oxygen (or helium for a white dwarf of low mass). Although the star no longer produces radiation pressure to counteract the inward pull of gravity, it is prevented from total collapse by electron degeneracy pressure. This quantum mechanical effect is a consequence of the uncertainty principle, described in Chapter 4, which states that particles have quantum uncertainty in their properties. This uncertainty is equivalent to infinitesimal motions, and these motions cause a pressure to be exerted. In a white dwarf, this pressure is exerted by electrons that are crowded together as the collapse of the star squeezes atoms together until they overlap. It is the pressure of these "degenerate" electrons that supports the white dwarf against the force of gravity. The more massive a white dwarf, the smaller its size. The maximum mass possible for a white dwarf is 1.4 $M_\odot$, the Chandrasekhar mass. An object more massive than this cannot be supported by electron degeneracy pressure.

The atomic nuclei occupy a much smaller volume than the electrons and are not degenerate at these densities. There are no further nuclear reactions in a white dwarf, which shines only by virtue of its accumulated heat supply. Once the star's energy has radiated away, after about 10 billion years, the white dwarf becomes a black dwarf. This is a compact star of about the earth's size, but with very low temperature and very little luminosity. Such objects are possible forms of dark matter.

A star that has a mass less than about 8 $M_\odot$ ends up forming a white dwarf of mass less than 1.4 $M_\odot$, the maximum for a star of this type. We see white dwarfs in the Hyades cluster, where stars below 6 $M_\odot$ are still on the hydrogen-burning main sequence. The Hyades white dwarfs must have formed from precursor stars of more than 6 $M_\odot$.

A star of initial mass more than 8 $M_\odot$ does not lose enough mass from stellar winds to become a white dwarf. Indeed, a massive star is defined to be one that is too massive to end its life in that form. It has another fate, which provides another possible form of baryonic dark matter.

The Formation of a Neutron Star

Massive stars spend relatively little time on the main sequence. A 15 $M_\odot$ star spends 10 million years on the main sequence, and a 30 $M_\odot$ star only one million years. Because a massive star evolves so rapidly, there is little time for outer shells to burn hydrogen once the helium core contracts and ignites again as the star becomes a red giant. After the helium is exhausted, the core again contracts and undergoes triple alpha burning, with three helium nuclei fusing into a carbon nucleus. The core eventually becomes hot enough for carbon to burn to oxygen; at this stage there is also a helium-burning shell, and the envelope expands to form a red supergiant. The core continues burning until it reaches a temperature of 1 billion degrees Kelvin, as thermonuclear fusion burns heavier and heavier elements, finally creating a core of iron. Iron is the final stage of fusion: the synthesis of any heavier element from lighter elements does not release energy; rather, energy must be used up. The iron core is said to be endothermic (energy absorbing) in fusion or fission reactions: no more nuclear processing can occur unless energy is supplied.

As heat escapes, the core contracts, and the temperature rises above 1 billion degrees Kelvin. When the mass of the core exceeds 1.4 $M_\odot$, even

The evolution of a high-mass star is traced on the Hertzsprung-Russell diagram, from its leaving the main sequence to finishing as a supernova.

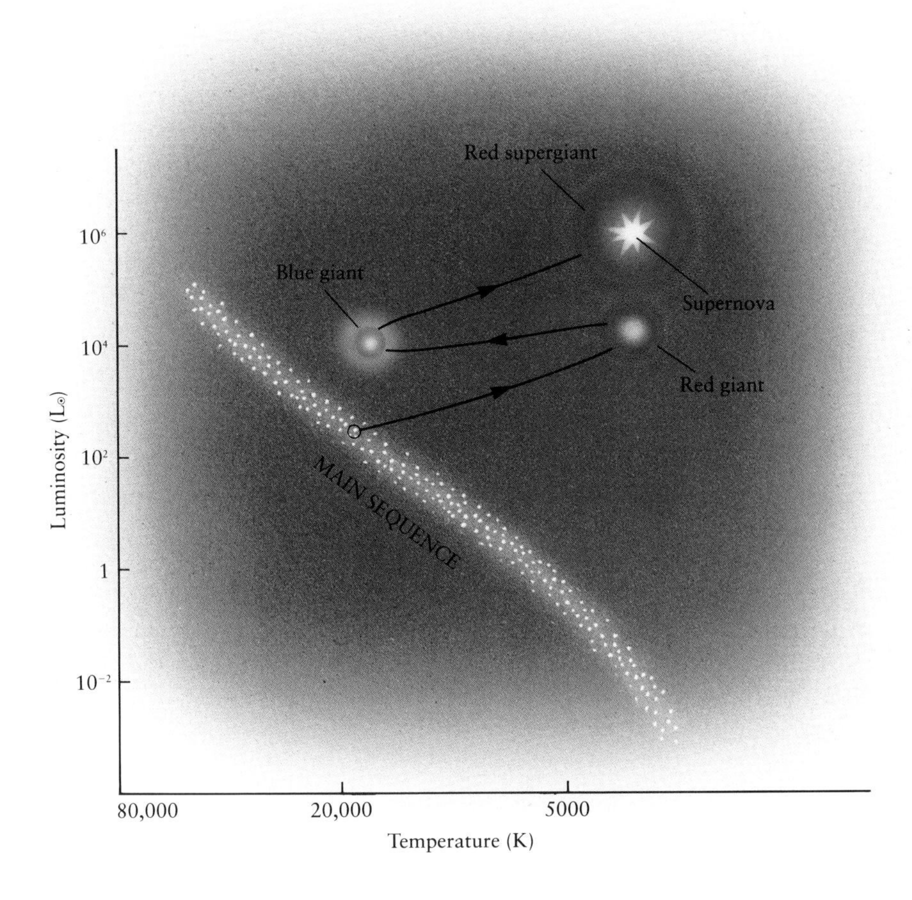

electron degeneracy pressure can no longer support it. The core collapses to a much denser state of matter in which the atomic nuclei, rather than the atoms, are squeezed together. In this state, protons are able to capture electrons to form neutrons. Simultaneously, energy is radiated away in the form of weakly interacting antineutrinos, which escape unimpeded from the star. The loss of energy hastens the formation of what amounts to one giant nucleus containing only neutrons. A neutron star is a sphere of gas compressed to the density of a nucleus and supported by neutron degeneracy pressure, the quantum mechanical pressure that arises when neutrons are squeezed to near contact.

The resulting neutron star has a radius of about 1 kilometer and a density of about 1 billion tons per cubic centimeter. The collapse of the stellar core drives a blast wave that expels the outer layers of the red supergiant in a huge explosion. This is a supernova. The relic neutron star is formed hot, and glows in x-rays. It has no source of energy to maintain its temperature, however, and it slowly cools down. After a few million years, it is invisible, at least through its thermal emission. It becomes dark matter.

Supernovae are prolific sources of neutrinos. These particles are the unambiguous signature of the formation of a neutron star. Indeed, 99 percent of the energy in the explosion is emitted in neutrinos and accompanying antineutrinos. These weakly interacting particles are notoriously difficult to detect; fortuitously, two neutrino "telescopes" were operating on 23 February 1987, when the supernova dubbed SN 1987A exploded. Indeed, with

Supernova SN 1987A appears in the photograph on the right, taken a few days after the explosion, as a bright spot in the upper right where before there was none. The supernova erupted on February 23, 1987, near the Tarantula nebula, from a 20 $M_\odot$ blue supergiant star in the Large Magellanic Cloud.

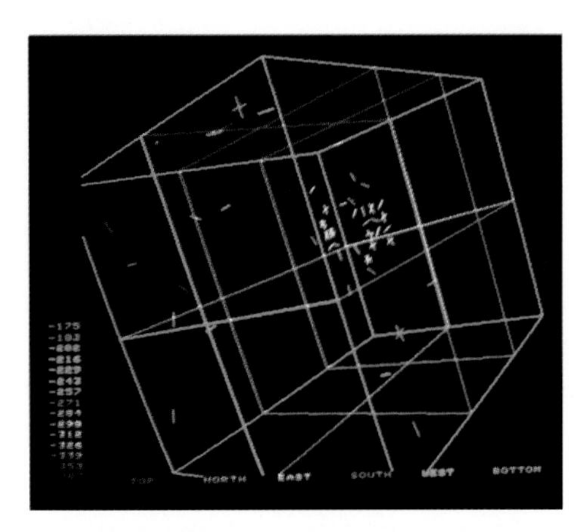

On February 23, 1987, about three hours before supernova SN 1987A was first photographed, 12 neutrinos interacted with electrons in the 8000-ton underground liquid detector in the Kamioka zinc mine near Tokyo, and 8 neutrinos were captured in a similar detector in the Morton salt mine near Cleveland. This photograph records the latter event.

hindsight, the neutrinos from the supernova reached the earth hours before the light signal was detected. Of course, the supernova really exploded about 50,000 years ago, but the Large Magellanic Cloud, where SN 1987A is located, is at a distance of 50,000 light-years from the sun.

The neutrino detection experiments are huge underground vats containing thousands of tons of water monitored by a vast array of phototubes. When neutrinos and antineutrinos reach the vats of water, the neutrinos are scattered in a forward direction by electrons in the water molecules, whereas the antineutrinos are absorbed by protons in those same molecules, and generate neutrons and randomly moving positrons. The fast electrons and positrons slow down in the water and emit a glow of light (called Cerenkov radiation) as they decelerate. On 23 February 1987, 20 neutrino-triggered scatterings were detected from the direction of the Large Magellanic Cloud over a 13-second period, about four hours before the first report from Chile of the optical supernova. Detection of the neutrinos and antineutrinos confirmed that a neutron star was being created. The handful of events detected implied that about 10^{57} neutrinos and antineutrinos per second were emitted as the neutron star formed.

Observing Neutron Stars

In A.D. 1054, Chinese historians of the Sung dynasty recorded a new star in the constellation of Taurus, now recognized to have been a supernova. Although the star faded over the course of the following year, astronomers have identified the remnant of that supernova at the location reported by the historical record. This remnant is the Crab nebula, which consists of filaments of gas expanding at a speed of about 1000 kilometers per second. The proper motion of these filaments confirms that they originated in the explosion of A.D. 1054. The Crab nebula is an intense source of polarized radio waves and optical light, as well as of x-rays. It was soon realized that an energy source was required in order to explain this radiation.

The puzzle was solved in 1968 with the discovery of a neutron star in the center of the Crab. An object inside the nebula was observed to emit regular pulses of light every 0.033 second, although the frequency of the pulses appeared to be decreasing very gradually. The slowing of the pulses was attributed to the slowing spin of a rapidly rotating star, called a pulsar, which was slowing down as it lost energy through the emission of electromagnetic radiation, similar to the radiation that was observed coming from the remnant. The source of the pulses had to be much more compact than a rotating or pulsating white dwarf; only a neutron star fit the requirements. The pulses of light are from radiation beamed out along

The Crab nebula is the expanding remnant of a star that exploded in the year 1054.

north and south magnetic poles of the spinning, highly magnetized neutron star, analogous to a lighthouse beacon. The rate of slowing fixed the neutron star's date of formation, as expected, in 1054, the year of the supernova.

Pulsars had been noticed as early as 1967, but the original pulsars were simply identified as starlike sources of radio signals, with highly regular periods. The discovery of a pulsar in the Crab nebula confirmed their identification as neutron stars that are formed in a supernova explosion. Hundreds of pulsars have now been observed, having periods ranging from milliseconds for the fastest pulsars up to seconds for the slowest. The pulsars are mostly located within the Milky Way, in the galactic plane. Two pulsars, one in the Crab supernova remnant and one in the Vela supernova remnant, formed as hot neutron stars in explosions about 1000 and 10,000 years ago, respectively. The pulsars identified in young supernova remnants appear to have had a surprisingly fast period of about 10 to 20 milliseconds when first formed.

The fastest pulsars known have millisecond periods. The stability of their periods compares favorably in accuracy of timekeeping with the best devices made by man. Pulsars keep time to a few parts in 10^{14} over a year

The moving magnetic field of a spinning pulsar generates a tight beam of electromagnetic radiation, directed out along the star's north and south magnetic poles. The beam revolves as the star spins; if the earth is in the way, we observe a flash of light.

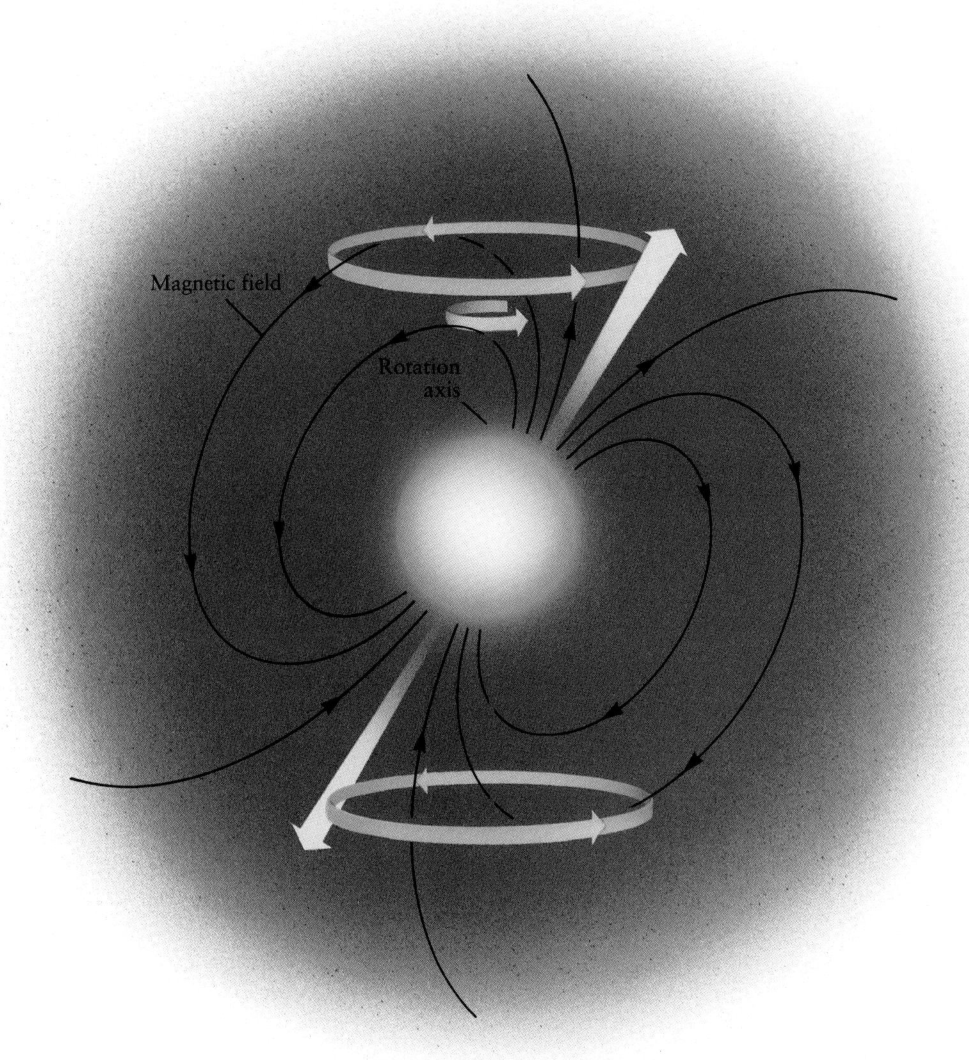

or longer, comparable to the accuracy of the best terrestrial atomic clocks. In addition to those in young supernova remnants, many millisecond pulsars have been discovered in globular clusters, systems of old stars that are at least 10 billion years old. It is believed that these millisecond pulsars have been caused to rotate faster as a result of accreting mass from a companion star that is evolving into a red supergiant. The close companion swells up and loses mass onto the neutron star.

Pulsars resemble spinning magnets; with time, they spin down as the electromagnetic radiation takes its toll and become invisible even at radio frequencies. Our galaxy is teeming with neutron stars that are long-dead pulsars.

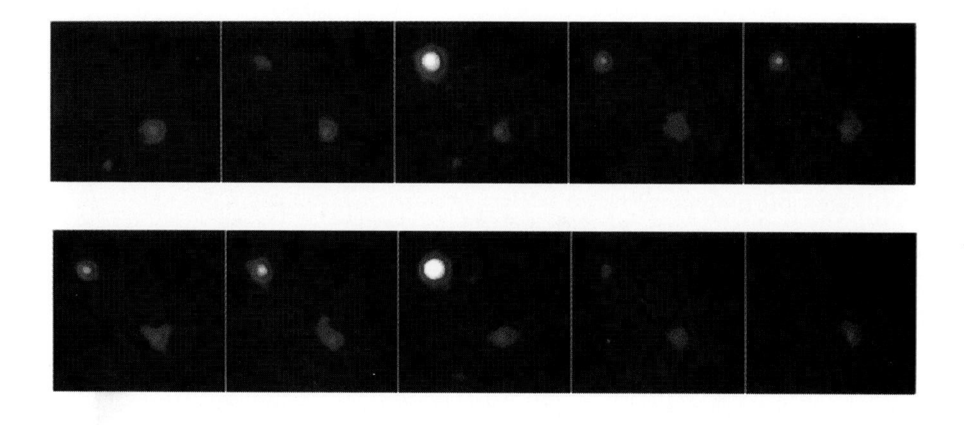

Geminga is a gamma ray pulsar, a spinning neutron star that beams gamma rays from its poles, like a lighthouse beacon, as it spins on its axis. Geminga is the closest known neutron star, about 100 parsecs away, and generates about four gamma ray flashes every second.

Brown Dwarfs

If some baryonic matter takes the form of dead stars, other baryonic dark matter may be objects that never managed to become stars. Brown dwarfs are objects so low in mass that their central temperatures never became high enough to ignite nuclear fusion reactions. Objects smaller than 0.08 $M_{\odot}$, and these include giant planets, generate no nuclear energy, and so are not luminous. However, they produce a small amount of gravitational energy since they are very slowly contracting. The core of Jupiter, for example, contracts a few millimeters per year. The planet actually radiates in the infrared about 50 percent more energy than it receives from the sun.

Astronomers search for brown dwarfs in two types of locations. In a binary system, the presence of a low-mass companion will cause the primary, or more massive, star to execute an orbital motion around the system's center of mass. This motion would be detectable either as a periodic Doppler shift of a few meters per second or as a periodic displacement of the center of light, as large as a fraction of an arc-second in a year for nearby systems. However, both spectroscopic and astrometric surveys of nearby stars have failed to turn up any brown dwarf companions.

Another approach is to search nearby, relatively young stellar associations for free-floating brown dwarfs that may still be visible as infrared sources. The small amount of energy generated by a brown dwarf as it settles should be visible only in the infrared because of the object's low surface temperature. There are other types of infrared-emitting stars, however, and the telltale signature of a brown dwarf would be its location above the lower end of the hydrogen-burning main sequence in the Hertzsprung-Russell diagram. To date, no strong candidates have emerged for objects

Karl Schwarzschild, the astrophysicist who discovered, from Einstein's theory of general relativity, the mathematical expression for the gravitational field around a black hole.

of substellar mass, outside the solar system, apart from two planetary mass objects inferred to be orbiting a pulsar. However, a modicum of theoretical faith suggests that despite the failure of our best efforts at detection, brown dwarfs should surely exist in considerable number. After all, why should a fragmenting interstellar cloud recognize the minimum mass of a hydrogen-burning star?

Black Holes

Black holes are the darkest form one can imagine for dark matter. Although they cannot be directly observed, the astronomer has indirect strategies for determining their presence. Because of their invisibility, they constitute a logical candidate for dark matter. Whether there are indeed enough black holes to resolve the dark matter problem is unknown.

What Is a Black Hole?

Black holes are predicted by the theory of general relativity, one of the theories of relativity created by Albert Einstein. The theory of special relativity describes the structure of space and time, whereas the theory of general relativity describes space, time, and gravity. The first of these theories tells us that space and time are different aspects of four-dimensional space-time. Thus a point in space usually has a past and a future. An explosion is an example of a point in space-time: the explosion is both at a specific point in space and at a specific time. Any observer, or any light signal, can be said to move toward the future. If it travels far enough and long enough, the observer, or light signal, is in theory eventually able to reach any other point in space at some future time. The one exception to this rule is experienced by an object near a black hole, which is defined to be a region in space-time where there are events from which there is no escape, even for light signals. A black hole is a *trapped surface:* cross this surface, and there is no going back!

The radius of a black hole, named after the astronomer Karl Schwarzschild, is proportional to the mass of the hole. For a black hole of a solar mass, one finds that the Schwarzschild radius is only 3 kilometers; for supermassive black holes, which are thought to contain up to one billion solar masses, the Schwarzschild radius would only be one ten-thousandth of a parsec, exceedingly small by astronomical standards. Planets would *not* be sucked in by a black hole: the earth would just as happily orbit if the sun were replaced with a black hole of the same mass, since the gravity field at this distance would be unchanged. A rocket ship visiting a black hole would feel that gravity was no different from that around a star, until the spacecraft approached within a few Schwarzschild radii of the hole. Then tidal forces would begin to affect the spacecraft.

Gravity is measured by observing its tidal effects: it acts on an extended body and pulls from ever so slightly different directions, causing the body to develop internal stresses. In the case of the rotating earth, covered with a deep layer of liquid water, ocean tides develop as a manifestation of the gravitational pull of the sun and moon. However, all forms of matter in a gravitational field behave like the tides in displaying this tendency of gravity to generate shear. There is indeed a subtlety about tides: they demonstrate that gravity acts not just at a single point in space, but over an extended region: it tugs and pulls.

A famous thought experiment suggested by Einstein illustrates how gravity can be measured by the inevitable force gradient that arises when it acts on anything other than a single idealized point mass. Consider a freely falling elevator in space near the earth. Now place an apple, an example of our idealized point mass, in the center. There is no motion of the apple relative to the elevator, no way of telling from inside the elevator if the apple is falling toward the earth or is stationary in deep space far from any gravity. But place a second apple alongside the first apple and the effects of the earth's gravity can be detected: the two apples are each falling toward the center of the earth, and so are slowly approaching each other. Tidal forces are very strong near a black hole, and no material object of finite size can withstand them: such an object would be ripped apart.

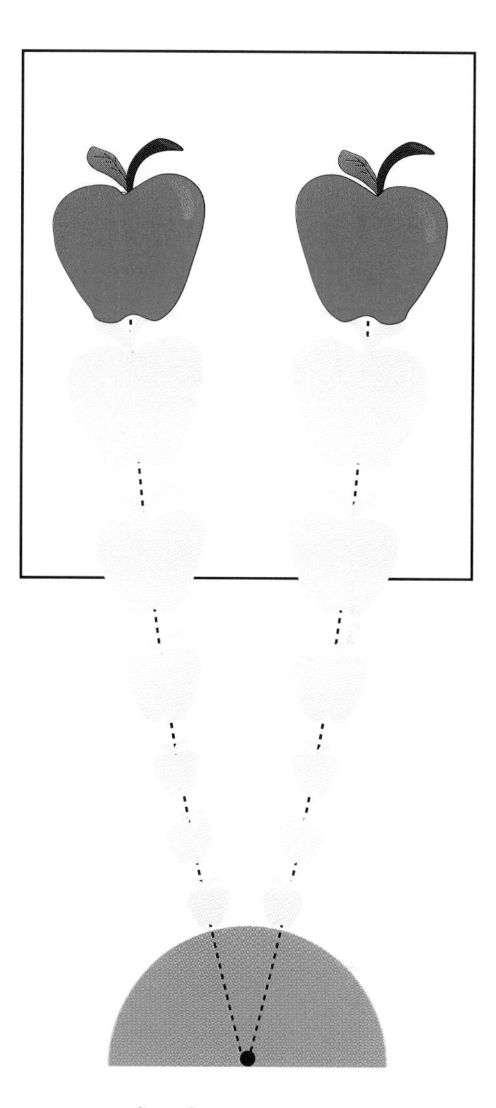

Einstein's thought experiment. Two apples are in an elevator falling toward the earth. To an observer inside the elevator, the only sign that they are falling is the narrowing of the distance between the apples as they approach the earth.

One of the most dramatic tests of the theory of general relativity probed the tidal forces in a strong gravitational field, in this case near a binary pair of neutron stars, observed as a pair of pulsars. If the theory is correct, the tidal forces should cause the orbital energy of the pulsar pair to be slowly radiated away as gravitational waves. As a consequence, the neutron stars, separated initially by about 1 million kilometers, should spiral together at the rate of about a few centimeters per year. Because the remarkably stable spinning pulsars act as clocks of superb accuracy, their orbits can be timed with high precision. Einstein's prediction is verified to an accuracy of at least 99.7 percent.

We have seen in Chapter 6 that space is distorted by gravity. General relativity predicts that time, and more generally space-time, is also distorted by gravity. The ticking of a clock is slowed down in a high gravity field. This effect has been measured in atomic crystals in laboratories on the earth: certain types of crystals, when excited, undergo very stable oscillations. The frequency of the oscillations is found to depend on the crystal's height above the earth's surface. As the height is changed, the frequency varies by about

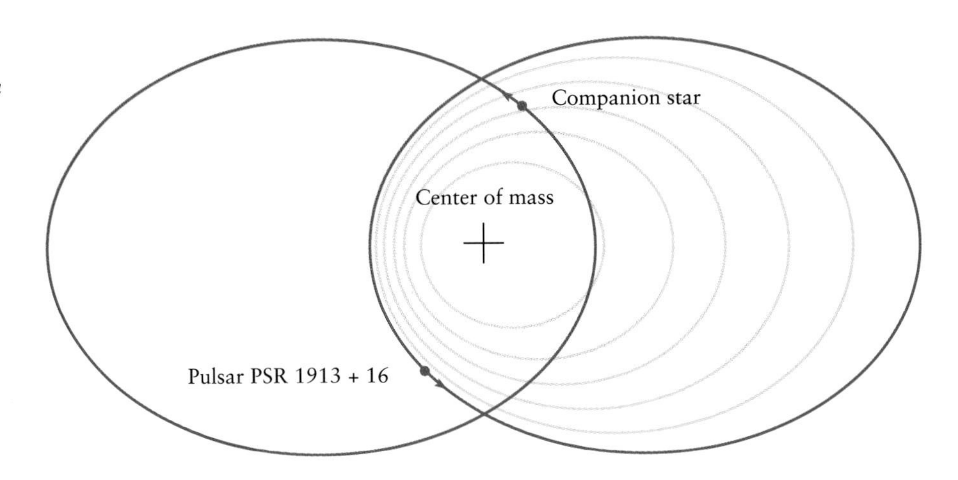

The orbit of pulsar PSR 1913 + 16 will become smaller and smaller as it spirals in toward its partner star. The timing of the orbits is a test of Einstein's theory of gravitation.

one part in a billion. This slowing down of time due to gravity is called a *gravitational redshift,* since the net effect is a redshift in light as it escapes from a strong gravity field.

Gravitational redshifts have been detected in white dwarf stars. These stars are about 100 times smaller than the sun, so the gravitational potential energy is 100 times larger. This means that the gravitational redshift near a white dwarf is 100 times greater than the gravitational redshift of the sun, or about 1 part in 10,000. Near a black hole, the slowing down of time is a far more dramatic phenomenon, however. Indeed, as an infalling object approaches the Schwarzschild radius, an outside observer would perceive a clock on board ticking ever more slowly. Any radiation emitted by the object would be more and more redshifted, until it eventually disappeared from view.

Where Do Black Holes Come From?

Black holes are the endstates of very massive stars. Perhaps stars of initial mass greater than about 50 $M_\odot$ are destined to end up as black holes. There is no known pressure, not even a pressure as extreme as electron degeneracy or neutron degeneracy, that can stop the collapse of a stellar core if its mass is more than a few solar masses.

Extremely massive black holes, with masses equal to thousands, millions, or even billions of solar masses, are believed to form when the core of a star cluster or the nucleus of a galaxy becomes unstable and collapses. When stars are packed together at a sufficiently high density, they some-

The spiral galaxy M100 in the Virgo cluster has a bright nucleus, within which a massive black hole may be embedded.

times collide with one another, and the resulting debris accumulates into a giant black hole, according to theoretical speculations. The density of matter when such a supermassive black hole forms need not be great. The reason is that the density is proportional to the mass divided by the volume (M/R^3), but the Schwarzschild radius of the black hole is $R = 2GM/c^2$: hence the density is proportional to $1/M^2$. This means that the density of matter in a black hole of a million solar masses is relatively low, about the density of air, when the black hole forms (and is at a size equal to the

Schwarzschild radius). It is just that so much mass is present in such a small region of space that the gravity field is large enough to prevent light from escaping. In a low-mass hole, the density is of course far higher.

Properties of Black Holes

The simplest black hole is defined only by its mass: fix this, and the resulting black hole is unique as observed from the outside. In fact, for simple black holes, the mass (which determines the Schwarzschild radius) is the *only* measurable quantity. The most complicated black holes that can exist are characterized by two additional quantities as well as mass: angular momentum and charge. These quantities can be measured by observing the orbits of particles around the black hole. Composition is irrelevant. It does not matter what collapses into a black hole, since the outside observer loses all information about the detailed composition. This conclusion is expressed as the *no hair* theorem: "Black holes have no hair." More specifically, the three quantities—mass, angular momentum, and charge—provide a complete description of any black hole.

Black holes have some exotic properties. Energy can be excavated from a spinning black hole, rather like bouncing a ball off a spinning disk. The center of a black hole is a singular point of space-time, where the theory of general relativity allows the possibility of a bridge to another space-time, known as a wormhole. While the theory of general relativity says that black holes are forever (once trapped, you can never escape), the theory of quantum gravity suggests a bizarre fate for mini–black holes: if sufficiently small, a black hole can evaporate. We described earlier how energy can effectively be liberated from space-time at the boundary of a black hole and ejected in the form of fast particles. According to quantum theory, pairs of particles are produced temporarily for a short period of time even in a vacuum; when such a pair comes into existence, the strong tidal force near a mini–black hole causes the pair to split up. One particle is trapped, freeing the other particle to escape, and thus liberating energy.

Thus the ordinary matter brought into being by the big bang may have been in the form of tiny black holes, in addition to being in ordinary forms such as protons and electrons. A black hole of about the mass of a small mountain (roughly 10^{15} grams) would evaporate in 10 billion years, and smaller black holes would evaporate more rapidly. These smaller black holes would have disappeared already, but the legacy of the early universe might be a proliferation of black holes, each with the mass of a mountain but the size of a proton. If ever a mini–black hole were discovered, it could only have come from the big bang, since stars cannot form black holes with masses less than 3 $M_{\odot}$.

Observing Black Holes

At the moment when mini–black holes completely evaporate, they would produce bursts of particles and photons lasting less than a millisecond. Bursts of gamma rays have been detected: their origin is unknown, and it is possible that they are the death signatures of mini–black holes produced in the early universe.

Although stellar black holes are invisible, they can sometimes be recognized when they are paired with a companion star, since in this circumstance they can become vigorous emitters of x-rays. The black hole may be close enough to its companion star so that when the star evolves to become a giant or a supergiant, matter from the star's atmosphere spews onto the black hole. The accreted gas spirals onto the black hole, forming a disk that slowly falls into the black hole itself. As the gas falls in, it heats up and emits x-rays.

Material from a normal star flows onto its companion black hole, forming an accretion disk of gas around the hole. The gas grows hot as it spirals toward the center of the disk and emits torrents of x-rays before being swallowed.

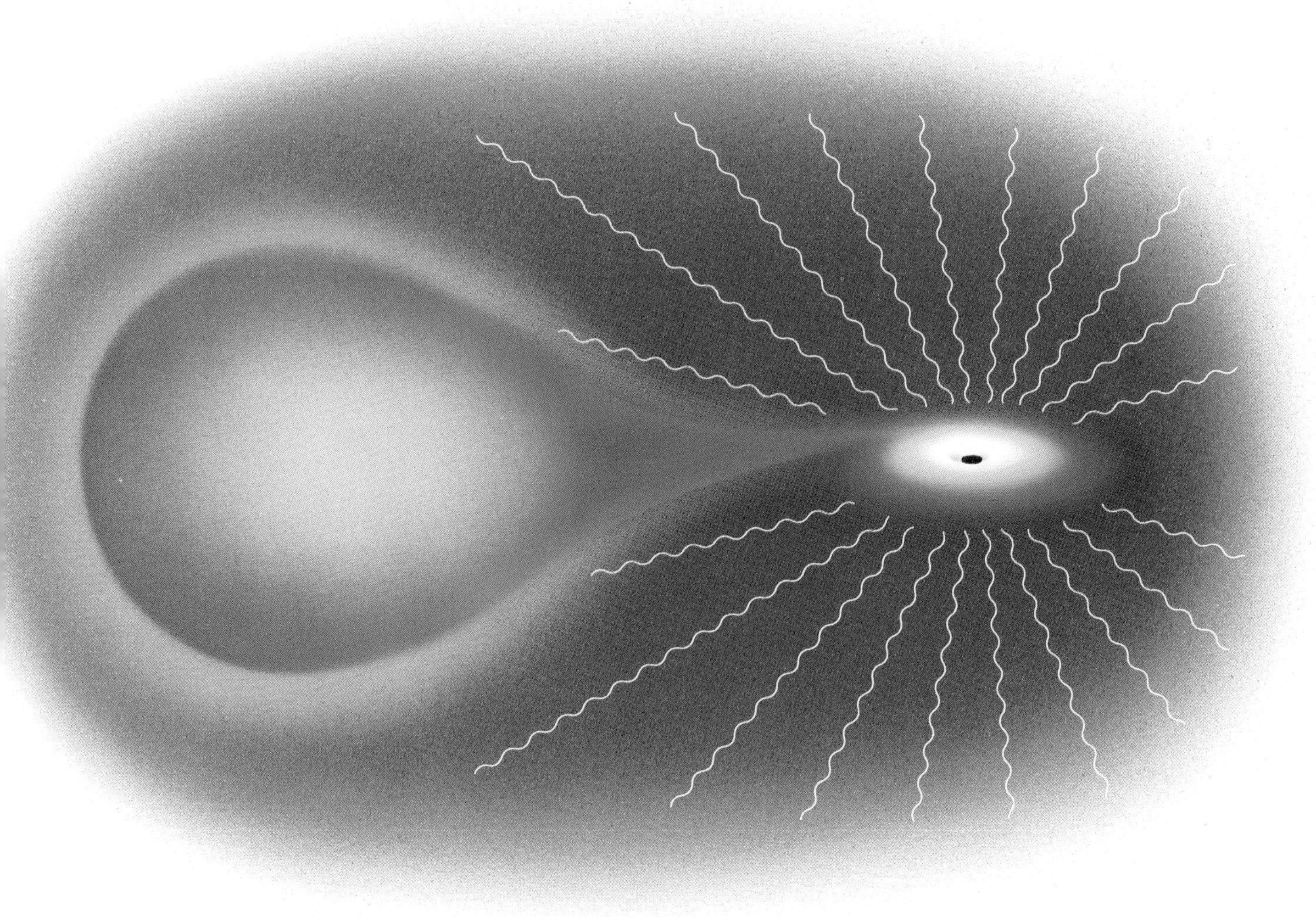

The best-known black hole candidate is Cygnus X-1, which is an invisible source of x-rays in orbit around a luminous blue giant star known as HDE 226868. The orbital period is 5.6 days, and the mass of the unseen component exceeds 3 $M_{\odot}$. Cygnus X-1 is five times brighter at x-ray frequencies than in the optical band: it is predominantly a producer of x-rays. The invisible object has an x-ray luminosity of about 5000 $L_{\odot}$, and the blue giant companion has an optical luminosity of about 1000 $L_{\odot}$. Moreover, the x-ray emission fluctuates rapidly every few milliseconds or less. These are precisely the characteristics expected for an accretion disk around a black hole. The x-rays are emitted as blobs of matter torn off the accretion disk fall into the black hole, while their rapid flickering is a result of the short time it takes for the blobs to fall through the Schwarzschild radius—about 10^{-4} second. There are several other massive x-ray binary systems believed to contain black holes.

Supermassive black holes are theoretically conjectured to power energetic phenomena emanating from the centers of certain galaxies. Seyfert galaxies, named after the astronomer Carl Seyfert, constitute about 1 percent of all spiral galaxies and are characterized by having extremely bright nuclei, called active galactic nuclei. The fact that active nuclei are observed in only 1 percent of otherwise normal galaxies suggests that the active phase may be short-lived, lasting perhaps 1 percent of the present age of the universe. Most galaxies may have undergone an active phase in their past and should still possess massive but inert black holes in their nuclei. Once the monster in the center is no longer fed, the accretion disk dies away and even a supermassive black hole might leave little trace of its presence in a galaxy.

In some cases, the spectra of active galactic nuclei reveal the presence of enormous velocity flows, up to one-thirtieth of the speed of light, in a compact region less than a light-year across. We can infer the diameter of the region of these flows from the variability of the light emission: if the luminosity of an object varies on a timescale of, say, a year, the region producing the light cannot be more than a light-year across. Gas shed by evolving stars in the inner galaxy is believed to be fueling a central massive black hole. If matter is falling toward the center on a radial orbit, it would be accelerated to near the speed of light before reaching the Schwarzschild radius. Because the galaxy is rotating, the gas has angular momentum, and for the most part it is expected to spiral in, forming an accretion disk of far greater size than predicted for stellar black holes, and heating up before disappearing across the event horizon of the black hole. Collisions between the rapidly moving gas atoms can efficiently release vast amounts of energy, if the supply of accreting material is adequate. Stars may collide in the closely packed nuclei of galaxies, and the debris released as the stars are disrupted is likely to provide an additional source of fuel for the black hole.

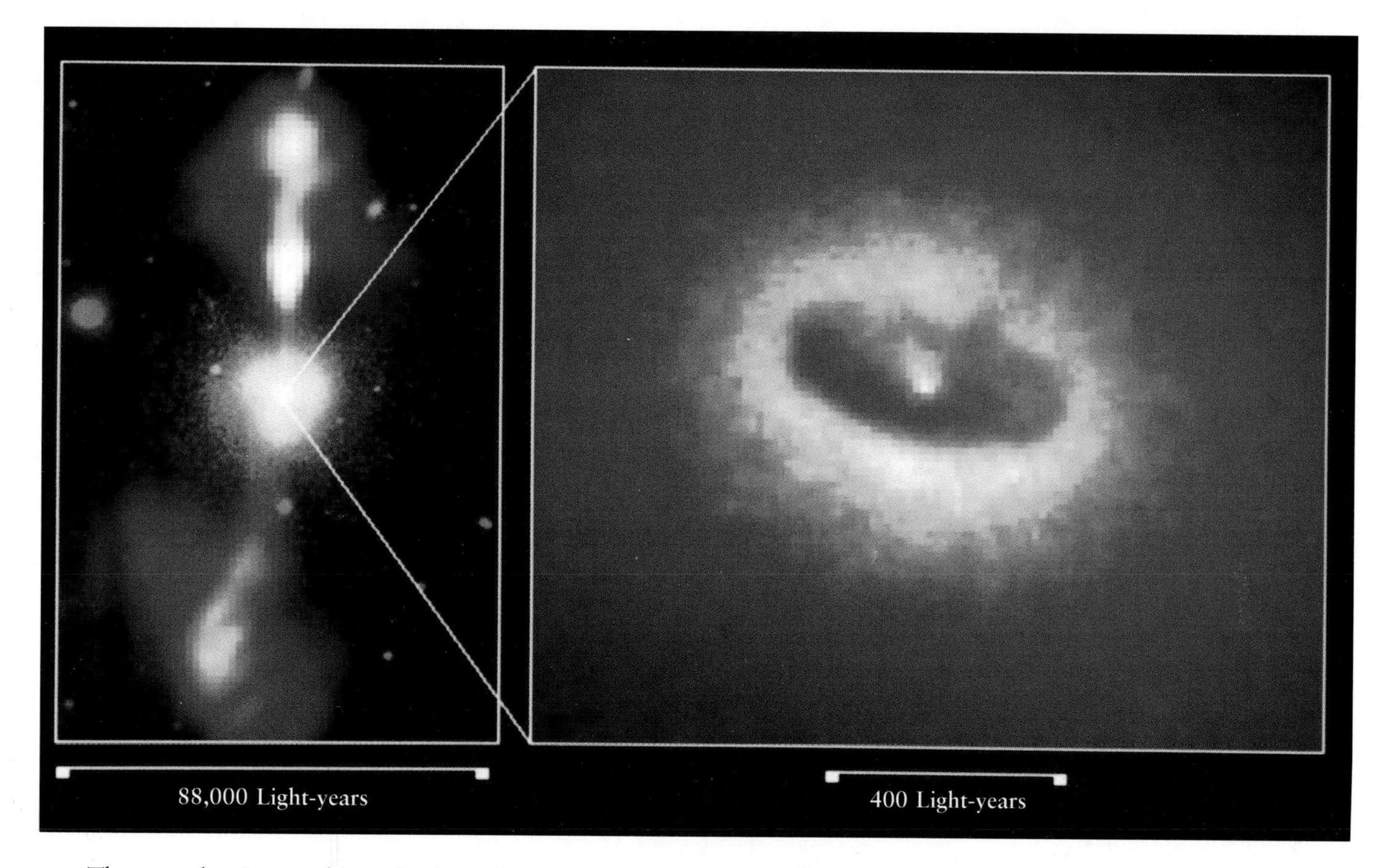

NGC 4261 is an active galaxy known to be the source of a jet of intense radio emission (left). The Hubble Space Telescope's high-resolution imaging of the galaxy nucleus reveals an accretion disk surrounding the energy source, presumed to be a massive black hole (right).

The most luminous objects in the universe are quasars, quasistellar radio sources, and these starlike objects are believed to be galaxies with such vigorous, luminous nuclei that the surrounding galaxy can barely be discerned, except in very few nearby cases. Quasars are commonly strong emitters of radiation at radio, infrared, x-ray, and gamma ray wavelengths; indeed, more energy is emitted in x-rays than at any other frequency. Tens of solar masses per year of fuel are required to supply the supermassive black holes that are believed to power quasars. It is possible that such a large supply of gas may be available during the formation phase of a galaxy, before many stars have formed. Indeed, nearby quasars are rare, and the number increases steeply at high redshift. Most quasars are at a redshift between 2 and 3, more than three-quarters of the way back in time to the big bang. This epoch, perhaps not coincidentally, is also the epoch in the universe when most of the galaxies we see may have formed.

The best places to look for evidence of supermassive black holes are therefore in the nuclei of nearby galaxies. The best-studied candidates are at the centers of our own galaxy, the Andromeda galaxy, and the nearby ultraluminous elliptical galaxy Messier 87. Astronomers probe the gravitational field close to the center of a galaxy in two ways. They can look for a central peak in the galaxy's light profile that would indicate an unusual concentration of mass at the center. Or, they can study the dispersion of velocities of gas clouds and stars in the central few parsecs. If there is a sufficiently massive black hole, the enhanced gravity will cause these velocities to be high. In our own galaxy, the center is highly obscured by interstellar dust along the line of sight, yet there is still some evidence for a central black hole: the mass estimates range from 10,000 to 1 million $M_{\odot}$. While the evidence remains ambiguous for the Milky Way, somewhat more convincing evidence has been found for a black hole of 30 million $M_{\odot}$ in the center of Andromeda, and for a black hole of 100 million or even 1 billion $M_{\odot}$ in the center of M87. However, higher-resolution observations of the best candidates are needed to provide truly convincing evidence for the existence of supermassive black holes.

The case for black holes powering active galactic nuclei is a strong one, although largely circumstantial: no other mechanism works nearly as well to explain all of the observed phenomena. Our closest view of the central activity is provided by jets of radio emission observed by radio astronomers

The Hubble Space Telescope has photographed a rotating disk of hot gas in the nucleus of the giant elliptical radio galaxy M87. The spectra of redshifts and blueshifts of light from the disk show the disk to be rotating at a velocity of about 550 kilometers per second. The likely reason for the high velocity is the presence of a central supermassive black hole of mass equal to 1 billion suns.

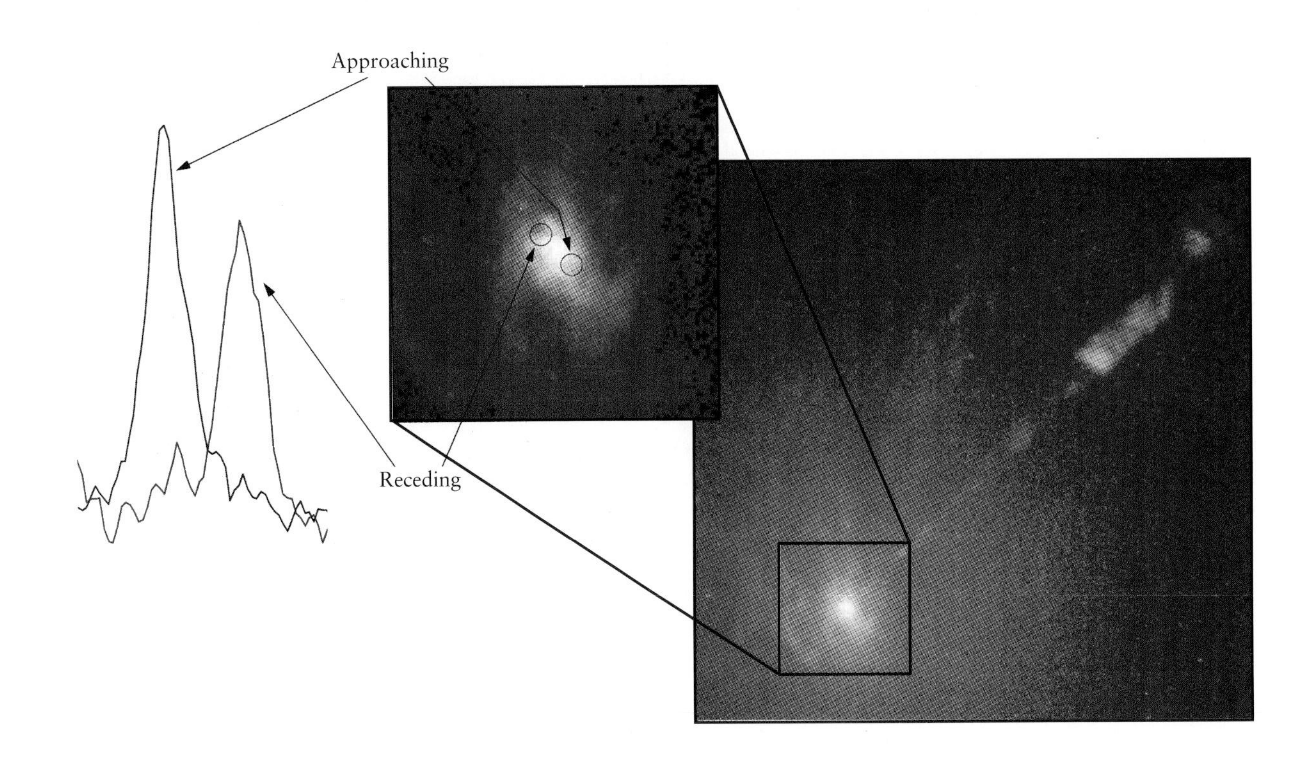

to emanate from the central nucleus. The origin of the jets is most simply explained if we assume the presence of a central supermassive black hole: the accretion disk provides a natural mechanism for channeling the energy emitted from the central source into narrow beams perpendicular to the disk.

To examine the jets more closely, astronomers turn to arrays of radio telescopes that collectively have far greater resolving power than a single telescope has. Many radio telescopes can observe simultaneously over a baseline of hundreds of miles or even in different continents, to effectively simulate part of a giant telescope with a diameter equal to the maximum separation between antennae. In this way, one can attain a resolution at radio wavelengths more than a thousand times superior to that reached at optical wavelengths even by the Hubble Space Telescope. These so-called very large baseline arrays even reveal radio-wave emitting blobs ejected at near light speed from active galaxy centers. It is difficult to imagine anything but a supermassive black hole providing all this energy.

It is unlikely, however, that supermassive, or even massive, black holes could be the halo dark matter. If black holes were orbiting the halo, they would frequently encounter globular star clusters, and their strong tidal forces, exerted during close passages, would have completely disrupted these clusters over the age of the galaxy. One can infer that the maximum mass of any black holes in the halo is about 1000 $M_\odot$. The possibility is certainly still open that halos are baryonic, even if black holes are absent, since halo dark matter could still consist of the remnants of massive, and even supermassive, stars.

Whether there are enough black holes to resolve the dark matter problem in general is unknown. Black holes of stellar mass would most likely have formed in supernova explosions, which we may recall are also the source of all the heavy elements. If the dark matter were to consist of black holes, there would have been so many supernovae early in the galaxy that we would have found a higher abundance of heavy elements than seems to exist. After all, one finds metal-poor stars in the halo that formed early in the history of the galaxy, yet not so early that they could not have picked up debris, as they formed, from even earlier, short-lived stars.

A more likely possibility may be massive black holes of thousands of solar masses apiece. These are also created by the deaths of very massive stars, but the black holes are so massive and their gravity so powerful that they would swallow the enriched layers of the imploding massive star before any explosion could have occurred. An interesting consequence would be a brilliant display of cosmic fireworks from the formation of these objects early in the universe. Although we cannot detect individual objects at great distance, the cumulative radiation should provide a highly redshifted, diffuse background of extragalactic light that would be visible in the in-

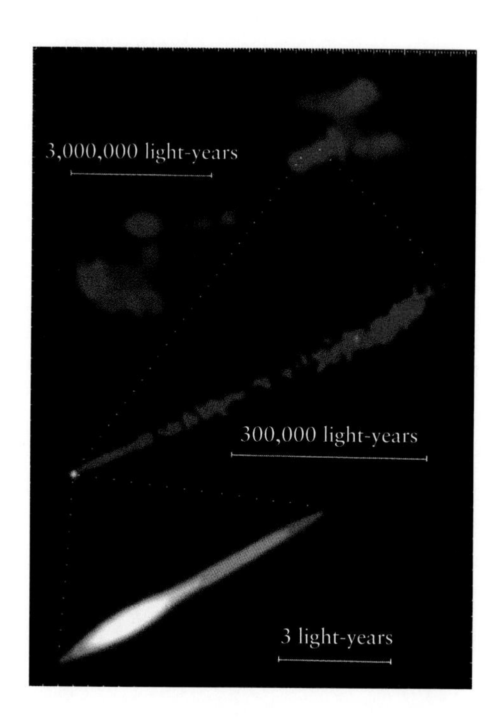

A radio galaxy viewed with progressively finer resolution. On a scale of 3 million light-years, giant lobes composed of extended radio emission are seen. The radio lobes are fed by a radio jet viewed below at one-tenth of this scale. A close-up of the central nucleus reveals a central jet on a scale of 3 light-years that is feeding the larger jet and radio lobes. It is presumed to emanate from a supermassive black hole.

frared spectral region. This hypothesis remains untested, since astronomers have yet to complete adequate experimental searches for such diffuse backgrounds.

But black holes, as we have seen, are far from being the only baryonic candidate for dark matter.

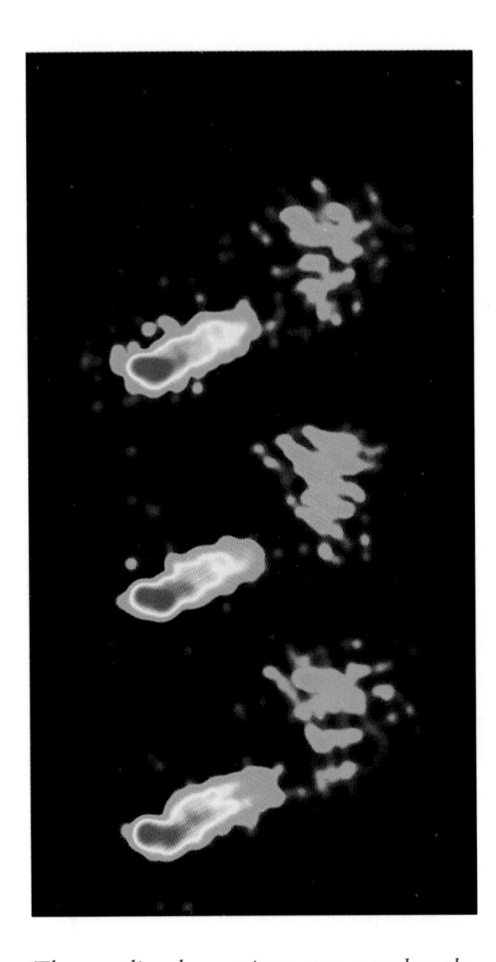

These radio observations at a wavelength of 6 centimeters were taken of the quasar 3C 345 at three epochs: April 1989, March 1990, and September 1991. They show a "jet" emerging from the nucleus of the quasar, on the left. The two red blobs in the jet start close together and move apart at an apparent speed several times the speed of light. In actuality, the jet is directed at a small inclination to the line of sight; therefore, blob velocities at near light speed can give the illusion of moving faster than the speed of light.

Are Halos Baryonic?

The most likely site for baryonic dark matter, given that there is most likely a surfeit of it, is in the dark halos of galaxies. For whether it consists of stellar relics or planetlike objects, one would expect to find the dark baryonic matter in proximity to the observed baryonic matter. Thus halos are a more obvious site than the depths of intergalactic space. Even the dark matter in galaxy clusters could plausibly be baryonic.

Some provocative clues suggest that galaxy halos may perhaps be mostly baryonic. The key to finding dark halo matter is to measure the velocity of galactic rotation within the halo. As we have seen, that rotation velocity is constant outside the region of the luminous matter, and its constancy provides direct evidence that dominant amounts of dark matter are present in galaxies. The higher the rotation velocity of stars and gas clouds orbiting the galaxy, and the farther out the measurements extend, the greater is the mass of dark matter that is required to confine the orbits of the stars. The curious fact is that the shape of the rotation curve (the distribution of rotation velocities outward from the galaxy center), as well as the maximum rotation velocity, seems to depend on the amount of matter in the disk, which of course consists largely of stars. This would be quite a cosmic conspiracy if halos consisted of exotic particles rather than baryonic matter.

The shapes of some dark halos are another indication that they are made of baryonic matter. Although the shapes of dark halos cannot be observed directly, they can sometimes be inferred from observing the distribution of stars. A case in point comes from studying several polar ring galaxies. These systems are the product of a merger between two galaxies that has left a ring of stars around the minor axis of the larger galaxy. The ring of stars enables one to study the orbits of stars along both the minor and the major axes, and to thereby probe the dark matter distribution. The galaxy halos are often found to be flattened, like oblate, or squashed, spheroids.

Another commonly seen sign of a flattened dark halo is a warped galactic disk. The warp is formed as the galaxy oscillates, like a flag in the wind. Because the oscillations would otherwise rapidly die away, the disk must be embedded in either a flattened or an elongated dark halo. Elliptical galaxies also are flattened, although whether they are mostly oblate or prolate, that is, football-shaped, is uncertain. A prolate system in projection on the sky, although intrinsically elongated, appears to be flattened.

Flattening, especially if oblate, is easy to understand if halos are baryonic. The gas loses thermal energy as it cools but retains its angular momentum, and so a flattened spheroid is the natural endstate of a rotating, collapsing gas cloud. In contrast, a prolate halo would be possible if baryonic clumps formed long before the disks and clustered to form dark halos. But this is exactly how nonbaryonic matter aggregates into halos. The extent and shape of dark halos will need to be better mapped to properly sort out this issue.

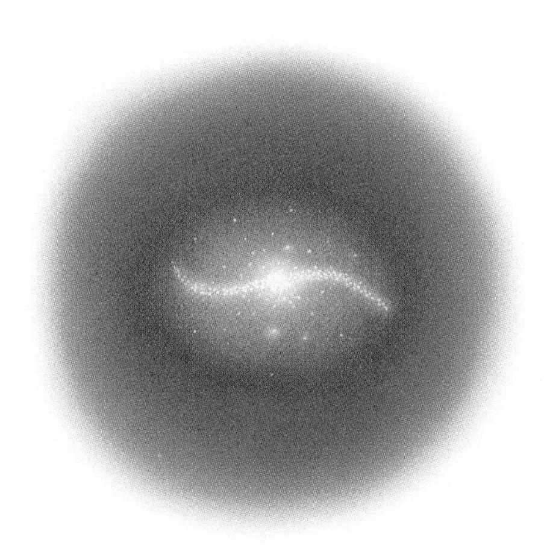

A warped galactic disk must be embedded in a flattened dark halo, either oblate (as shown here) or prolate.

MACHOs: Astrophysical Dark Matter

Among the possible astrophysical objects contained in the halo are the relics of stars, neutron stars, dim stars such as white dwarfs, or even black holes, as well as objects that have never quite fulfilled themselves as stars because of their low mass. Because these objects are invisible, or almost so, they are excellent candidates for dark matter. Moreover, MACHOs are more natural candidates for the halo dark matter than WIMPs, because they are already known to exist.

Two experiments reported in 1993 have found strong evidence for the existence of MACHOs. The technique used is gravitational microlensing. If a MACHO in our galaxy's halo passes very close to the line of sight from earth to a distant star, the gravity of the otherwise invisible MACHO acts as a lens that bends the starlight. The star splits into multiple images that are separated by a milliarc-second, far too small to observe from the ground. However, the background star temporarily brightens as the MACHO moves across the line of sight in the course of its orbit around the Milky Way halo.

The idea is to measure the brightening effect in background stars. There are two major difficulties, however. First, the microlensing events are very rare. Only about one background star in two million will be microlensed at a given time. Second, many stars are intrinsically variable and will also show temporary brightening. Fortunately, the microlensing event has some unique signatures that distinguish it from a variable star. It should be symmetrical in time, be achromatic, and occur only once for a given star.

To overcome the low probability of observing a microlensing event, the experiments were designed to monitor several million stars in the Large Magellanic Cloud. Each star was observed hundreds of times over the course of a year. A preliminary analysis of the data, taken with both red and blue filters, revealed several events that displayed the characteristic microlensing signatures. The event durations were between 30 and 50 days.

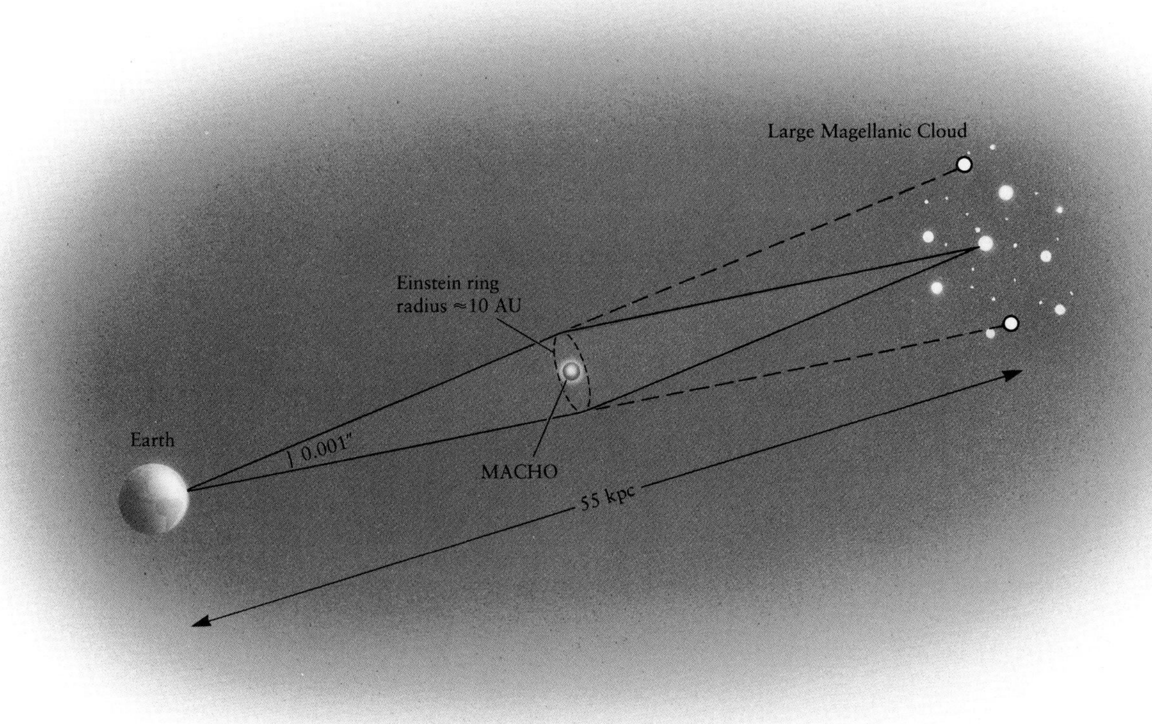

The gravitational lensing of light from a star in the Large Magellanic Cloud by a clump of dark matter (a MACHO). The light is bent slightly as it passes through the Einstein ring radius, generating double images of a single bright object. The image splitting is too small to be observable, but the associated amplification of light from the background star has been detected.

The duration of the microlensing event directly measures the mass of the MACHO, although there is some uncertainty because of the unknown distance and transverse velocity of the MACHO across the line of sight. The event duration is simply the time for the MACHO to cross the effective size of the gravitational lens, known as the Einstein ring radius. The radius of the Einstein ring is approximately equal to the geometric mean of the Schwarzschild radius of the MACHO and the distance to the MACHO. For a MACHO halfway to the Large Magellanic Cloud, that distance is half of 55 kiloparsecs. The Einstein ring radius is about equal to 1 astronomical unit, or the earth–sun distance. In order to be lensed, the MACHOs must be objects that are smaller than the lens, so they must be smaller than an astronomical unit, roughly the radius of a red giant star. The events detected are, to within a factor of a few, as the MACHO model of dark halo matter predicts, and the event durations suggest a typical mass of around 0.1 $M_\odot$; however, there is at least a factor of 3 uncertainty in either direction.

The microlensing studies continue, and the MACHO interpretation, if correct, predicts a definite pattern of results. There should be many more events of shorter duration, and there should be many weaker, low-amplification events. The microlensed stars are randomly selected, and so there should be no preferential microlensing of peculiar stars, which might be intrinsically variable stars that are fooling the observers. All of this will become clearer as more data is collected. For now, one can say that these results are certainly the strongest evidence yet of dark matter detection. Unless there are perverse types of rare variable stars, MACHOS are likely to constitute a significant fraction, at least 10 percent, of the dark halo. While baryonic dark matter probably exists in the halo, it is still unclear what fraction still may be in the form of the nonbaryonic WIMPs.

Hot gas in a small galaxy group. Superimposed on an optical photograph of the galaxy group NGC 2300, 50 megaparsecs away, is an x-ray photograph revealing a huge cloud of glowing intergalactic gas 1 million light-years across.

9

Intergalactic Matter

The next step in our cosmic saga of the evolution of the universe is the formation of galaxies; that step takes us to the creation of diffuse intergalactic gas clouds, the raw material out of which the luminous regions of galaxies are formed. Intergalactic gas plays a dual role: not only is it the progenitor of the galaxies, but it may also persist to provide a significant contribution to dark matter.

Recombination and Decoupling

For the first few hundred thousand years, the matter in the universe largely consisted of ionized hydrogen, protons, and electrons, with a 10 percent admixture of helium nuclei. At high temperature, energetic photons knock electrons away from protons, preventing the two particles from uniting to form hydrogen atoms, and the natural state of hydrogen gas is the plasma or ionized form. There are so many ionizing photons that if an atom does form, it is immediately re-ionized. However, as the universe expands and cools, the average energy of a photon falls until eventually hydrogen atoms are able to persist. This is the epoch of the "recombination" of hydrogen gas. (I suppose a more accurate term would be "combination.") From then on, the universe is almost entirely atomic.

The universe made the transition from the plasma to the atomic phase when it was one-thousandth of its present size. Then the cosmic blackbody radiation was at one thousand times the temperature of the microwave background today, 2.73 degrees Kelvin. Before the epoch of recombination, the radiation temperature was precisely equal to the gas temperature, because the radiation scattering off the free electrons remained closely coupled to the matter. Once the matter became atomic, however, the electrons are all bound into atoms. Now the radiation travels freely, propagating in a straight line without scattering as it has continued to do until the present epoch of the universe. We say that matter and radiation have decoupled, and the matter and radiation temperatures are no longer precisely the same.

Whereas the radiation ends up at 2.73 degrees Kelvin today, the matter must have had a very different history. The radiation can be viewed as an expanding relativistic gas consisting of photons with a dispersion of velocities equal to 170,000 kilometers per second, or about 60 percent of the speed of light. Such a gas cools more slowly than a nonrelativistic gas of hydrogen atoms. The atoms in a nonrelativistic gas travel on average at the speed of sound, which after recombination is only a few kilometers per second. While the radiation temperature drops by a factor of 1000 between recombination and the present epoch, the matter temperature would have fallen far more, by a factor of 1 million. As a result, the present-day temperature of intergalactic gas would be only about a thousandth of a degree above absolute zero.

Is our galaxy surrounded by such cold intergalactic gas? If present in sufficient quantity, such an intergalactic medium could contribute enough

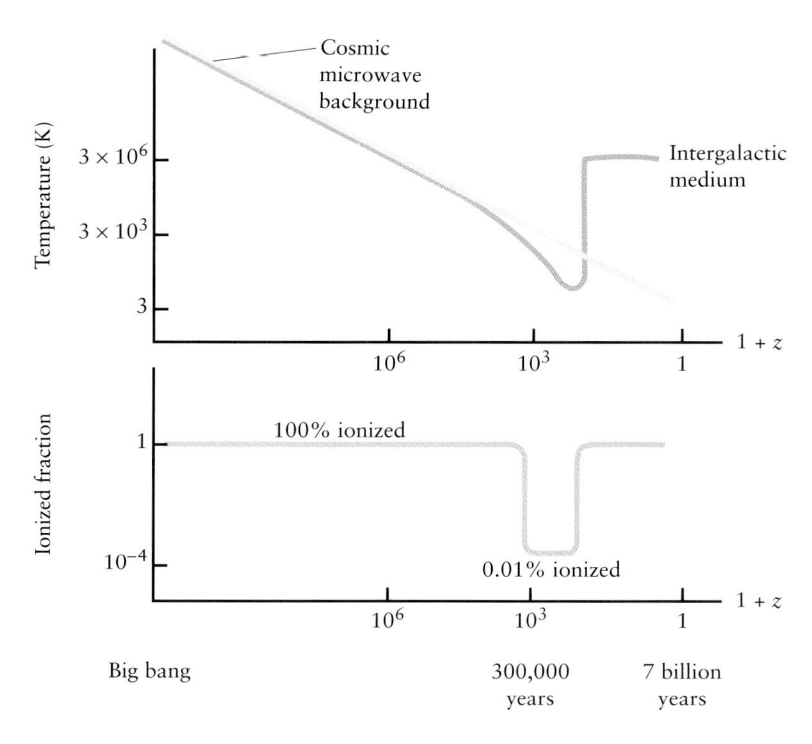

The continuously cooling intergalactic medium would have become cool enough to leave the plasma, or ionized, state after the epoch of last scattering. If it still exists, it must have been reheated and reionized.

mass to account for all of the dark baryonic matter whose existence is implied by the early nucleosynthesis of helium, deuterium, and lithium. It could even contribute substantially toward closing the universe. There are strong arguments, however, that any surviving intergalactic gas must have been reheated. Although intergalactic gas is "dark" in the sense that it emits little optical light, its presence would be revealed by its emission in other wavebands as well as by its absorption of the light from distant quasars.

The Search for the Intergalactic Medium

Much of the hydrogen gas created in the early universe was used up in the formation of galaxies, where it condensed to create stars. It is not possible to predict how much intergalactic matter remained after galaxies formed. Rather, along a given line of sight, we appeal to observation. An intergalactic medium extending throughout space, consisting of cold, atomic hydrogen, would easily be observable from its spectroscopic signature, since atomic hydrogen absorbs and emits radiation at a wavelength of 21 centimeters in the radio band. However, spectroscopic searches have not succeeded in finding much atomic hydrogen in intergalactic space. A diffuse medium has not been seen at a density anywhere approaching the critical density for closure, about 10^{-5} atoms per cubic centimeter. The most severe limit, however, comes from looking at another part of the spectrum, observed in the light from distant quasars.

Quasar light passes through billions of light-years of intergalactic space on its way to us. Were clouds of atomic hydrogen in the path of that light, the hydrogen would absorb the light at 1216 angstroms, the Lyman alpha line, in the far ultraviolet spectral region. Photons of this wavelength are absorbed by hydrogen in its ground state, and the electron then jumps to the first excited state. If the quasar is at high redshift, the far ultraviolet spectral region is shifted to longer wavelengths in the optical band, which are readily accessible from ground-based telescopes. Lyman alpha absorption from a uniformly distributed intergalactic medium has not been seen. We conclude that at redshift 4 the density of neutral atoms is less than

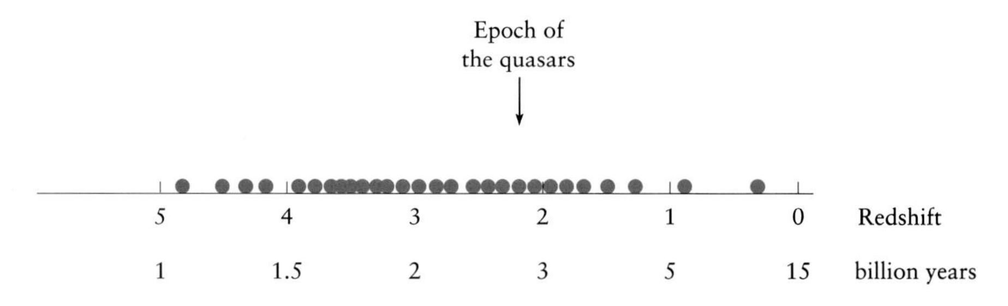

More quasars are found at high redshift.

10^{-11} per cubic centimeter, or one-ten-millionth of the closure value. Only a negligible abundance of atomic hydrogen is present in the form of a uniform intergalactic medium.

If there is an intergalactic gaseous medium, it must be ionized, and for it to be ionized, the gas must have been reheated. To heat the gas, there are ample supplies of energy available in the form of ionizing radiation from hot stars in newly forming galaxies and cosmic ray particles from supernovae and radio galaxies. However, there is no direct, unambiguous evidence for the presence of a pervasive, uniform intergalactic medium in ionized form. If the gas were very hot, at a temperature in excess of 1 million degrees Kelvin, it would emit x-rays in detectable quantity, unless the temperature of the medium were very carefully adjusted so that the bulk of the emission occurred in the far ultraviolet. If the intergalactic gas is at a temperature between 10,000 and 100,000 degrees Kelvin, it will produce most of its radiation in a wavelength regime that cannot be adequately observed at present.

Intergalactic Clouds

While a uniform intergalactic medium has not been discovered, gas has been detected in intergalactic space. Quasars have provided the tool for studying intergalactic gas. Any diffuse intergalactic medium would cover all of space between the quasar and our galaxy. Near the quasar, its absorption lines would be strongly redshifted, but light absorbed by the medium closer and closer to our galaxy would be redshifted less and less. The result would be a continuous band of absorption. In contrast, discrete clouds of intergalactic atomic gas would produce narrow spectral lines. Such narrow lines at the Lyman alpha wavelength are indeed found in the spectra of distant quasars; they are produced in numerous hydrogen gas clouds along the sight to those quasars, just as interstellar gas clouds produce absorption lines in nearby stars. Absorption lines from heavier elements are observed as well, but the advantage of the Lyman alpha line is that it is produced by the most abundant element in the universe, hydrogen.

Quasars contain ionized gas that emits radiation at the Lyman alpha wavelength. The emission is particularly strong if hot, massive stars or an active galactic nucleus is present, since these objects are prolific sources of ionizing photons. These photons ionize hydrogen, which recombines and produces intense Lyman alpha emission. It is difficult to detect Lyman alpha emission in nearby galaxies because our own interstellar medium absorbs radiation so strongly at that wavelength. The redshift helps, however, by shifting the Lyman alpha line from distant galaxies to a more transparent part of the spectrum.

A uniform intergalactic medium absorbing the light from a quasar would produce a continuous absorption band (top). In this case, the flux of light received from the quasar across a wide range of wavelengths would be uniformly less than that expected were there no medium (dashed line). In contrast, an intergalactic medium formed of distinct clouds would produce many narrow absorption lines (center). The actual spectrum is shown (bottom) of a high redshift quasar.

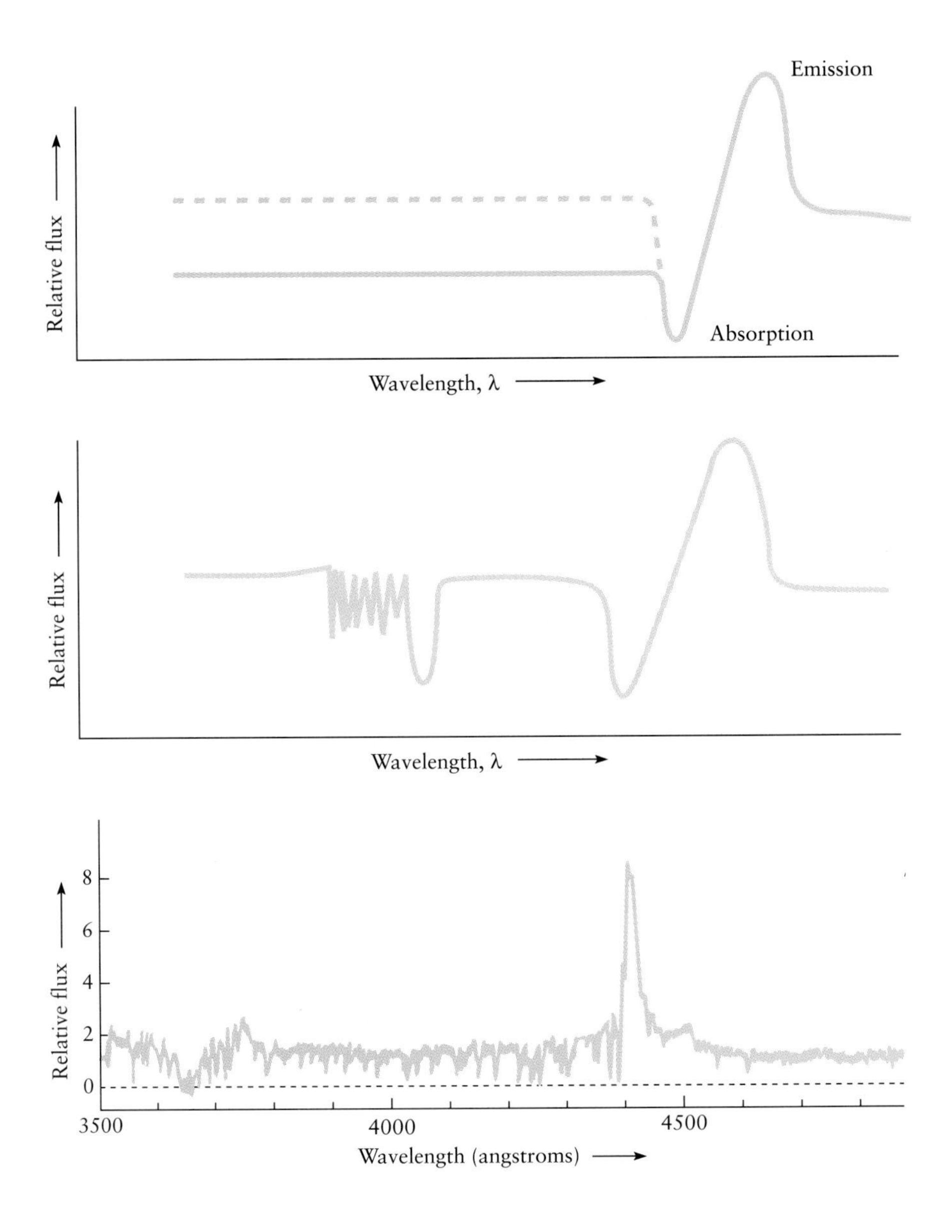

As a consequence, one can study Lyman alpha lines produced in the ultraviolet by both emission and absorption, but shifted toward the red, into the optical, by a factor of 3 or 4. In this case, the gas clouds are said to be at a redshift of 2 or 3 (recall that redshift z is the fractional wavelength shift, $\Delta\lambda/\lambda$, so that the total shift $(\lambda + \Delta\lambda)/\lambda$ is equal to $1 + z$). At a redshift of 2, one is looking perhaps three-quarters of the way back in time to the beginning of the big bang, some 10 billion light-years distant from us.

Astronomers have a special reason to search intensely for highly redshifted Lyman alpha emission; they hope that such emission is the sign of a galaxy in the process of forming from an intergalactic cloud. Although their searches have had little success to date, Lyman alpha absorption lines have provided an intriguing glimpse of the gaseous content of galaxies often too distant to be imaged directly, and some of these may be in a state of very early evolution. Why do we suspect that Lyman alpha absorption lines toward quasars are the sign of a galaxy in its early phase, when it was far more gas-rich than a typical galaxy is today? The expected probability of running into a galaxy en route to a quasar is small, only about 1 percent. Thus the absorbing clouds, if as abundant as galaxies, must be much larger than typical nearby galaxies. Alternatively, they could be much more abundant than galaxies. Curiously, both possibilities seem to be true for different types of clouds. In either case, one must be studying systems that are more gas-rich than those customarily found at the present epoch.

There is an even stronger clue that the absorbing clouds are much less evolved than typical interstellar gas clouds: many clouds that produce Lyman alpha absorption lines contain no or very few detectable heavy elements. Although other clouds are found that contain carbon, nitrogen, oxygen, iron, silicon, and even zinc and chromium, as well as hydrogen, there is little doubt that the abundance of heavy elements is greatly below the level found in, say, the sun. An extreme deficiency of heavy elements is the characteristic signature of objects formed early in galactic history. Three fairly distinct classes of absorbing clouds have in fact been detected.

The so-called pristine Lyman alpha clouds, which contain no heavy elements, are very weak in line strength. The strength of the absorption line measures the number of absorbing atoms in the cloud, so the line weakness tells us that these clouds have a relatively small proportion of hydrogen in atomic form. The number of atoms is far lower than is seen in any interstellar clouds, which are primarily atomic. This suggests that the pristine Lyman alpha clouds may consist of highly ionized hydrogen at very low density.

The size of the actual cloud can be estimated from the Lyman alpha forest, the name given to the weak lines because they occur in great numbers as though from a forest of intervening clouds. Astronomers compare the Lyman alpha forest along adjacent lines of sight to a gravitationally lensed quasar. A lensed quasar is really a single quasar that appears double: it has two identical images, separated by a few arc-seconds, produced as the path traversed by light from the quasar is bent by an intervening galaxy. The two quasar images are typically separated at the redshift of the absorbing clouds by only 30 kiloparsecs or so. The spectra of the two adjacent quasars are identical, except that lines from all clouds are not found in both. We conclude that the clouds must be smaller than 30 kiloparsecs

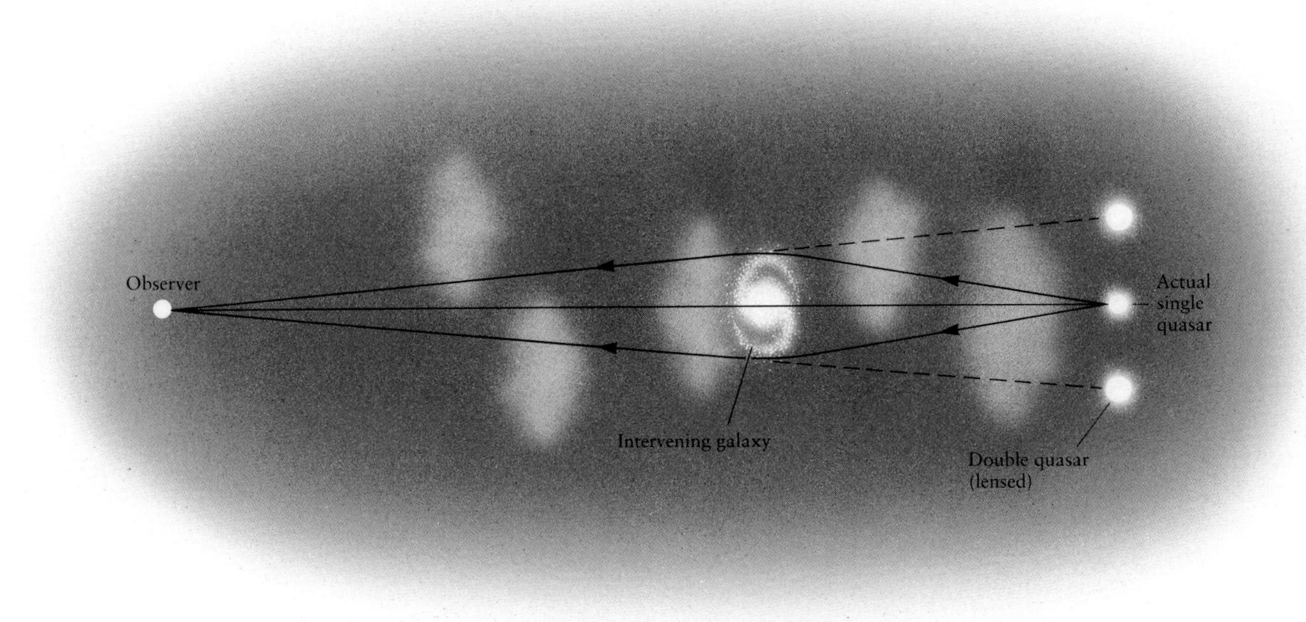

Hydrogen clouds appear frequently along the line of sight to a distant quasar. An intervening galaxy acts as a gravitational lens that bends light from the quasar, causing two images of the quasar to appear. The hydrogen clouds produce absorption lines in the quasar spectra. However, the pattern of absorption features is not identical for the two images. Some clouds will absorb the light from one image, and some from the other, proving that most clouds are not large enough to block both images.

in diameter, in the range of 10 to 20 kiloparsecs. The clouds are found to occur very frequently along a given line of sight, as compared to the number of galaxies like the Milky Way, and are thought to be primitive gas clouds that are destined to eventually form a numerous population of dwarf galaxies.

The metal-rich absorbing clouds, on the other hand, are probably interstellar clouds in distant galaxies. These galaxies are so distant and faint that it is usually difficult to see them against the glare of the relatively bright quasar. Yet some of the closer ones can be detected: sensitive images of some absorbing clouds at low redshift, taken over long exposure times, often reveal the presence of a nearby galaxy at the same redshift. We deduce that the absorbing gas is in the halo of the galaxy, and can infer that it extends out 100 kiloparsecs from the galaxy center, like a typical halo.

The most massive clouds are identified by their very strong Lyman alpha lines. The number of hydrogen atoms in a single such cloud is comparable to that in a gas-rich disk galaxy, viewed at a random angle. Such a cloud might correspond to the Milky Way before it had formed many stars and was still predominantly gaseous. The spectra of these relatively massive clouds also reveal the presence of heavy elements, although at an abundance of only about 1 or 10 percent that in the sun. We conjecture that these clouds are galactic disks in formation. We see them mostly at a redshift between 2 and 3, when the universe was less than a third of its present age.

Intracluster Gas

A much hotter form of intergalactic gas is detected in galaxy groups and in the even larger and more massive clusters. These are aggregates of hundreds or thousands of galaxies, bound together by their own gravity and that of dark matter. We have already noted that dark matter must be present simply to account for these stable groupings of galaxies.

The large mass of a galaxy cluster means that the typical galaxy is orbiting the cluster at about 1000 kilometers per second. Any hydrogen atoms immersed in this cluster would move at a similar speed. The large energy of atomic motions alone would make such a gas ionized and very hot. In fact, its temperature would be 100 million degrees Kelvin. At this temperature, the rapid motions of the electrons and protons should result in the gas prolifically emitting x-rays.

Galaxy clusters have indeed been found to be strong emitters of x-rays. These x-ray signals can only be detected by a telescope sited above much of the earth's atmosphere, which, fortunately for us, absorbs x-rays and far ultraviolet radiation. The cluster x-rays have therefore been mapped with special telescopes flown in rockets and in satellites, and are found to pervade the entire volume of the cluster. We learn from these studies that galaxy clusters contain about 10 percent of their mass in the form of intergalactic gas, in addition to the 10 percent or so in ordinary stars. The remaining 80 percent is dark matter; we can only speculate as to its nature.

An x-ray map of the Coma cluster of galaxies. A giant cloud of hot gas, at a temperature of 50 million degrees Kelvin, permeates the Coma cluster over a diameter in excess of 1 megaparsec. X-ray emission from several individual galaxies can also be seen. The diffuse blob in the lower right is diffuse x-ray emission from a group of galaxies that is falling into the Coma cluster.

One notable and unexpected discovery was that ionized iron exists in the intracluster gas. At 100 million degrees Kelvin, iron still retains one or two electrons, which are excited by collisions with the fast-moving free electrons in the cluster gas to emit spectral lines at x-ray energies in the analog of Lyman alpha emission. The amount of iron seen is considerable: the abundance of iron relative to hydrogen is about one-third that in the sun. This means that most of the intracluster gas must have been processed by stars and subsequently ejected from the galaxies. The ejected gas is driven out by exploding supernovae, which heat the interstellar gas and propel a galactic wind.

The Puzzle of the Baryon Fraction

Clusters of galaxies contain a surprisingly large fraction of baryons in the form of hot, x-ray–emitting intracluster gas; the hot gas accounts for at least 10 percent of a cluster's mass. A few clusters have been mapped in exceptional depth, and in these one can trace the gas to well outside the cluster's optical radius. In fact, the fraction of the cluster mass in the form of hot gas appears to rise with increasing radius, reaching a value of 30 or

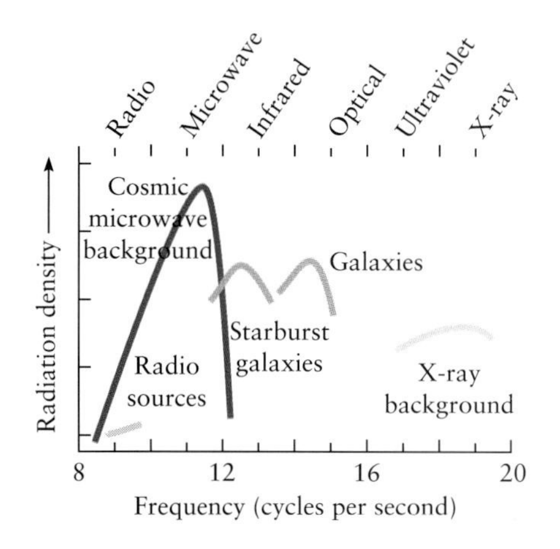

An observer looking toward the galactic poles, or out of the plane of the Milky Way, would measure a diffuse background radiation, at a range of frequencies from radio to x-ray. The dominant contribution to the radiation density arises from the cosmic microwave background.

even 50 percent. Compare this result with the prediction from primordial nucleosynthesis that the baryon fraction, relative to the closure value, is about 3 percent. The obvious interpretation is that we do not live in a universe at critical density. If the mean density of the universe were 10 percent of the critical value, however, one could easily reconcile the gas fraction with primordial nucleosynthesis.

Could the value determined for the gas mass be in error? Or could clusters of galaxies be sampling a volume of the universe exceptionally rich in baryons, by a factor of 10? If so, then we could still be living in a universe at closure density and dominated by cold dark matter. These questions will only be resolved when we have improved observations of clusters at x-ray frequencies and improved simulations of cluster formation.

The amount of gas in intergalactic clouds and in diffuse form, inside clusters and out, adds up to a negligible fraction of the density needed to close the universe. For clouds with Lyman alpha absorption signatures, one can estimate the gas fraction seen directly as one probes the universe along various lines of sight to the many different quasars in different parts of the sky. The baryon fraction in these gas clouds is about 1 percent of the closure value. For clusters, the numbers are less certain, but unlikely to exceed this. The gas fraction in galaxy groups is poorly known, since they emit x-rays at lower energies that are more difficult to observe.

Two arguments strongly suggest that the intergalactic medium could not be ionized and dense enough to close the universe. The strongest of these arguments is that the big bang would have produced negligible deuterium in such a baryon-dominated universe. The observed deuterium abundance cannot be explained unless the baryon density, and therefore the maximum density of an intergalactic medium, does not exceed 10 or 20 percent of the critical closure value. But perhaps we should strongly discount what may have happened in the first three minutes and ask whether, if there exists a dense intergalactic medium, we would have detected it.

Indeed, a dense intergalactic medium probably would have been detected. If it is at a temperature of more than a million degrees Kelvin, it would be emitting x-rays observable as a diffuse, isotropic glow. Such a background of diffuse x-rays has actually been seen. However, at least three-quarters of the x-ray background can be traced not to a hot, diffuse intergalactic medium, but to many very distant quasars and active galaxies, too far away for the most part to be individually resolved. Still, there is a residual diffuse flux that is unaccounted for, amounting to about 25 percent of the x-ray background. Some astronomers have attempted to attribute this residual flux to an intergalactic medium at about one-quarter of the critical density (although the density could be somewhat less if the medium were clumpy) and at an exceedingly high temperature. If the residual diffuse x-ray background does come from a hot intergalactic gas, its x-ray spectrum provides a measure of the gas temperature, and its x-ray intensity a measure of the average gas density. That gas would have to be at a temperature of about 400 million degrees Kelvin.

There cannot be enough of this hot gas to close the universe however, otherwise far more x-rays would be produced than are observed. Moreover, a gas at 400 million degrees Kelvin would reveal itself, not only by its x-ray emission, but also by an even more constraining signature. Photons in the cosmic microwave background are scattered by the hot electrons in such a gas. Each photon scattering transfers a small amount of energy from the gas to the radiation. The result would be to distort the blackbody spectrum at frequencies near its peak intensity, even though the probability of scattering is small at the low density of the intergalactic medium. The COBE satellite measured no such distortion, however, and so any distortion is at a level below the ability of COBE to detect. We conclude that any distortion arising from this electron scattering effect, called Comptonization, is less than 2 parts in 100,000. Comptonization distortions are indeed measured when the microwave background is studied in directions toward rich, x-ray–emitting clusters of galaxies, but these are from the scattering of background photons by the hot intracluster gas. The distortion measured toward galaxy clusters amounts to about 1 part in 10,000.

We conclude that a dense, hot intergalactic medium at the critical density, outside the great clusters of galaxies, with a temperature of tens or hundreds of millions of degrees in the nearby universe, would have been detectable from its x-ray emission. Nor could there have been such a hot intergalactic medium in the early universe, at the epoch, say, of galaxy formation. Although the x-ray emission from such a medium might have been redshifted and rendered undetectable, it would still have excessively distorted the spectrum of the cosmic microwave background.

There remains a possible loophole. There could exist a uniform intergalactic medium over a narrow temperature range, chosen to just evade detection. The intergalactic gas would still need to be heated to avoid cooling and condensing into galaxies. It must, however, be sufficiently cool, less than a million degrees, to avoid distorting the microwave background or emitting excessive numbers of x-rays. This scenario seems unduly contrived, and reminiscent of epicycles. Such a fix would require the input of vast amounts of energy to maintain the intergalactic medium in a delicate balance at a "just-so" temperature at which cooling is rapid and fragmentation of the gas is likely.

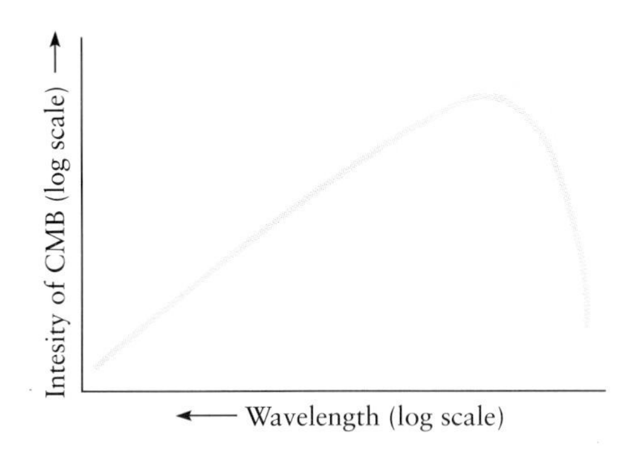

Occasional cosmic microwave background photons traversing hot intervening gas are heated, and shifted from the long wavelength side of the blackbody curve to the short wavelength side.

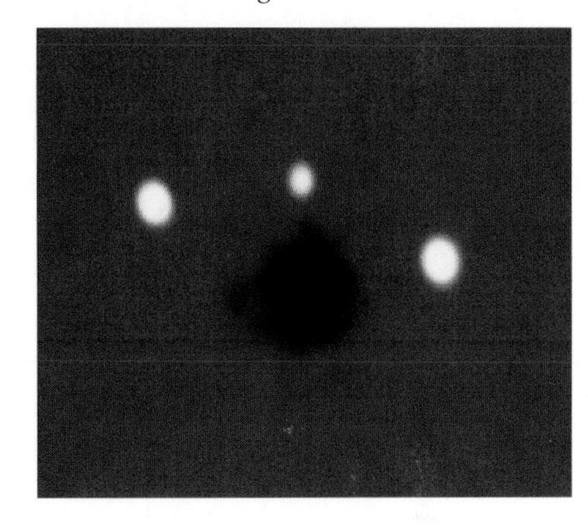

A radio map of the galaxy cluster Abell 2218, showing a dark "hole" in the microwave background radiation. Hot gas in the cluster adds energy to photons on the long wavelength (radio) side of the peak in cosmic blackbody radiation. As a result, there is a deficiency of radio photons in the direction of the cluster, tracing the hot gas distribution. The three yellow spots are radio sources in the background.

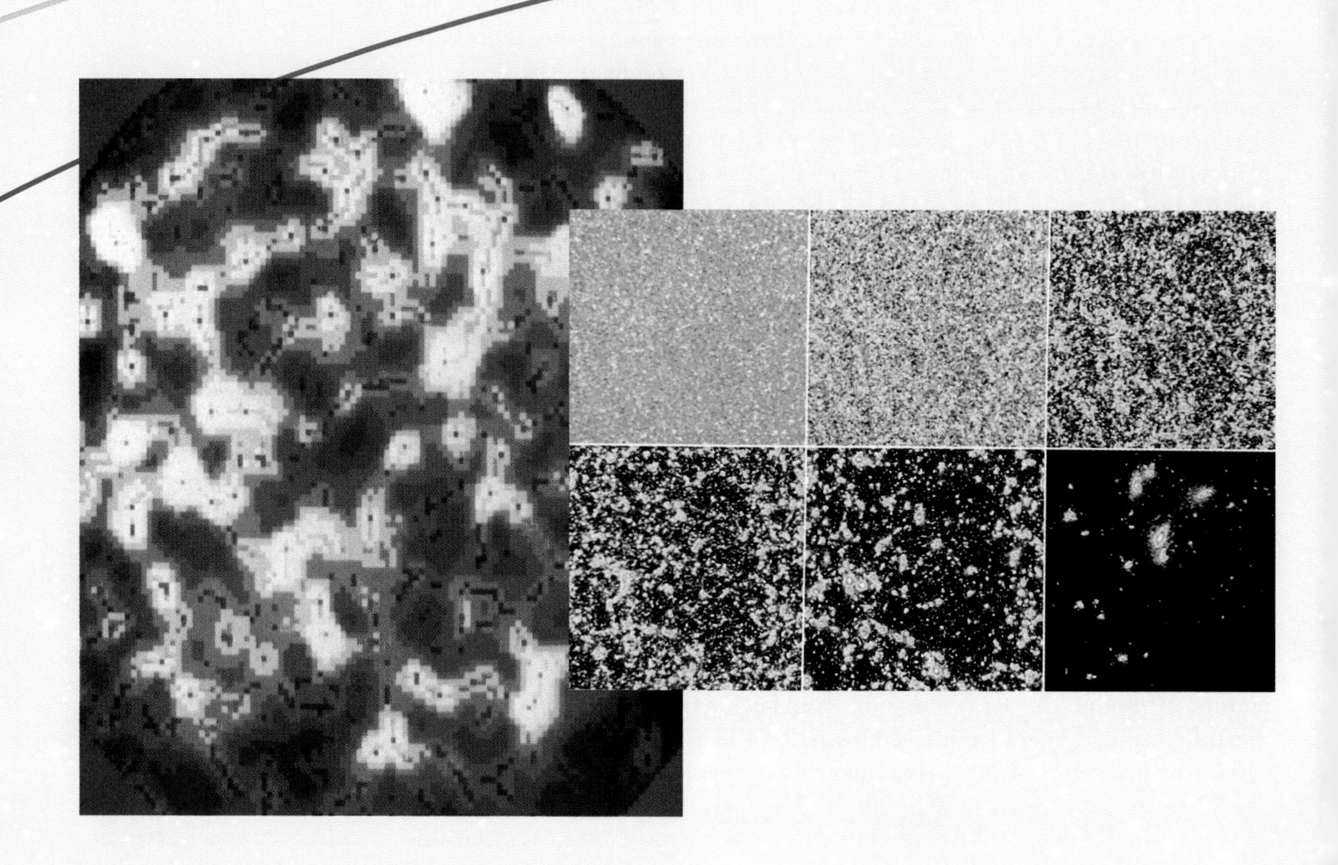

Left: *The large-scale structure of the observed universe, as revealed in the distribution of 400,000 galaxies across 100 degrees of the sky, mapped by a computer program designed to enhance the filamentary structure. Black pixels represent areas having the least number of galaxies; green and red pixels correspond to local peaks in the distribution. The green filamentary ridges may correspond to superclusters of galaxies in space.* ***Right:*** *A computer simulation shows how individual galaxies and galaxy groups would inevitably have developed by gravitational instability from tiny fluctuations. The simulation tracks 1 million particles laid down at random and shows the projection of a comoving cube of the universe viewed, from top left, at expansion factors 8, 23, 66, 125, and 250 after the start of the simulation. The bottom right panel, a 10-fold magnification of the adjacent panel, represents the present-day universe over a scale of roughly 20 megaparsecs.*

Origin of Structure

Working forward in time, we pass successively through the epochs of inflation, baryon genesis, nucleosynthesis, and radiation decoupling. The culmination is formation of structure. How structure originated from tiny fluctuations in density is one of the burning questions in cosmology.

All structure in the universe, according to big bang gospel, evolved from infinitesimal fluctuations in density imprinted at the dawn of time. These fluctuations were the seeds from which inexorably developed galaxies, clusters of galaxies, and even larger structures. We observe today structure on scales of up to 100 megaparsecs in the nearby universe. Yet its origin from density fluctuations has proved difficult to confirm. The reason is simply that the connection is tenuous between what was and what is present.

Small Fluctuations

The uniformity of the cosmic microwave background to better than a few thousandths of a percent tells us that the early universe, at an age of about 300,000 years, and at a redshift of about 1000, was exceedingly homogeneous. It could not have been completely homogeneous, however; otherwise galaxies would never have formed. There must have been small fluctuations in density.

Fluctuations are simply local variations in density relative to the background level: some fluctuations are upward in density, some downward. The degree of excess density defines the strength, or amplitude, of a fluctuation. These fluctuations are tiny at the beginning, but grow more pronounced over time by the process of gravitational instability. An upward fluctuation represents a slight local excess of gravity that attracts, and tends to pull in, nearby regions of average density. Although the matter is expanding everywhere, it is expanding at a slightly decelerated rate near the upward fluctuation. The effect is weak but proceeds inevitably, and the original density fluctuation grows larger and larger in strength. Once the density locally exceeds more than about twice the average value, the local effect of gravity is now strong enough to overpower the effects of the universal expansion. A cloud of matter begins to collapse. It is on its way to forming a galaxy or, if sufficiently massive, a cluster of galaxies.

In a static universe, infinitesimal fluctuations in density would grow at an exponentially rapid rate. Such a universe is said to be gravitationally unstable. Indeed, the entire universe, if static, would either spontaneously collapse or expand: this was an early theoretical objection to a static universe that had provoked Einstein into introducing a hypothetical force of cosmic repulsion to prop up the universe and stabilize it, at least against collapse. In a continuously expanding universe, however, the matter in upward fluctuations is being pulled apart by expansion and by pressure, especially from the radiation, at the same time as it is being pushed together by the pull of gravity. The decreasing density severely weakens the rate at which fluctuations can grow. In such a universe, structure takes time to de-

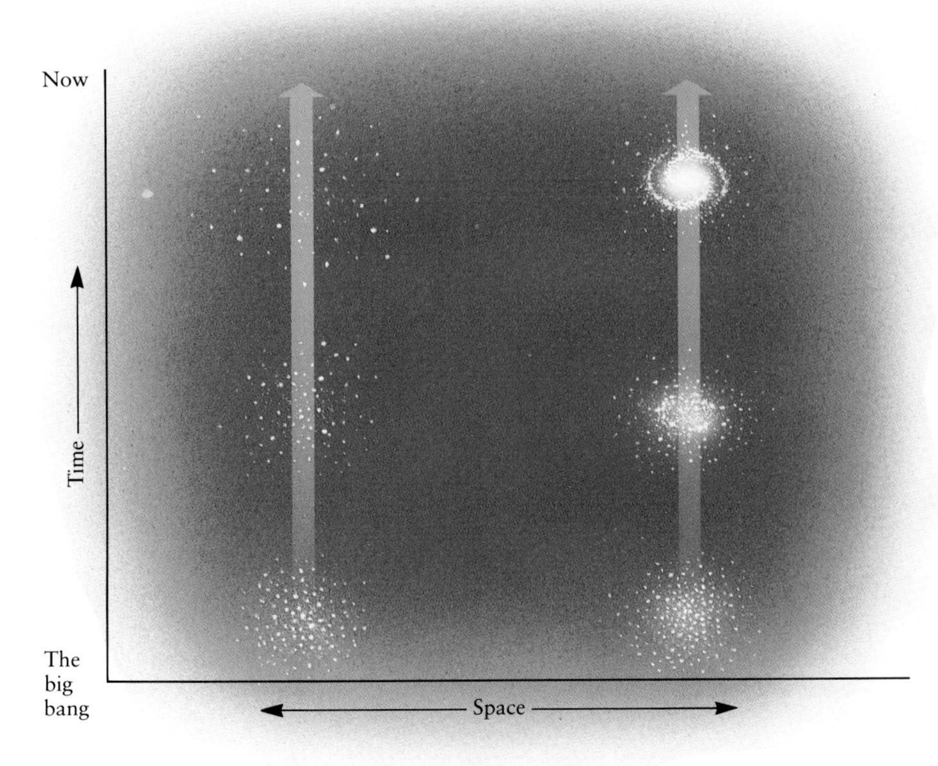

Small-scale density fluctuations, of a relatively low mass, disperse over time (left). For the largest fluctuations, however, the attractive force of gravity dominates over the dispersive tendency of pressure. Fluctuations of the mass of a galaxy gradually become more concentrated and eventually contract to form a galaxy (right).

velop, so small fluctuations must have been present from the beginning to act as seeds. If these fluctuations were too small, structure would not have formed by now. If they were too large, the universe would be too lumpy compared to what we see in the galaxy distribution. Theory predicts that primeval fluctuations in density of about 1 part in 100,000 were just what was required in order for the observed galaxy clustering to develop.

The growth of the primordial density fluctuations is largely controlled by the radiation content of the universe. There are approximately 3 billion photons for every proton, and sufficiently early in the universe, the energy density in the radiation was larger than the equivalent mass density in protons. This radiation-dominated phase lasted for the first 10,000 years of the big bang, during which fluctuation growth was suppressed. Only when ordinary matter became the dominant source of mass density in the universe did gravity become effective in stimulating the growth of fluctuations from the infinitesimal level originally laid down at the inflationary epoch to eventually develop into distinct cosmic structures.

Structure could have formed in one of two distinct sequences: either large structures the size of a galaxy cluster formed first, and galaxies formed when these massive structures fragmented, or else dwarf galaxies were the first objects to form, and these merged to produce progressively more and more massive structures, culminating in the formation of galaxy clusters. I discuss each of these scenarios in turn.

As radiation "leaks out," it drags matter with it, smoothing out fluctuations in the density of matter.

The Top-Down Scenario and Cosmic Pancakes

Let us first consider a universe in which the predominant form of matter is ordinary baryons. In such a case, not all fluctuations in density survive the hostile, hot environment of the early universe. In fact, during the first 10,000 years while the universe is dominated by radiation, fluctuations do not grow at all. At that epoch, the fluctuations comprise matter and radiation compressed together; such fluctuations are called *adiabatic* fluctuations, as opposed to *isothermal* fluctuations, in which only the matter is non-uniform.

It is very difficult to "bottle up" radiation: it tends to "leak" out, dragging matter along. As matter and radiation leak away, adiabatic fluctuations are smoothed out. Only the largest-scale fluctuations survive this homogenizing process, since there is not enough time for them to be erased. The larger the fluctuations, the longer it takes for a universe containing mostly ordinary baryonic matter to erase them. The smoothing out of fluctuations continues after the first 10,000 years, as long as the radiation is still scattering off the free electrons. All possible smoothing ceases once the universe becomes transparent to the radiation at the decoupling epoch, after about 300,000 years. The surviving fluctuations are at least 10^{15} $M_\odot$, the mass of a cluster. Only fluctuations this size or larger persist into the matter era.

These galaxy cluster–sized perturbations accrete matter and grow, until finally they collapse and fragment into galaxies. We refer to such a scenario, in which large structures develop before smaller structures, as "top-

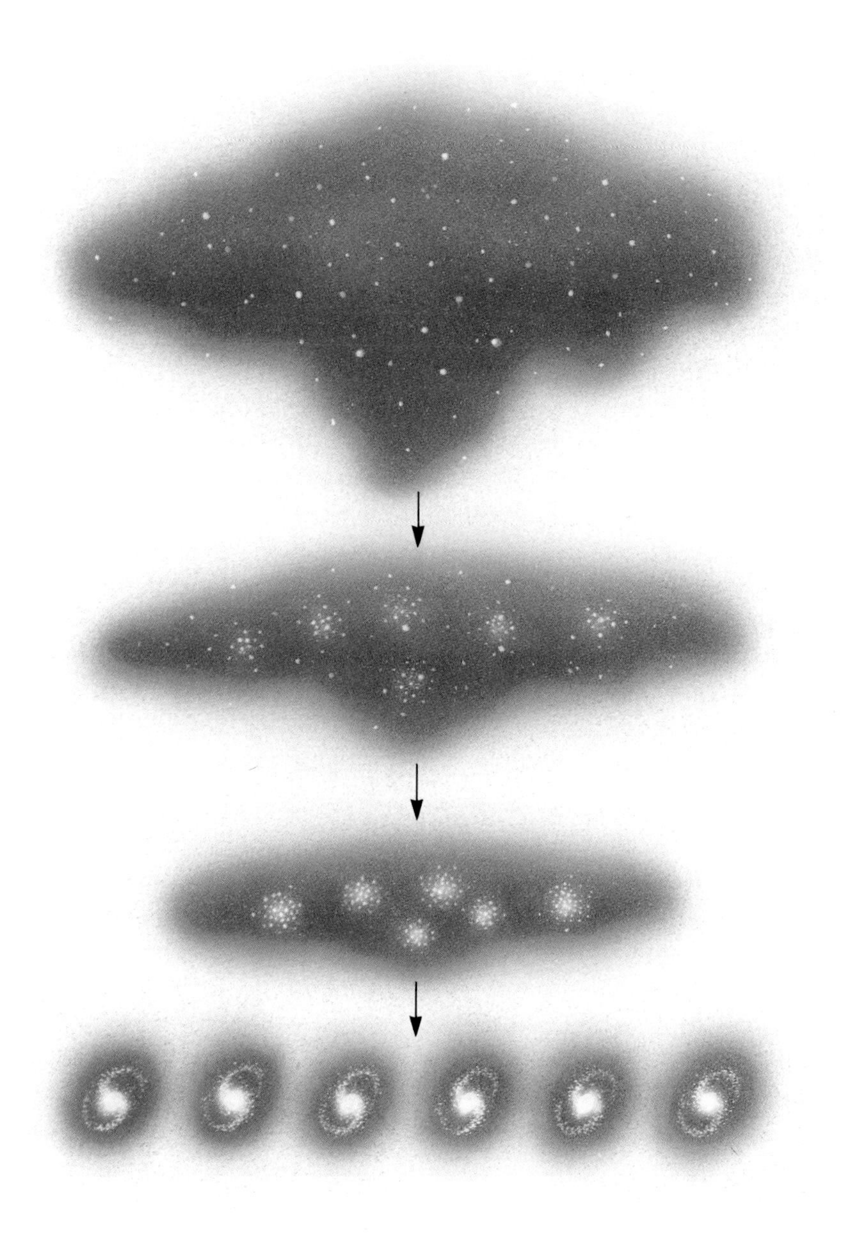

In the top-down scenario, large structures develop before smaller structures. A large cloud of matter typically collapses along two axes to form a thin, high-density sheet or pancake. Protogalaxies begin to coalesce in the sheet, and continue evolving into galaxies as the sheet fragments.

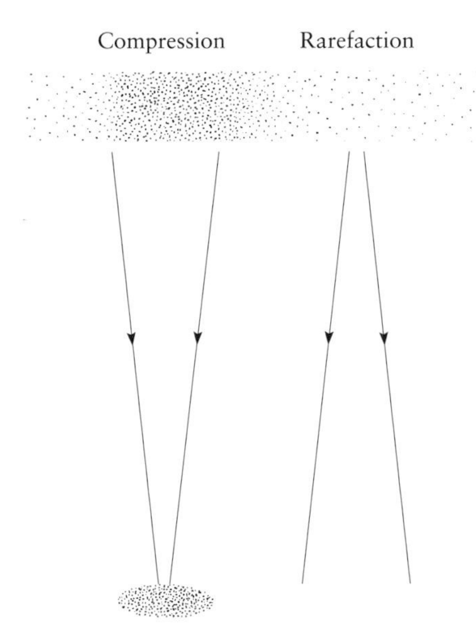

Overdensities pancake while rarefactions form voids, if primordial fluctuations on galaxy scales are smoothed out, as in the hot dark matter model.

down." It happens that a large cloud typically collapses to a thin sheet or pancake. The cloud is unlikely to collapse in synchrony along all three orthogonal axes to form a perfect sphere, or even along two orthogonal directions to form a cylinder-shaped cloud. One direction always wins, in general, over the other two, unless the initial fluctuation was precisely symmetrical, a rare and unlikely possibility. The result is a high-density pancake that breaks up into galaxy-sized chunks, creating large-scale structure in which galaxies lie mostly on sheets. There are giant clusters where sheets intersect and large voids between the sheets where the matter has been evacuated by the cloud's collapse. We shall see in Chapter 11 that there are large voids, sheets, and filaments in the large-scale distribution of galaxies that most likely have formed as a consequence of pancaking clouds.

Such a universe could not be at the critical density for closure and be dominated by baryons. I have already mentioned that the predictions for primordial nucleosynthesis would be violated if baryons provided the critical density for closure. Moreover, such a baryon-dominated universe results in excessive fluctuations in the cosmic microwave background for reasons explained on page 185. One needs to invoke a dark matter candidate that is nonbaryonic to rescue the top-down model.

The most successful version of a top-down model invokes dark matter in the form of a neutrino with a small mass of about 10 or 30 eV. The mass is easily arrived at: one knows the number of relic neutrinos from the big bang, which can be calculated from the temperature of the microwave background, and given this number, one designs a neutrino mass to provide a closure density in neutrinos.

Massive neutrinos would be moving at or near light speed when the universe became matter-dominated. The typical energy of a neutrino moving at that speed is about 10 eV, comparable to its mass. Only a much heavier particle would be moving slowly. Initially, the high-speed neutrinos move in random directions, and the effect is to create a high spread or dispersion in their velocities. It is this spread in velocities that enables neutrinos to smooth out all small-scale fluctuations as the particles stream away from any local region of excess density. Only the largest fluctuations, again above about 10^{15} $M_{\odot}$, survive this smearing, which eventually stops as the neutrinos slow down. The cluster-mass clouds that remain fragment into galaxies, thereby providing an alternative top-down scenario for the formation of structure.

The eminent Soviet cosmologist Yaakov Zel'dovich discovered cosmic pancakes and was a proponent of the top-down scenario for the formation of large-scale structure.

Cold Dark Matter and Bottom-Up Formation

The top-down scenarios have an important strike against them: there is considerable evidence that galaxies formed before clusters. Many clusters

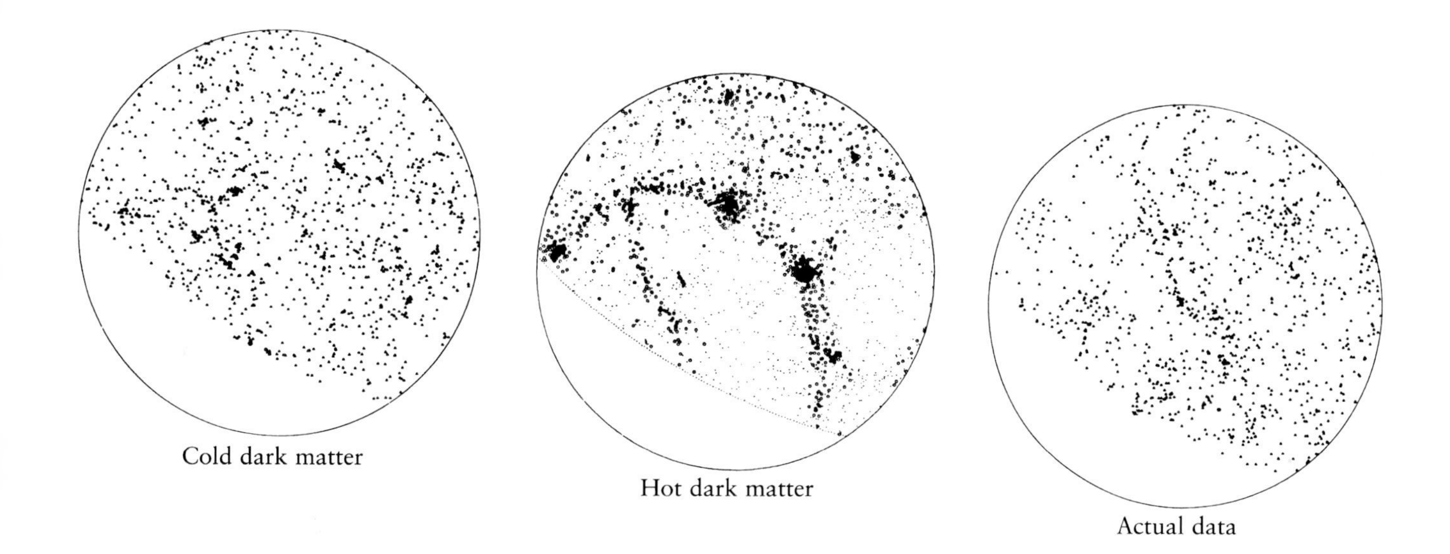

The two images at the left and center are computer simulations of a cold dark matter universe and a hot dark matter universe, displayed as projections on the sky, with the Milky Way outside the circumference and the north galactic pole at the center. Because structure develops in a top-down sequence in a hot dark matter model, that model leads to much more prominent large structures. At the right, the actual data describing the observed universe is closer to the predictions of the cold dark matter model.

appear still to be forming, whereas many, but not all, of the most distant galaxies appear to be relatively old, as judged by the range of ages in the stars present. An alternative approach leads to a "bottom-up" picture, in which small galaxies cluster and merge together, progressively forming more and more massive systems. Such a sequence best resembles the universe we observe, with small structures forming before large ones.

The bottom-up scenario is most simply enacted if the universe is dominated by cold dark matter, which is usually assumed to consist of weakly interacting massive particles (WIMPs), although it could also consist of condensed baryonic lumps (MACHOs) that formed sufficiently early before galaxies formed. As in the top-down scenario, the fluctuations begin to grow only when the universe is first dominated by matter, rather than radiation, at an epoch about 10,000 years after the big bang. Afterward, the radiation is still scattering off free electrons, and it remains tightly coupled to the baryons in the universe until hydrogen atoms form at an epoch of 300,000 years. In this scenario, however, the baryons do not control gravity. Rather, the cold dark matter is dominant, and other than by gravity, it does not interact with the radiation at all. Hence, the density fluctuations are not damped by "leaking" radiation carrying mass with it. Moreover, the dark matter, being cold, exerts no thermal pressure to counter the attractive force of gravity, so it cannot resist the clumping of density fluctuations by gravitational instability.

By contrast, in the previous example of a baryon- or neutrino-dominated universe, the smaller fluctuations were destroyed by the streaming of hot particles. We can think of the "bottled-up" radiation and baryonic or neutrino matter as being "hot," in contrast to the cold dark matter particles, which are too heavy and too unlikely to interact with photons to have any but negligible streaming motions.

Until very recently, our instruments were unable to detect fluctuations in the microwave background. We knew only that such fluctuations, if they

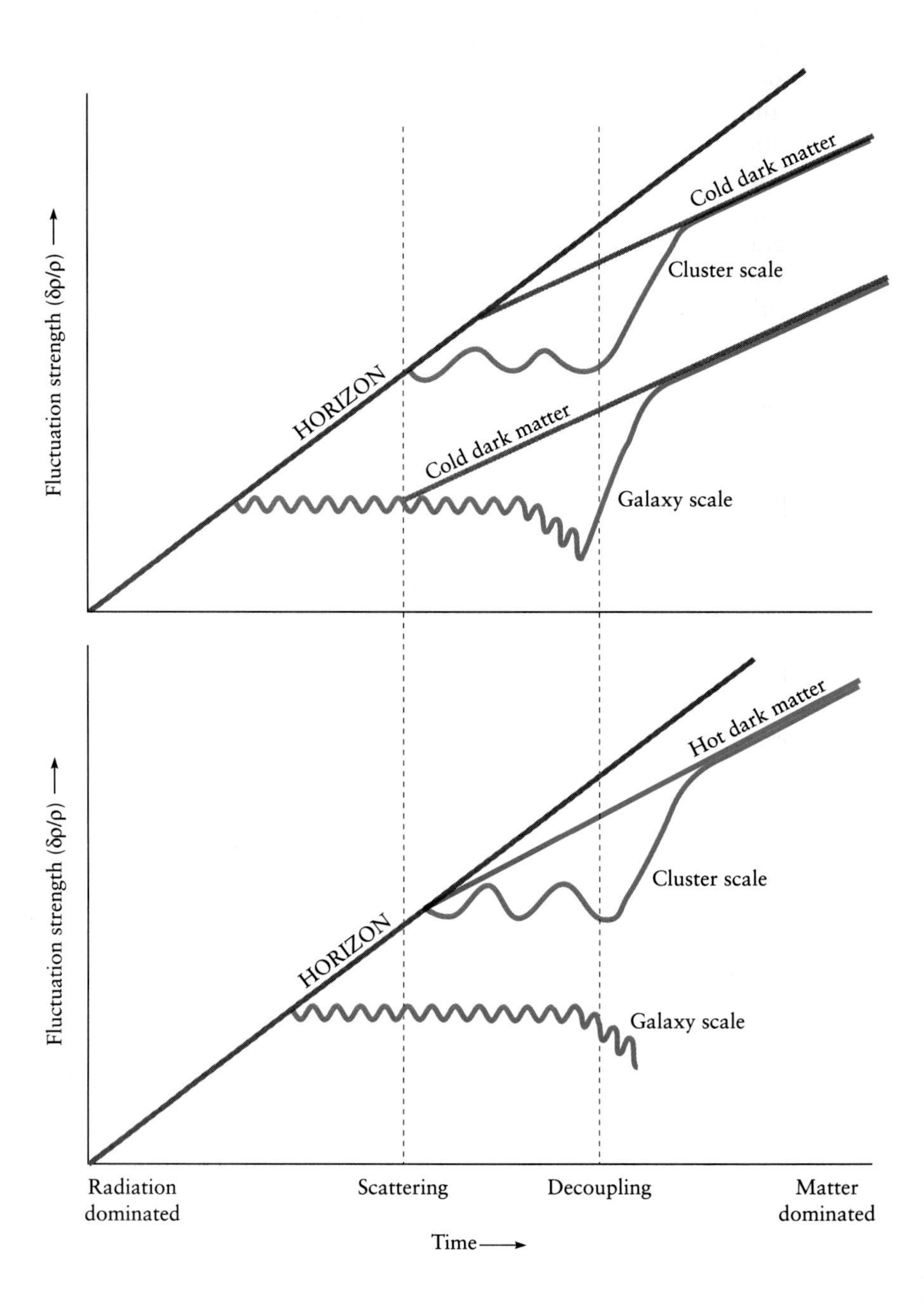

In a universe dominated by cold dark matter (top), density fluctuations in the cold dark matter grow on cluster and galaxy scales once the universe is matter-dominated. After decoupling, the baryon density fluctuations catch up with the dark matter fluctuations. However, in a universe dominated by hot dark matter (bottom), only cluster-scale fluctuations survive into the matter-dominated era.

existed, had to be weaker than the weakest fluctuations our instruments were capable of measuring. Thus we had a very low upper limit on fluctuation strength. And that meant that the structure we see today must have emerged from very weak fluctuations. This limit was enough to severely constrain models in which galaxies were made solely from baryonic, or ordinary, matter.

A strong rationale emerged for cold dark matter cosmologies. Fluctuations in the distribution of cold dark matter get a head start over purely baryonic matter fluctuations: they start growing while baryonic matter is still trapped by radiation. Collapse only occurs, however, after matter and radiation have decoupled. When baryons eventually break loose, they fall into the gravitational potential wells created by cold dark matter. Large-scale structure arises from much smaller initial fluctuations in the cold dark matter, with correspondingly smaller variations in the microwave background.

The smallest regions of the initial density fluctuations are expected to be the strongest in amplitude. This means that small-scale fluctuations are the first fluctuations to develop and that they have the most time to gain strength. Thus, as time progresses, larger and larger regions in turn develop fluctuations, which grow in amplitude as the smaller-scale fluctuations cluster together. The fluctuations grow from small to large scales in a hierarchical fashion. The smallest fluctuations are thus the first to collapse, because they have the longest period in which to grow in strength.

The first objects to form in a universe dominated by cold dark matter are of very low mass, corresponding to dwarf galaxies. Ordinary galaxies do not form until the universe is already about one billion years old, when it was about one-tenth of its present size. The forming galaxies consist of cold dark matter halos, within which baryons are able to settle. The baryons radiate away energy and cool down, thereby becoming more condensed than the dark matter. The cloud of matter is able to collapse into a galaxy only when gravity overcomes thermal pressure.

The size of the first clouds to collapse is dictated in part by the competition between the attractive tug of self-gravity and the opposing tendency created by the difference in thermal pressure between the center and surface of the cloud. That size can be estimated by comparing the gravitational potential energy of a cloud at this epoch with the thermal energy, that is, with the random motions (V_s) of gas atoms that tend to resist gravitational collapse. The gravitational potential energy of a cloud of mass M and radius R represents the energy released under gravitational free fall in contracting to its present size, and is GM^2/R. The thermal energy is the kinetic energy of all the gas atoms, and is $MV^2/2$. Gravity wins over pressure if the mass is sufficiently large and, in particular, exceeds a critical value

$$M_{\text{Jeans}} = V_s^3/G^{3/2}\rho^{1/2},$$

where ρ is the gas density and I have set $M \approx \rho R^3$ and ignored the factor of 2. This expression was first derived by the English astrophysicist James Jeans.

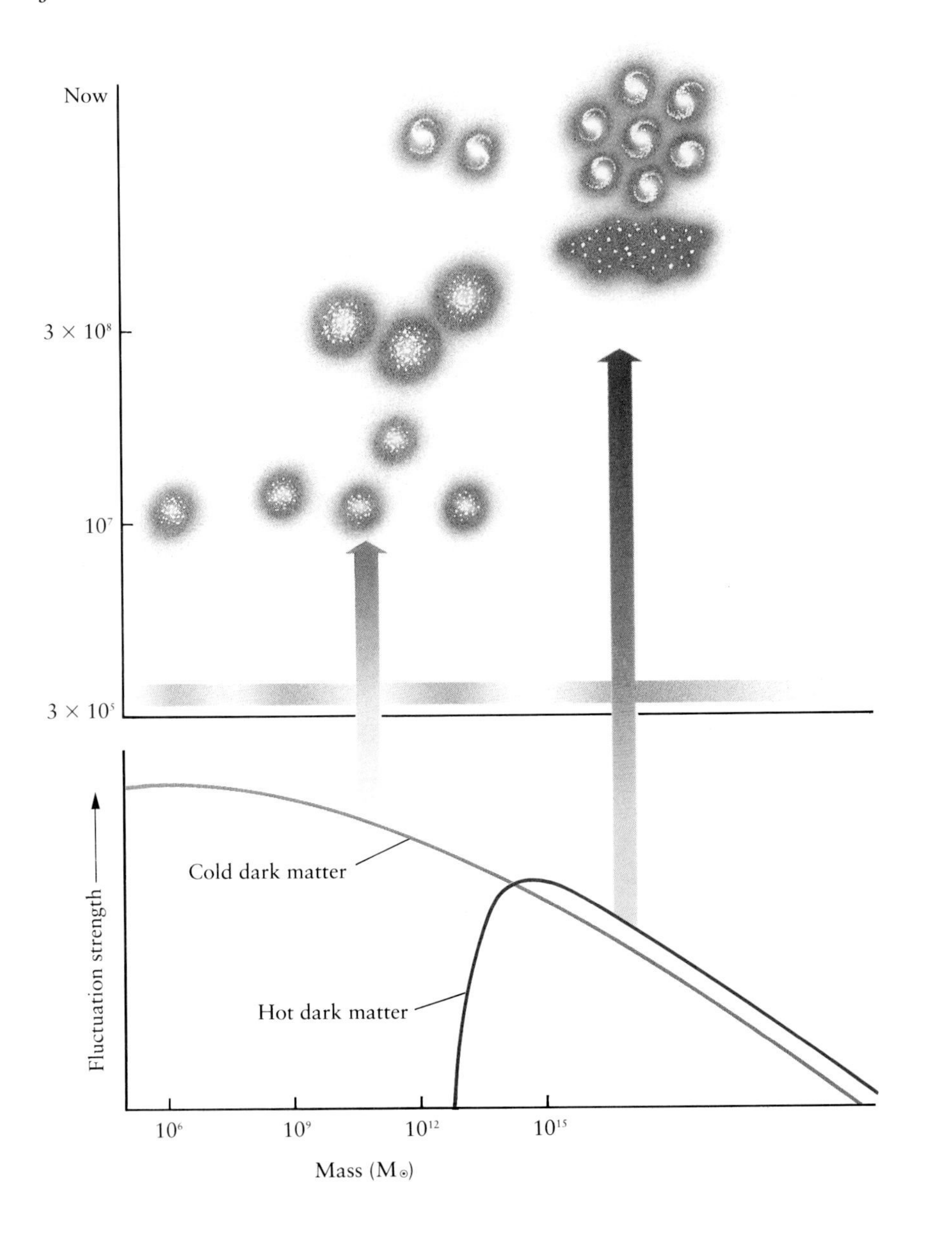

In the cold dark matter scenario, fluctuations are strongest at galaxy scales, and so galaxies form from early-evolving clouds. In the hot dark matter scenario, huge clouds of matter form first, at late times.

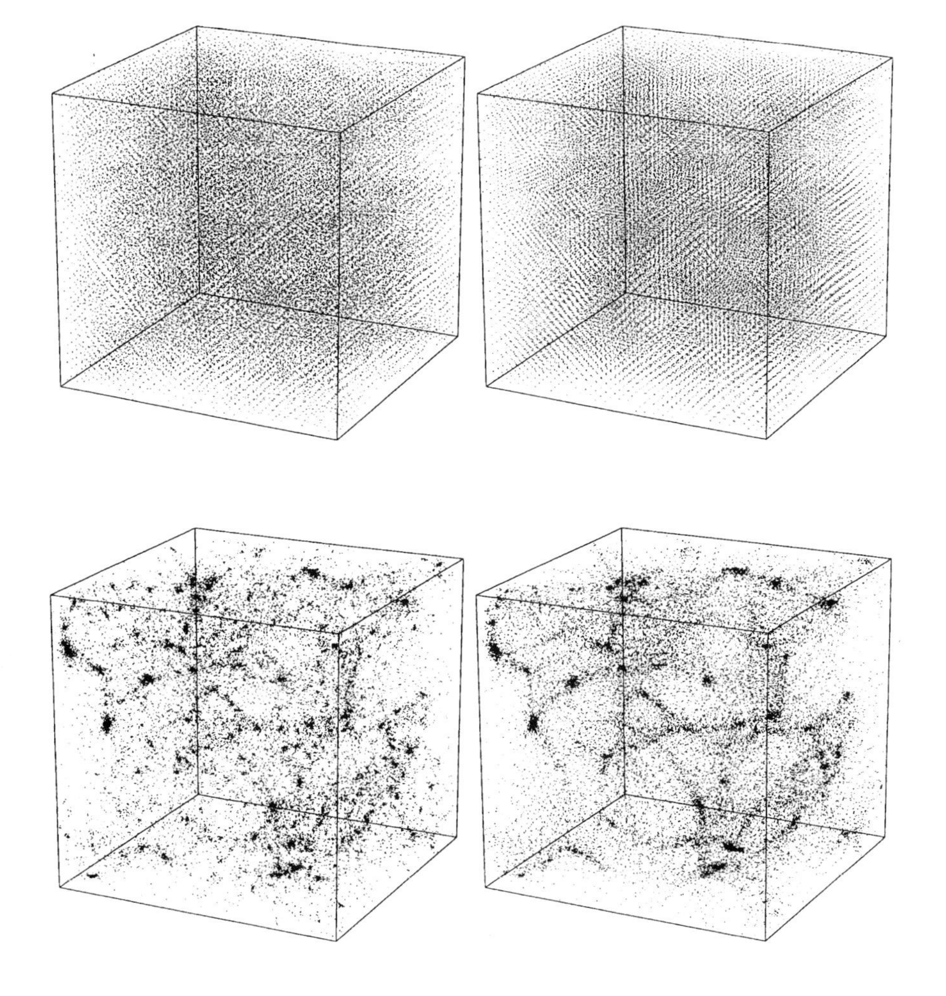

Three-dimensional views of simulations showing the evolution of a cold dark matter universe (left) and a hot dark matter universe (right), from the specified initial conditions (top) to the present epoch, when the universe has expanded by a factor of 10 (bottom). Each box is a comoving cube (the expansion of the universe is taken out) 64 megaparsecs on a side. The hot dark matter model develops a more pronounced network of large sheets, filaments, and superclusters than does the cold dark matter model.

We can now estimate the minimum mass scale to collapse, called the Jeans mass. At the epoch of decoupling, the universe was one-thousandth of its present size, and consequently one billion times denser than it is today. Similarly, the temperature of the background radiation was 1000 times higher than its present value. The temperature gives us the thermal velocity V_s. Once we have it and the density, we can evaluate the Jeans mass, which is found to be about 1 million $M_\odot$, the size of the smallest galaxies. Hence, in a bottom-up scenario, such dwarf galaxies are the smallest structures to form and collapse first, followed by larger and larger structures, until eventually galaxy clusters are formed.

Origin of the Fluctuations

The density fluctuations responsible for cosmic structure existed before radiation and matter decoupled, but where did they come from originally? A major breakthrough in our understanding of their origin came about with the invention of the inflationary theory of cosmology. The quantum vacuum, as we saw earlier, inevitably contains fluctuations, although they are on an infinitesimal length scale. Inflation captures these fluctuations and amplifies them up to scales that correspond to those of galaxies, galaxy clusters, and beyond. All of this happens during the first 10^{-35} second of the universe. As the expansion continues, and the horizon expands, by definition of course, at the speed of light, larger and larger fluctuations come into view. The fluctuations could grow in amplitude only much later, when the universe was dominated by matter. Inflationary fluctuations have two noteworthy properties. The density fluctuations are adiabatic. And they are of the same strength, whatever the scale on which they are sampled.

Without the magic of inflation, any physical origin for fluctuations fails dramatically. Random fluctuations are expected in any gas, but such fluctuations over galactic scales early in the universe were probably negligible. Take two identical boxes of the same gas: the number of molecules always differs by a number equal to the square root of the average number of molecules in the box. Now apply the same reasoning to the Milky Way: there are about $N = 10^{68}$ protons in the Milky Way, and the random fluctuations amount to $\sqrt{N}$ protons. We conclude that the fractional amount of mass in a fluctuation is only $\sqrt{N}/N$, or $1/\sqrt{N}$, or 10^{-34}.

This degree of fluctuation is indeed tiny, but why couldn't it have grown over time? It makes sense to consider such fluctuations on the scale of a galaxy only when the universe is at least a month old. Fluctuations at that scale certainly existed before one month, but in no sense could they grow. Only after a month does the light travel time since the big bang, or horizon, exceed the size of the region that will eventually form a galaxy. Earlier, when these fluctuations were larger than the horizon, different parts of the fluctuations evolved like separate bits of the big bang, and there could have been no coherent growth. Although galaxy-scale fluctuations could eventually begin to gain in strength, the obstacle is, there is simply not time enough for such tiny fluctuations to grow between one month and today.

Inflation solves this problem by commencing with a region so tiny and containing so few particles that the fluctuations in the gravitational field are large. These fluctuations have their origin in quantum uncertainty, which inevitably plays a role at the infinitesimal scales being considered. Inflation blows up tiny quantum fluctuations in the gravity field to scales

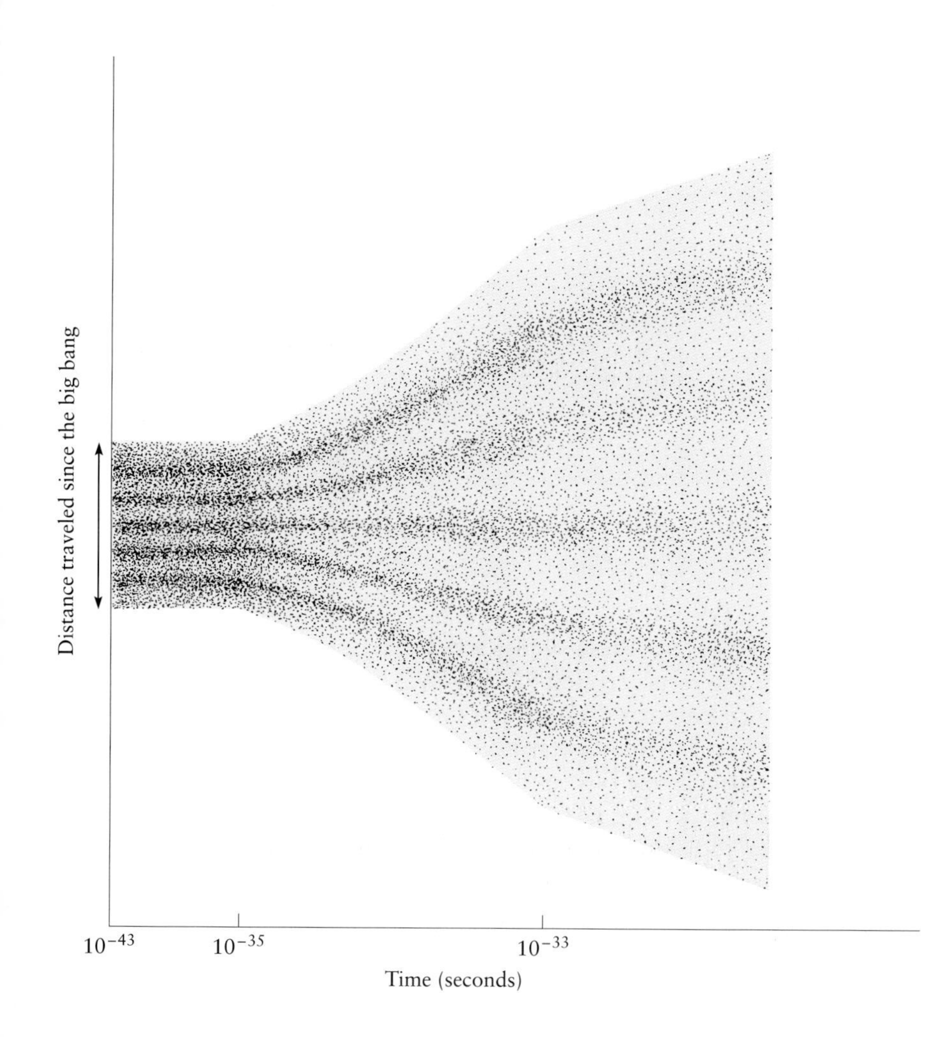

Inflation expands the scale of density fluctuations in matter. The fluctuations also become more pronounced as time passes.

that are macroscopic today. After inflation is over, the fluctuations exist on all scales up to the largest scale to which the universe has inflated.

Another noteworthy property of the fluctuations is that they have the same amplitude over all scales on which they initially are generated. This scale invariance, having the same strength regardless of scale, is an important prediction of the theory of inflation. However, the theory fails to make a prediction for the actual value of the fluctuation amplitude. For the structure we see today to have evolved by the gravitational instability of small fluctuations, the initial fluctuation level must have been about 10^{-5}. We know that the primordial inflationary fluctuations must have had this strength because temperature fluctuations at close to this level were mea-

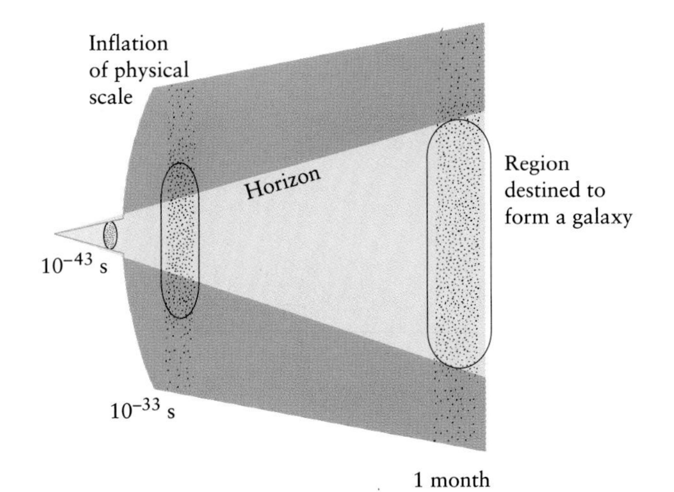

Galaxy-scale fluctuations in the density of matter first enter the horizon one month after the big bang.

sured by the COBE satellite, as described below. It is actually rather remarkable that the measured fluctuation level is so close to what is needed. This coincidence supports the idea that inflation, and indeed cold dark matter, cannot be too far from the truth.

One can actually use this result to constrain the very uncertain choices of various parameters that go into the inflation model. Inflation is far from being a unique prescription: seemingly arbitrary choices must be made for various numbers that are needed to prescribe, for example, the duration of inflation, the degree of acceleration, and the generation of emerging density fluctuations. The energy scale of inflation is so far removed from everyday physics that standard elementary particle theory offers little guidance.

When parameters are chosen to create an initial fluctuation level of 10^{-5}, we find that inflation is a perfectly viable theory. Admittedly, its parameters, specifically the source of energy that drives inflation, must be chosen with care. Inflation cannot be predicted a priori; rather, cosmologists try to infer the parameters that are necessary for inflation to have occurred by comparing observations with such predicted relics of inflation as the density fluctuations.

It would be attractive if inflation occurred regardless of the initial conditions, but this seems not to be the case. Our model of the big bang has a boundary in time, the Planck instant, before which we know essentially nothing. In setting up a model for inflation, we have to assume that the universe began in a certain way, perhaps with inhomogeneities and anisotropies in the expansion, and perhaps with pre-existing curvature of space. Even the properties of the various elementary particles and their interactions must be assumed, for they need not have been as described by current theories. Indeed, theories of quantum gravity advocate a beginning when even space was not three-dimensional but had 10 or more dimensions that somehow collapsed soon after the Planck time, to form the space we inhabit. All of these possibilities may be considered as initial conditions of the universe.

Under a great variety of possible initial conditions, inflationary models succeed in smoothing out most wrinkles in space curvature. However, there are some models, strongly curved, that simply do not inflate. As for the details of the inflationary particle physics, any general prescription for inflation seems unachievable. There is an infinity of initial conditions for inflation, almost all of which do not result in a universe compatible with what is observed.

The moral behind this is that we only have one universe to try to comprehend. It is often thought desirable in particle physics to avoid both small and large dimensionless numbers, since these are often considered unnatural, and, by the same token, to avoid any seemingly special initial conditions. The attitude of cosmologists is somewhat different: for one thing, there is only one universe to study, which makes it difficult to argue about the desirability of "naturalness." Maybe apparent "fine-tuning" is not too bitter a pill to swallow. Ultimately, we may hope that a theory of quantum gravity will emerge to provide the possibly unique initial conditions that led to inflation.

A Baryon-Dominated Universe

There may be strong reasons for arguing that baryons are not the dominant constituent of dark matter, but they are, after all, the only dark matter candidate that we actually know to exist. Whether enough baryons exist to be all of the dark matter is extremely uncertain, but it is instructive to reconsider the evolution of large-scale structure in a universe dominated by baryons.

Suppose that the only dark matter is baryonic. One can live with the nucleosynthesis predictions only if the density is at most 10 percent of the critical value. To construct a viable model of fluctuations in this case, one can no longer appeal to primordial adiabatic fluctuations. There would have been insufficient time in a low-density universe to grow galaxies from adiabatic density fluctuations, without requiring large initial amplitudes. These would consequently have generated excessive fluctuations in the cosmic microwave background.

The alternative approach resorts to primordial entropy fluctuations. Blackbody photons represent the ultimate in disorder, and entropy may be thought of as a measure of their number, relative to the number of protons, in the universe. As entropy increases, we will all end up in a soup of blackbody photons: even massive black holes, into which every surviving galactic star eventually will fall, are destined in the almost, but not quite, infinite future to evaporate into radiation. There are about 3 billion blackbody photons in the cosmic microwave background radiation for every proton in the universe if the baryon number is taken from the nucleosynthesis prediction. The contribution of photons from starlight is negligible. Normally the ratio of photons to baryons would be an invariant number, fixed once and for all time by the physics of baryon genesis.

However, we do not have a definitive understanding of the origin of baryon number. Perhaps as argued in an earlier chapter, it arose shortly after the end of inflation as a consequence of CP violation. In such a universe, the baryon number is invariant everywhere. If inflation is discounted, however, one can equally well imagine that the very early universe generated a baryon number that varied from place to place. These variations would be equivalent to primordial entropy fluctuations. One can think of an entropy fluctuation as being equivalent to keeping the radiation uniform but allowing the density of baryons to fluctuate in space.

The baryon-dominated universe has to be open. Inflation does not result in an open universe, however; therefore, the inflationary prediction that density fluctuations remain invariant over all scales does not apply. In an open universe, there is no guide to the distribution of amplitudes of primordial entropy fluctuations. Theorists cannot predict the distribution of these fluctuations a priori, but must adjust the distribution to fit whatever best reproduces the observations of large-scale structure. The procedure is to make different guesses as to the amplitude of the primordial fluctuations over the range of clusters to superclusters of galaxies, and then to check these guesses by simulating the formation of these clusters on a computer. The primordial entropy fluctuations always lead to very early galaxy formation.

Structure again forms from the bottom up in an open universe. In a low-density universe, however, gravity is unimportant at late times in aiding the growth of fluctuations. All the action occurs early. If the mean density of the universe is only 10 percent of the critical value for closure, large structures last formed at a redshift of 10, about 90 percent of the way back to the beginning. Subsequently, the kinetic energy of the universe dominates over the gravitational potential energy, whereas earlier the two types of energy were in approximate balance. In other words, an open and closed universe are identical at high redshift, but begin to drift apart at low redshift.

The best simulations imply that the small-scale fluctuations are initially much stronger than the large-scale fluctuations. Hence the first small structures may form soon after the recombination epoch at a redshift of 1000, a mere 300,000 years from the beginning. They are of scale close to that of the Jeans mass, or 1 million $M_{\odot}$, similar to the masses of globular clusters or the smallest dwarf galaxies. A hierarchy of structure develops, in which galaxies form when the universe has expanded by another factor of 10, and clusters of galaxies form when it has expanded by a factor of 100. The universe we see is thus in place at an early epoch, little more than a billion years after the beginning. This is a very different expectation from that of a cosmological model at critical density.

By undertaking surveys of the most distant galaxies, cosmologists may be able to look back in time sufficiently far to discriminate between the rival theories of structure formation. Their observations of the development of galaxy clustering, especially at an earlier epoch, may provide a potentially powerful cosmological probe.

Detection of Fluctuations

The most compelling versions of inflationary cosmology predict equal-strength fluctuations on all scales. This distribution had been previously advocated on grounds of simplicity, and is not unique to inflation. What is remarkable is that inflation predicts fluctuations of this form on scales so large that there has been insufficient time for any physical processes to have either generated or destroyed the fluctuations, simply because there was not enough time before the last scattering of the radiation for light to have traveled from one end of the fluctuation to the other. Light had traveled some 300,000 light-years between the big bang and the time of last scattering, a distance as measured today of about 300 megaparsecs. This distance translates into an angular scale on the sky of about 2 degrees. If one observes the cosmic microwave background on scales of a few degrees, one is looking directly at temperature differences on the sky that, although infinitesimal, must have been generated at the epoch of inflation.

Theorists had been calculating the strength of fluctuations in the fireball radiation ever since 1967, and these calculations were constantly refined to remain just below the available experimental upper limits. They knew that fluctuations could not begin growing until the transitional epoch when matter first dominated over radiation in density. Moreover, fluctuations grow only when within the horizon, the distance light has traveled since the big bang. Only fluctuations within the horizon are "causally connected," that is, are small enough in scale that a physical process like gravity has had time to operate across the entire fluctuation. Consequently, the smaller-scale fluctuations that first enter the horizon during radiation dom-

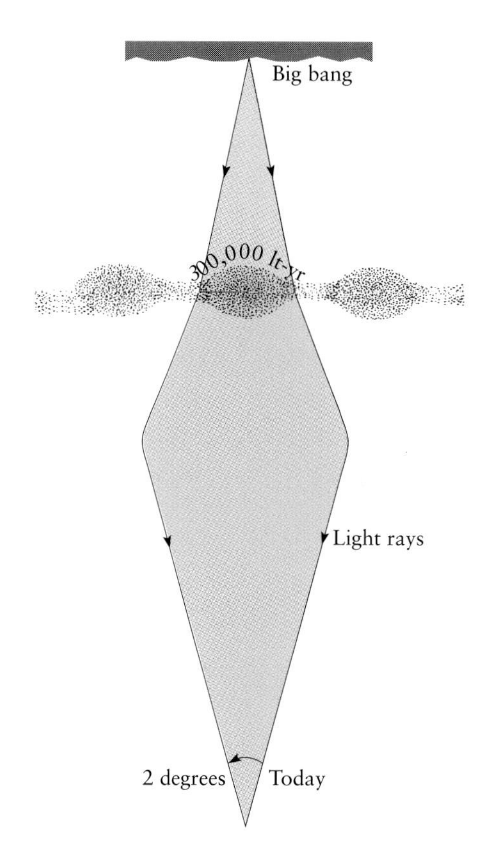

The horizon scale at last scattering subtends an angle of about 2 degrees on the sky.

ination suffer retarded growth relative to the larger-scale fluctuations that enter during matter domination.

The horizon scale at the transitional epoch, when the densities of matter and radiation are equal, corresponds to a distance today of about 10 megaparsecs. This scale is imprinted on the distribution of density fluctuations, and its imprint shows up in the distribution of the fluctuations observable today. For example, galaxy clustering is relatively strong on scales of a few megaparsecs, and it is thought that the only important force acting on such large scales is the force of gravity. This allows astronomers to

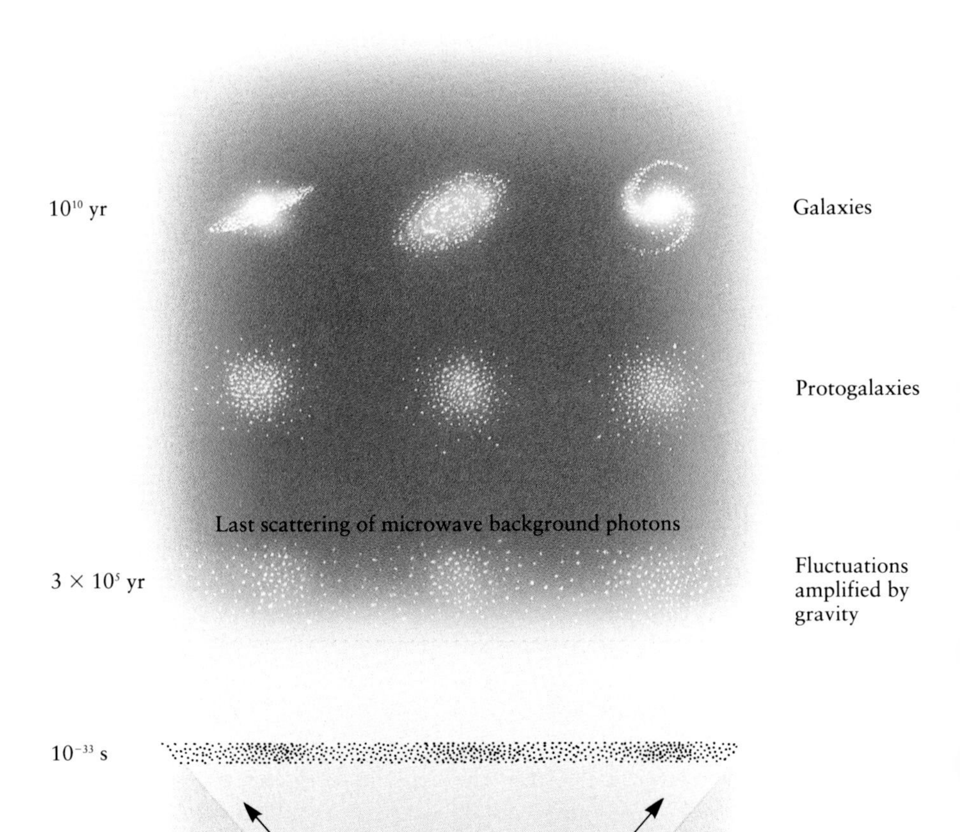

The evolution of density fluctuations, from tiny fluctuations in the density of matter 10^{-33} seconds after the big bang, to fully formed galaxies.

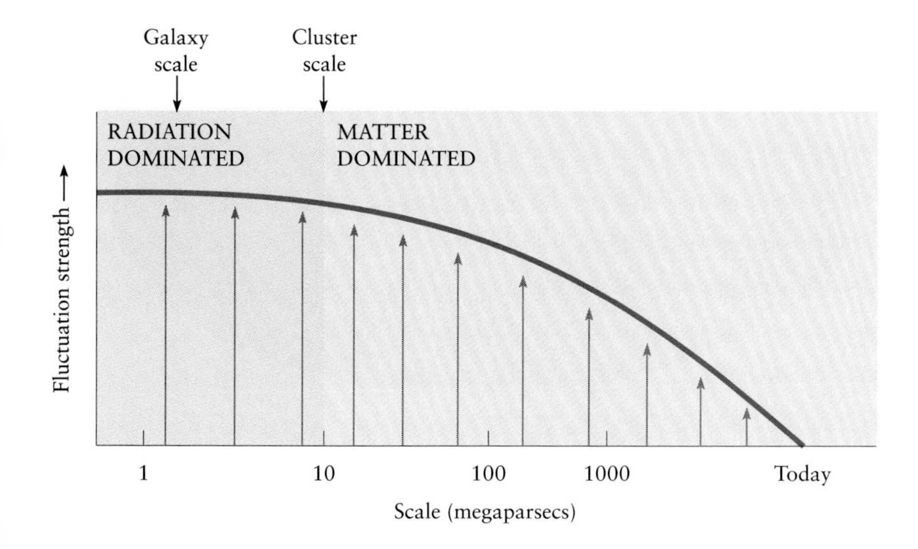

The size of the horizon at the transition from matter to radiation domination marks a natural division between fluctuations at different scales. Fluctuations at lesser scales have all undergone an identical growth in strength within the horizon until the present epoch. Fluctuations at larger and larger scales, however, have undergone progressively less growth within the horizon.

hope that they can safely compare theory to observation on a scale where the physics is thought to be simple.

Fluctuations at last scattering of the radiation leave a signature on the cosmic microwave background. Photons at that epoch are scattered in the slightly stronger gravity field of a fluctuation and lose energy in climbing out of the gravity field. They thereby acquire a small gravitationally induced redshift, which is measurable as a slightly colder spot on the sky, in the cosmic microwave background. Conversely, where there were holes in the matter distribution at last scattering there should now be slightly warmer spots on the microwave sky. Many of these cold and warm spots should be present, observable as infinitesimal temperature fluctuations on large angular scales. On smaller scales, additional complications arise because the matter that scatters the radiation is itself moving in the gravity field induced by the density fluctuations. The resulting Doppler shifts further enhance the temperature fluctuations.

With the cold dark matter theory, one can now predict the shape of the fluctuation distribution one might expect on a variety of angular scales. One can test this theory by comparing these predictions with the fluctuations actually observed in the cosmic microwave radiation. We have seen that the simplest inflationary theory predicts equal strength fluctuations to be initially present on all scales. Above an angular scale of a few degrees, the prediction refers to fluctuations that have undergone no causal interactions throughout their history, coming to us directly from inflation. Near an angular scale of 1 degree, one can make a direct connection between fluctuations and the structure we observe in the universe around us; the

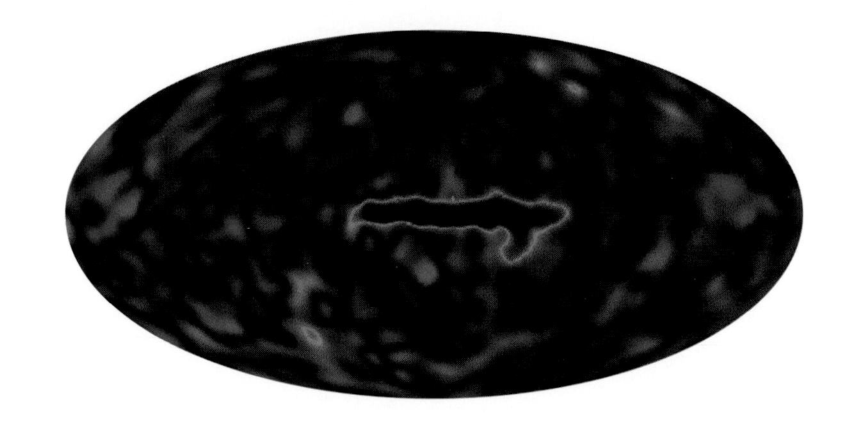

A map of the microwave sky at a frequency of 53 gigahertz, created by the COBE satellite's DMR experiment. The map shows hot (blue) and cold (red) spots at a level of about 1 part in 100,000 of the microwave background intensity. The dipole signal, due to the motion of our galaxy, and foreground galactic emission have been subtracted in constructing this map, although the imperfect subtraction of the galaxy has left the central blue streak. Individual fluctuations are likely to be real only about 50 percent of the time, but statistical analysis of the map confirms their detection.

fluctuations can be mapped on the same 100-megaparsec scales as the galaxy distribution into which they are destined to develop. The simple, "no preferred scale" spectrum leads to the prediction of temperature fluctuations of about 1 part in 100,000 over angular scales from 1 to 90 degrees.

Even if one takes the viewpoint of a skeptic and discards all inflationary concepts, one still has to explain the galaxy clustering, superclustering, and large-scale flows we see around us. If these grew from tiny, long-ago density fluctuations according to the gravitational instability picture, then there must inevitably be slightly smaller fluctuations in the microwave background.

Cosmologists breathed great sighs of relief when the discovery of the fluctuations was first announced in 1992, at more or less the expected level. The observed amplitude indeed comes very close to matching the amplitude of the density fluctuations required to account for large-scale structure, although, as we will see in the following chapter, a detailed comparison does not quite succeed in the case at least of the simplest version of cold dark matter.

The presence of fluctuations was confirmed within a year of the COBE announcement by several other groups searching for temperature anisotropies. Observations from the South Pole and balloon-borne telescopes revealed fluctuations at a level of between 1 and 3 parts in 100,000 over angular scales between $\frac{1}{2}$ and 10 degrees. In fact, 1993 saw the announcement of no fewer than eight independent experimental detections of fluctuations. These fluctuations were mostly found on angular scales smaller than the resolution of the COBE sky map. It is only on scales of a degree or less that one finds the fluctuations that are direct precursors of the largest observed structures in the nearby galaxy distribution.

A note of caution deserves to be added. Our galaxy also emits radiation having low-level fluctuations; its sources are interstellar synchrotron emission (radio waves emitted by ultrafast electrons spiraling around the interstellar magnetic field), bremsstrahlung, and dust emission. This foreground noise is especially severe at a few degrees or less, and mostly affects experiments that only cover a small patch of sky. The large-scale microwave background fluctuations are surely of cosmic origin. It is likely that some of the smaller-scale experiments are also measuring true cosmic microwave background fluctuations, but some are undoubtedly contaminated by galactic emission. It remains unresolved to what extent any of the temperature variations on smaller angular scales constitute a residual signal from the cosmic background.

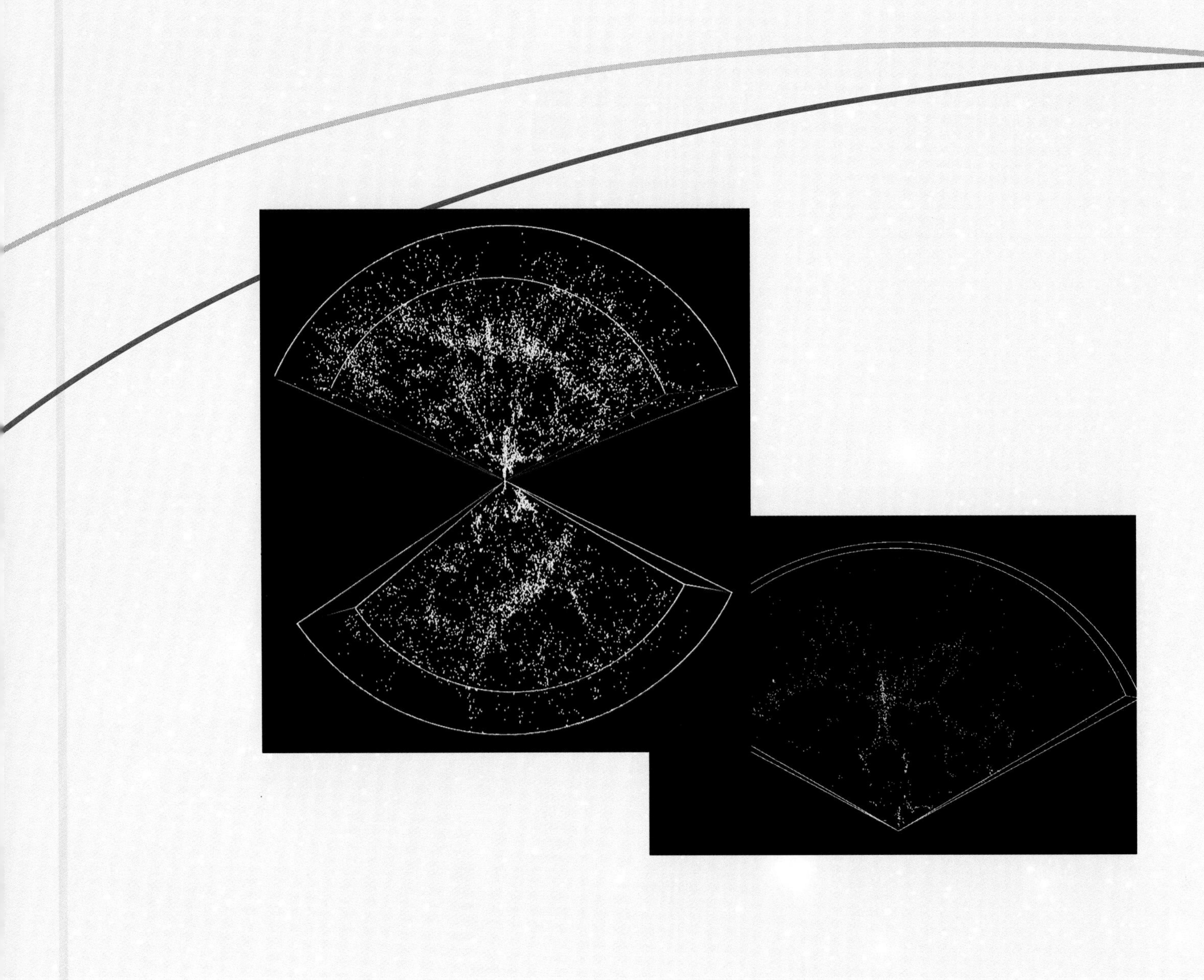

Left: *A three-dimensional map of the universe, showing the positions of some 5000 galaxies to magnitude 15.5 in two slices of the sky that are about 70 megaparsecs deep and 20 megaparsecs thick. We are at the center, and the lack of any correction for the local velocities of galaxies in clusters accounts for the apparent "finger of God" effect, really due to the presence of rich clusters of galaxies. The region in the outer arc is inadequately sampled, but within the inner arc one can clearly see giant voids and elongated structures such as the "Great Wall," a prominent nearby horizontal feature in the top slice.*
Right: *Another, shallower slice of the sky shows the Coma cluster of galaxies and its surroundings, which in the map resemble a "stick man."*

11

Large-Scale Structure

Perhaps the most important observational evidence in cosmology comes from studying the large-scale structure of the universe. With the aid of large telescopes, we can examine galaxies and their spatial distribution around us back to when the universe was approximately half its present age. Our goal is to understand the evolution of structure. Our theories suggest that on large scales, structure develops in a relatively simple and straightforward fashion as long as the only important force is gravity.

The Greatest Structures: Galaxy Clusters

The breadth of a galaxy is measured in tens of thousands of parsecs, where one parsec equals 3.1 light-years. Galaxies in turn are typically separated by about five million parsecs. But statistics fail to describe why the Milky Way's nearest neighbor of comparable size, Andromeda, is a mere two million light-years away. In fact, our galaxy is a member of a group of galaxies called the Local Group. Some galaxies have nearest neighbors that are much closer, perhaps 10 times closer. This is the case in a cluster of galaxies.

Clusters are the largest self-gravitating aggregates of matter in the universe. In a rich cluster, there may be one thousand or more galaxies per cubic megaparsec, orbiting around one another in their mutual gravitational field. There are also superclusters, or clusters of clusters of galaxies. The Milky Way galaxy, for example, is on the outskirts of the Virgo supercluster. Although most ordinary matter in a galaxy is in stars, the known matter in a cluster is predominantly in the form of diffuse, hot, x-ray–emitting intergalactic gas.

The orbital velocities, whether of stars in a galaxy or galaxies in a cluster, probe the gravity field. We infer that for the galaxies not to fly apart,

A mixture of spiral and elliptical galaxies is clearly visible in this remote cluster, viewed at a redshift of 0.4 by the Hubble Space Telescope. In addition, there are several systems that appear to be merging galaxies in the throes of formation.

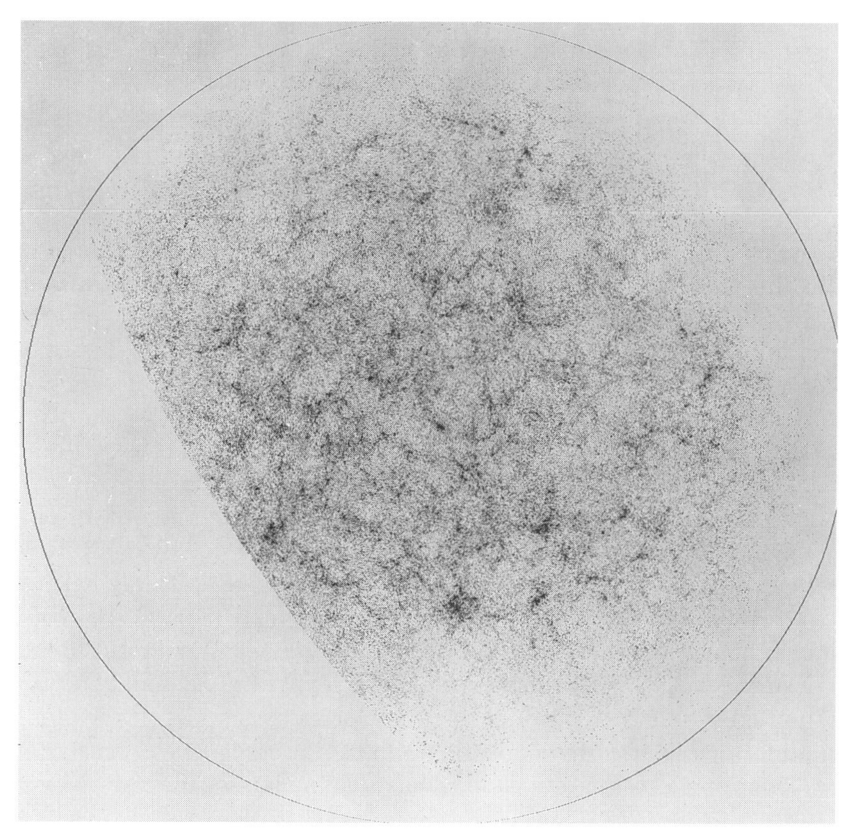

Bottom right: Approximately 10,000 galaxies brighter than magnitude 15.5 are plotted in this equal-area map of the northern sky: the north galactic pole is at the center and the Milky Way is outside the circumference of the circle. The Coma cluster of galaxies is visible near the north galactic pole, and the nearby Virgo supercluster is the largest visible galaxy concentration. Bottom left: Many other clusters of richness comparable to the Coma cluster become visible when one million galaxies, all brighter than magnitude 19, are mapped in the northern sky. Again, the Coma cluster is visible near the north galactic pole at the center.

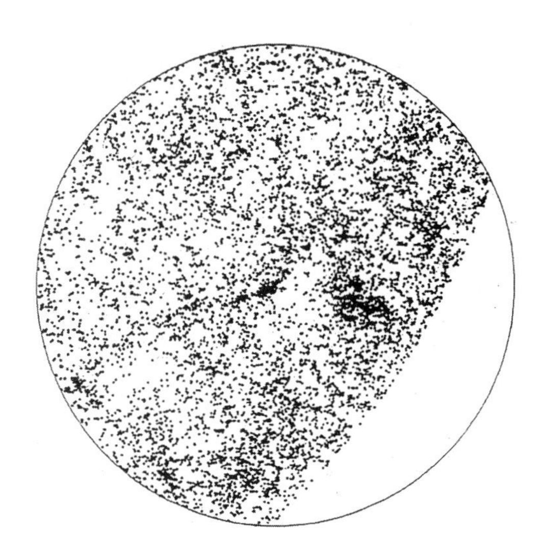

some 80 percent of a cluster's mass must be dark matter. This means that most of the mass in the universe is in invisible form, of a type unknown and undetected at any wavelength. Far more mass than is directly measured in the form of ordinary stars or diffuse gas remains unaccounted for.

Imagine a group of clumps, each of which is a newly formed galaxy, or better still, a protogalaxy. These are expanding away from each other, some one billion years or so after the big bang. We will discuss in Chapter 12 how a galaxy itself forms, but it is much simpler to understand how galaxy groupings develop on the very largest scales. The formation of clusters is relatively uncomplicated because, early on, only the comparatively simple force of gravity is involved.

To understand how easily structure develops, imagine that the clump of protogalaxies is destined to become a small cluster of, say, 100 galaxies. Such a region has a slightly higher density of protogalaxies relative to its surroundings, perhaps only by 10 percent or so. In another region of precisely the same size, there would probably be either fewer or more galaxies. Such discrepancies are inevitable even if the universe were, on the average, uniform, as it must have been early on to produce the smoothness of the cosmic microwave radiation. Even a uniform universe would contain statistical fluctuations, equal to the square root of the average number of galaxies in a region, so that one volume might have 110 galaxies while another would have 90. The region with the slight excess exerts an enhanced gravitational pull on its surroundings, relative to other regions, and it gradually accretes more galaxies. Its density grows from a 10 percent excess at 110 galaxies to a 200 percent excess at 200 galaxies, by the time the universe has expanded by a factor of 10.

At this stage, the local region is exerting a gravitational pull on itself that is strong enough to counteract the expansion of the universe, and the region stops expanding. It reaches a maximum radius when its density is about five times greater than the background density, which of course is continually decreasing.

At its radius of maximum expansion, the cloud of galaxies has only gravitational potential energy and no kinetic energy. As it begins to collapse, it loses potential energy and gains kinetic energy in the form of random motions of the galaxies. Once the kinetic energy equals one-half of the (negative) potential energy, the cloud reaches a stage of energy balance and there is a dynamic equilibrium: the gravitational field of the cloud no longer changes with time. The kinetic energy of the randomly moving galaxies acts like a pressure that balances the force of gravitational attraction, and the result is to produce a stable cluster of galaxies. If the cloud loses no further energy by cooling and radiating away its thermal energy, it stops collapsing. At this stage, the cloud has collapsed in radius by a factor of 2.

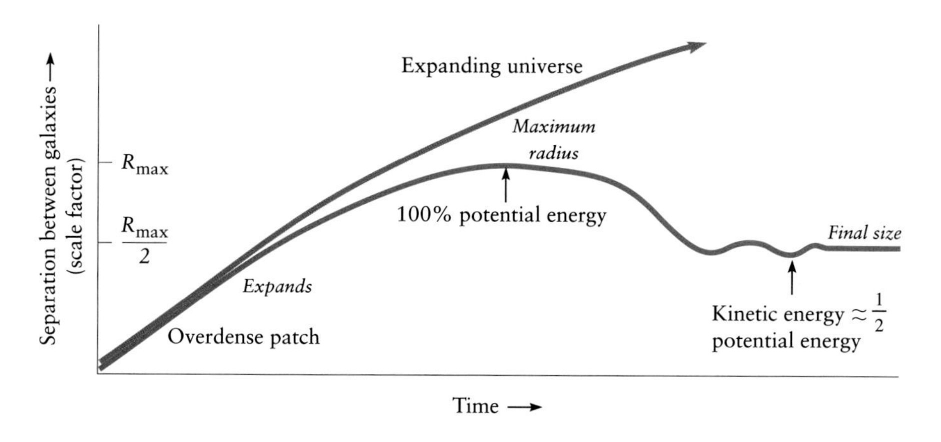

An overdense patch of the universe is destined to form a cluster of galaxies. That patch gradually breaks away from the expanding background, reaches a maximum size, and finally attains an equilibrium at half the maximum radius, when its kinetic and gravitational potential energies are in balance.

While this argument describes a region collapsing to form an aggregation of galaxies, the physics initially is identical for weakly interacting dark matter collapsing to form a halo, and very similar for a gas cloud collapsing to form a single galaxy. The only difference is that a single galaxy continues to collapse because the gas atoms are able to cool by radiating away their thermal energy before the gas eventually fragments to form stars.

The Milky Way galaxy is in such a state of equilibrium. It is not expanding like the universe, nor is it contracting like a collapsing interstellar gas cloud. Its stability can be explained in two ways. The density of the Milky Way greatly exceeds the mean density of the universe, and so its enhanced gravitational pull keeps the galaxy from expanding. Moreover, the Milky Way consists mostly of stars, which act like collisionless particles, conserving energy in their orbits. Consequently, a galaxy like the Milky Way is in a state of equilibrium determined by the balance between its own self-gravity and the kinetic energy of the stellar motions.

Voids and Sheets

To learn precisely how galaxies are distributed, astronomers have been constructing three-dimensional maps of the universe. Redshifts are measured for thousands of galaxies in a well-defined region of the sky; these surveys map nearly every galaxy above a chosen brightness, which is as low as the availability of telescopes and observing time will allow. After converting redshift to distance, the astronomer can reconstruct the three-dimensional distribution of the galaxies.

The first galaxy maps were a surprise. True, the expected great clusters of galaxies were found, but on larger scales the galaxy distribution did not immediately settle down to the uniform universe beloved by the

Astronomers produced this "slice" of the sky, 6 degrees wide and 30 degrees across, by measuring the redshifts of all galaxies to magnitude 15.5, then plotting those redshifts as radial distances from our position at the vertex. The "stick man" is the Coma cluster of galaxies and its surroundings, with the radial extension being an artifact of the plot due to the peculiar velocities of the Coma galaxies. The map extends to a depth of 100 megaparsecs.

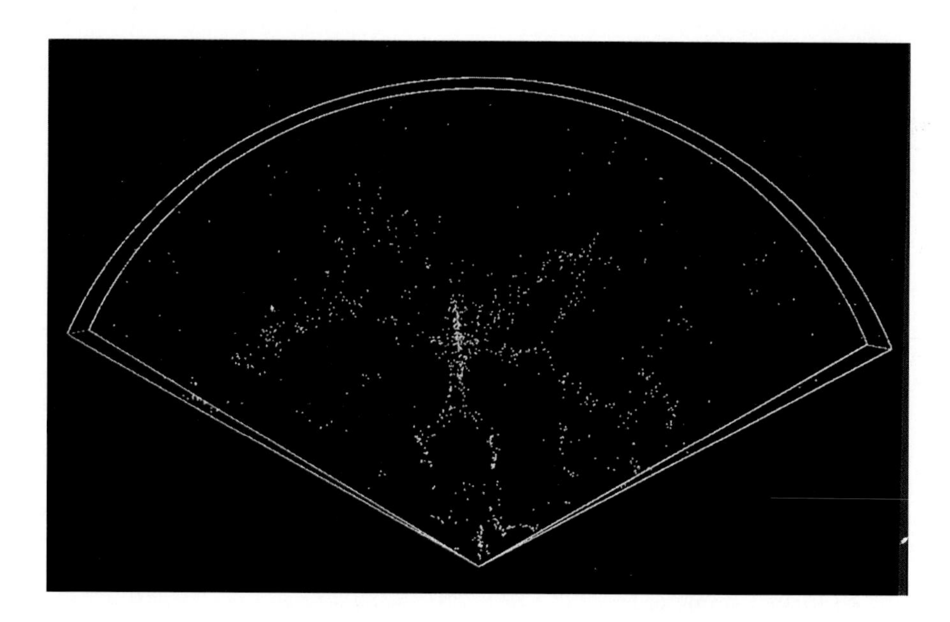

]cosmologists. There appear to be prominent holes or voids, often several megaparsecs across. Between these voids are agglomerations of several galaxy clusters (superclusters), thin sheets of galaxies, and occasional filaments or chains of galaxies that stretch over 10 megaparsecs or more. The universe appears textured, almost spongelike, on scales of tens of megaparsecs. On much larger scales, above 100 megaparsecs, the galaxy distribution finally appears to attain uniformity.

Any cosmological theory has to be able to explain how this diversity of structure evolved from an almost uniform universe. Does any one theory do so more successfully than others? There are two main competing explanations. One is suggested by the cold dark matter scenario, in which density fluctuations are present on all scales. It proposes that galaxies form only at the extreme, and therefore rare, density peaks in the primordial fluctuation spectrum. This is known as the biasing hypothesis, because it suggests that galaxy formation is biased to rare density peaks.

The hypothesis has two important consequences. Even if the underlying fluctuations are completely random, rare peaks will be clustered together for much the same reason that the highest mountains occur together, as in the Himalayas. The average peaks, which resemble hills rather than mountains, are more uniformly distributed. Hence most of the cold dark matter is outside the concentrations of galaxies, rather than centered on them. This helps explain why one can have $\Omega = 1$, and yet only measure $\Omega = 0.1$ by studying the mass associated with galaxy halos, clusters, and superclusters. A second implication is that most of space is filled by large

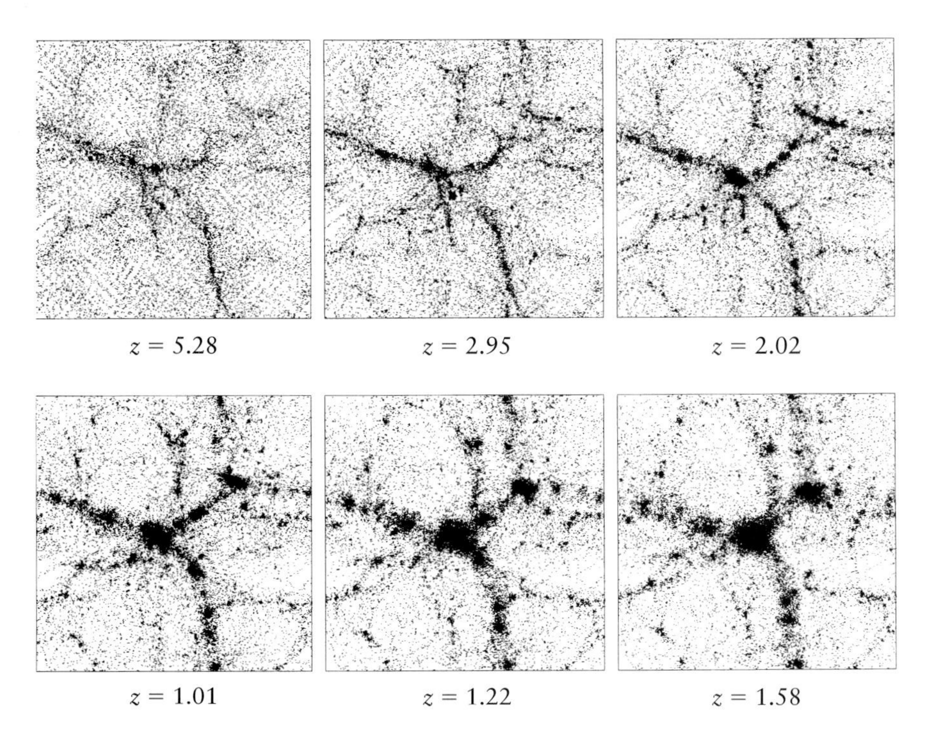

In this computer simulation, a cluster of galaxies grows as matter is accreted along filaments. Each square is the two-dimensional projection of a comoving cube and represents a step in the evolution of the universe, commencing at redshift 5.36, less than a billion years after the big bang, and stopping at redshift 1, about 3 billion years later.

regions, containing only average peaks, and devoid, by hypothesis, of galaxies. Occasionally, groupings of peaks occur that are filamentary or sheet-like in shape, but this is relatively rare. Most groupings tend to resemble elongated spheroids.

An alternative scenario, that of hot dark matter, results in a spectrum of fluctuations, the smallest of which in scale is characteristically a galaxy cluster in mass. Cluster-mass clouds collapse, forming structures that ini-

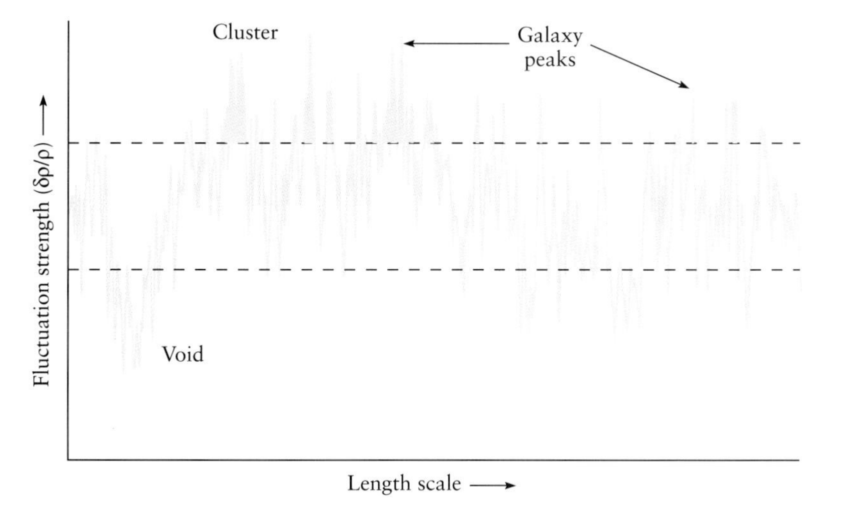

Only fluctuations above a certain strength can form galaxies in a universe dominated by cold dark matter. Thus galaxies form from peaks in the distribution of density fluctuations. These rare peaks are more frequent in the denser regions, destined to eventually form galaxy clusters, and rarer in the less dense regions, which eventually form voids.

tially are sheets or pancakes but later fragment into smaller clouds of galactic mass. The result is the formation of sheets of galaxies. Clusters form at the nodes where the randomly oriented pancakes intersect one another, while large voids are evacuated regions between the pancakes.

This would be an attractive picture of large-scale structure, were it not for the fact that in a hot dark matter scenario most galaxies inevitably form late, within the past few billion years. The problem is that clusters appear to be still forming, and galaxies cannot form before clusters do in a universe exclusively dominated by hot dark matter. Yet some galaxies must have formed more than 10 billion years ago, when the universe was only 20 percent of its present age, since galaxies are seen beyond redshift 4, when the universe was one-fifth its present size, and quasars to a redshift of 5, when the universe was one-sixth its present size.

Cosmologists have had to resort to additional hypotheses to salvage hot dark matter. One way out is to postulate that the initial fluctuations contained some galactic-scale clumps that allowed galaxies to form before the epoch of pancaking.

Large-Scale Flows

One way to test how structure evolved is to search for direct evidence of primordial fluctuations on very large scales where collapse has not yet occurred. Such large-scale fluctuations could manifest themselves in several different ways. One might expect to see density fluctuations directly in the galaxy distribution on scales of 100 megaparsecs or larger. The task of mapping out the deviations from uniformity has been a daunting one, however: over these scales, one is looking for variations of a few percent in galaxy number. The astronomer must measure the redshifts of many thousands of galaxies and obtain accurate distances. Unfortunately, inordinate amounts of time on the largest telescopes are necessary in order to make much headway. Only recently have new instruments been developed that enable astronomers to simultaneously measure hundreds of galaxy redshifts in a single exposure at a large telescope, and one can anticipate that this field is on the verge of making substantial progress.

There is an alternative approach that can be pursued with fewer redshifts, although one needs data of even higher quality. Far-away galaxies are not seen to be simply receding from us; they also move to and fro under the influence of local gravitational fields. Galaxies move within clusters, and the clusters themselves acquire additional velocities by virtue of their being in or even near superclusters. These to-and-fro motions are superimposed on the expansion of the universe, much as the ocean waves are superimposed on the tides. These peculiar velocities, or deviations from the Hubble flow, can be measured with some accuracy.

The technique makes use of the fact that two internal properties of galaxies, the rotation speed and the luminosity, are correlated. We came across this correlation, named the Tully-Fisher relation, in Chapter 2, where it was used to measure the Hubble constant. Measure rotation speed, and the true luminosity can be deduced. Since rotation velocity does not depend on distance but apparent luminosity does, we obtain a measure of distance by comparing the two luminosities. The distance, when plugged into Hubble's law, yields the average velocity of expansion expected if there were no interference from density fluctuations, while redshift measures actual speed. Comparing the two velocities gives the peculiar velocity of the galaxy.

One can study deviations from the uniform Hubble flow by applying the Tully-Fisher correlation to large numbers of galaxies with measured redshifts in different directions. The peculiar velocities probe the effects of gravity both from the matter we see and also from any dark matter. When galaxy peculiar velocities are averaged over large scales, they provide a unique map of large-scale density fluctuations. Peculiar velocities have the advantage over redshift surveys of probing the dark matter as well as the luminous matter. Simply counting galaxies traces only the luminous component of the universe.

The strength, and even the reality, of these large-scale flows are still controversial because the origin of the Tully-Fisher correlation is not yet understood. For all we know, it may vary with environment. The only unambiguous measurement of peculiar velocities comes from a different technique; the idea is to measure the motion of our galaxy against the cosmic

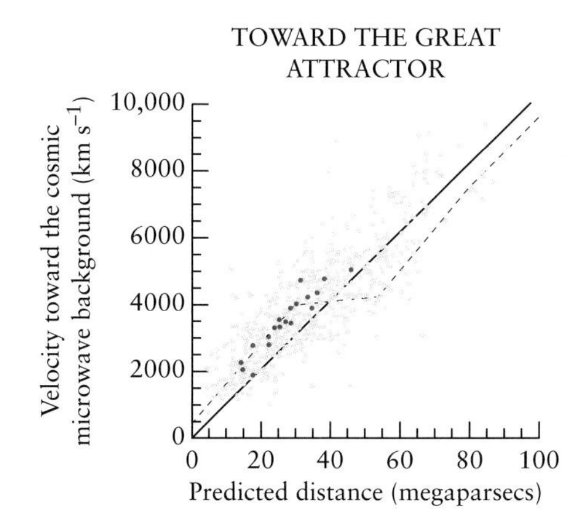

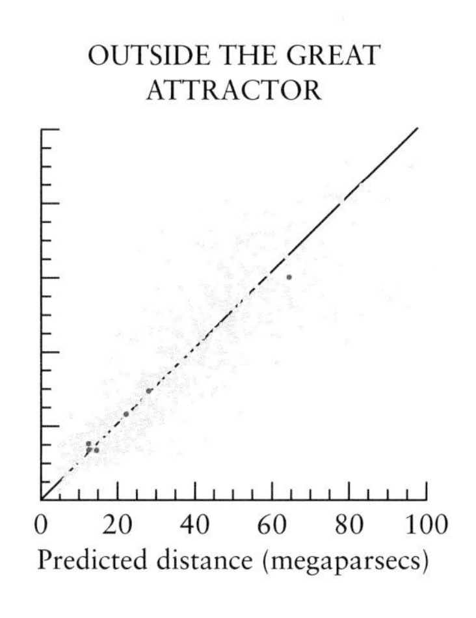

The Hubble diagram for two large samples of galaxies: it plots galaxy velocity (measured relative to the cosmic microwave background) against predicted galaxy distance as obtained from the Tully-Fisher relation. An unperturbed Hubble flow in a uniform universe corresponds to the solid line. In one direction (toward the Great Attractor), there is a deviation from the Hubble flow due to the gravitational attraction of the Great Attractor. The dashed line shows the expected distortion: galaxies fall toward the Great Attractor, reducing their apparent Hubble velocity on the far side but increasing it on the near side.

microwave background. The dipole anisotropy in the microwave background, described in Chapter 3, tells us that our galaxy is moving in the frame of the cosmic microwave background at a speed of 600 kilometers per second in a direction about 45 degrees away from the Virgo cluster. Indeed, the entire Local Group is moving at about this speed, since the motion of the Milky Way relative to Andromeda is only 50 kilometers per second. Some studies of galaxy peculiar velocities conclude that aggregations of galaxies extending over distances of between 10 and 100 megaparsecs may also be moving at a similar speed. Of course, one expects that over a sufficiently large distance, a region should be at rest with respect to the cosmic microwave background. Precisely how far one has to go in mapping the galaxy distribution of peculiar motions to find this cosmic restframe has yet to be determined.

The peculiar motions, taken at face value, challenge many of the theories of structure formation. For example, in an open universe, the expansion energy dominates today over gravitational potential energy, and the peculiar velocities of galaxies are expected to be much lower than in a universe at critical density, such as that predicted by inflation. We have seen that in such a universe, there is a delicate balance between expansion energy and gravitational energy so long as the universe is uniform. Thus any sufficiently large-scale density fluctuations, no matter how weak, can help tilt the balance toward gravitational effects and thereby help drive peculiar velocities of galaxies. To account for the observations, one needs more than just a critical density, one also needs fluctuations in density of sufficient strength. We now consider the successes and failures of the rival theories.

The Failure of Cold Dark Matter?

The cold dark matter scenario makes a unique prediction for the strength of fluctuations destined to evolve into galaxy clusters and galaxies. Cold dark matter cannot resist developing clumpiness in a universe at critical density, so these systems should be vast in number. Unfortunately, excessive fluctuations are present in the theory: one sees far fewer clusters of galaxies and massive galaxies than predicted. Nor is the low number of these systems the only problem. The theory predicts that some galaxy clusters should show a spread of relative galaxy velocities two or three times larger than is observed, as should galaxy pairs, and that some massive galaxies should be rotating at speeds that are also two or three times larger than is observed.

These predictions are accurate only if we have correctly identified which primordial fluctuations form galaxies. If density peaks eventually turn into galaxies, then the power in the cold dark matter scenario remains exces-

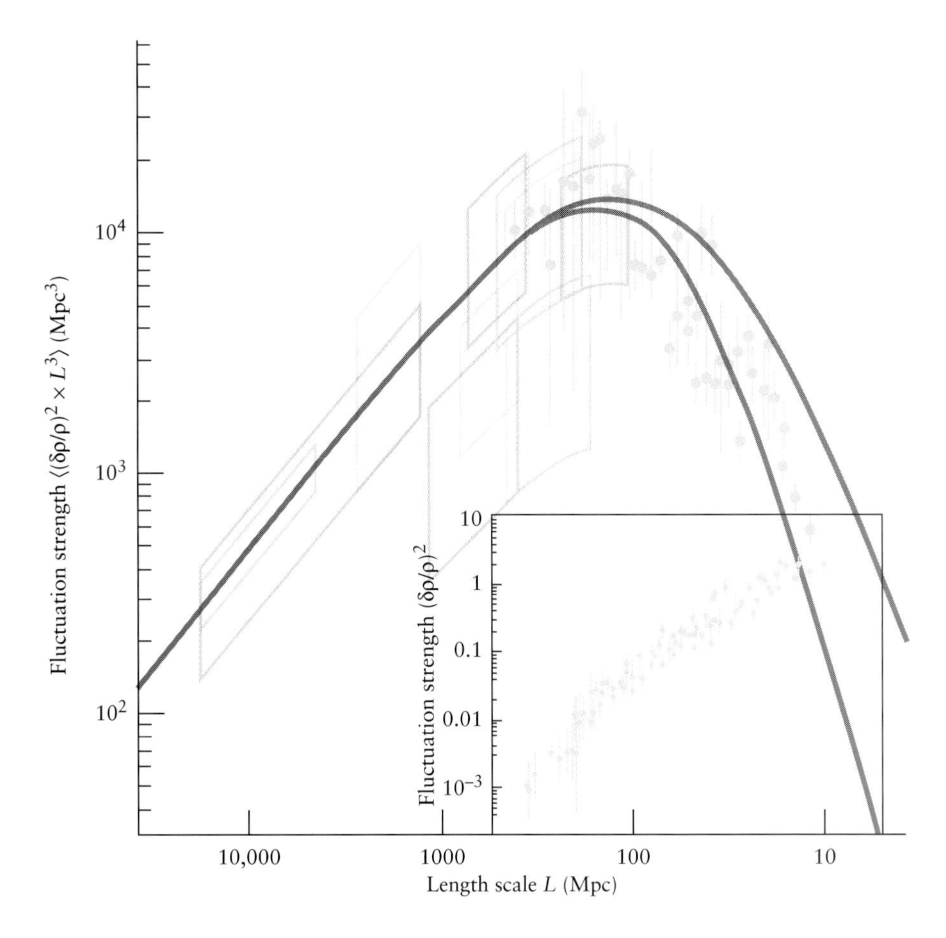

The power spectrum of density fluctuations, defined to be the square of the density fluctuation strength multiplied by the volume over which the fluctuation is being sampled, is plotted against length scale. The data points (and vertical bars representing uncertainties) are taken from several surveys of galaxies and galaxy redshifts, and are also shown directly in terms of the fluctuation strength in the insert. The boxes are the strengths of density fluctuations inferred from various measurements of fluctuations in the cosmic microwave background. The smallest box on the left is the result from the COBE experiment; other boxes are from ground-based and balloon-borne experiments. The solid lines are predicted power spectra for the inflationary cold dark matter-dominated universe (CDM) and for a universe containing a mixture (two-thirds cold and one-third hot [MDM] dark matter). Both types of dark matter have power spectra compatible with fluctuations in the cosmic microwave background, but at small scales a universe of cold dark matter predicts fluctuations of greater strength than have been observed.

sive. If, to take the opposite point of view, only the weakest fluctuations in the most quiescent regions of the universe eventually form galaxies, then the number of galaxies and galaxy clusters formed is still large, but many would be unobservable. There would be dark clusters or dark galaxies, objects with high primordial velocities and large velocity dispersion, that have no optical counterparts. We may be able to test this hypothesis: interstellar gas falling toward an invisible massive cluster would heat up and radiate x-rays, enough to act as a useful tracer of these dark galaxies.

One concludes that short of some rather radical measures, the cold dark matter universe is in serious difficulty. A compromise is necessary. The simplest solutions on offer adjust the properties of the dark matter so that it modifies the inflationary fluctuations. Not all survive, depending, for example, on whether the dark matter is sufficiently hot. There would be far fewer fluctuations on galaxy scales in such a picture. Another, more radical approach tinkers with inflation itself. One can try to custom design a

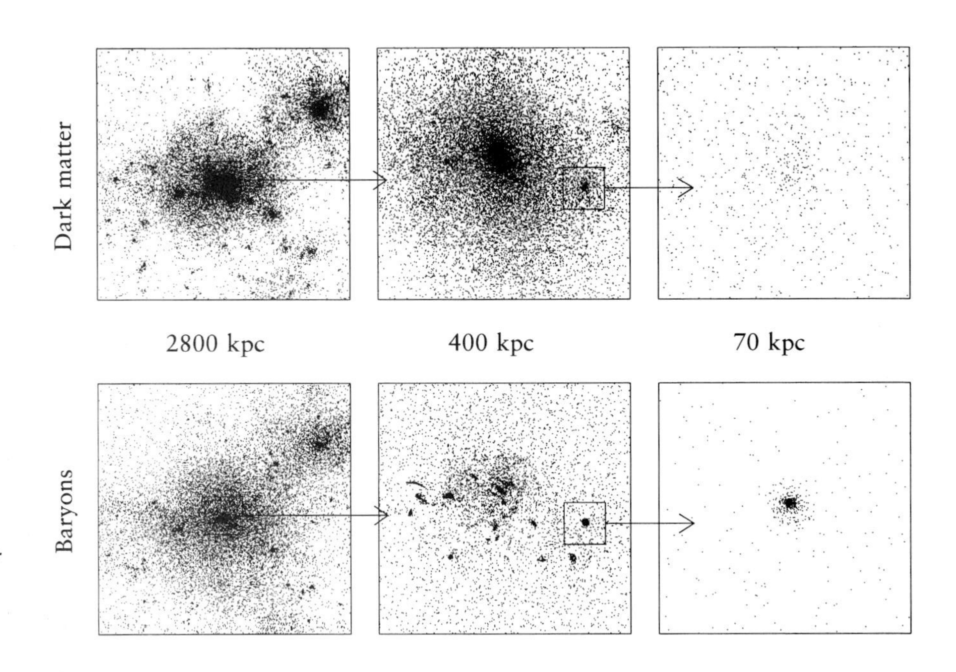

In this computer simulation, the tendency of cold dark matter to clump produces a massive, collective dark halo that is the size of a group of galaxies. The zoom sequence shows the structure in the dark matter distribution at a redshift of 1, two-thirds of the way back to the big bang (top); each dot represents 10^9 solar masses of cold dark matter. The lower sequence shows the corresponding structure of the baryonic component; here each dot represents 10^8 solar masses of ordinary matter. The baryonic gas loses energy by cooling and is therefore able to form discrete clumps of galaxy mass. The massive concentration of cold dark matter (top, center) results in relative galaxy velocities as large as 1000 kilometers per second, about twice as large as is observed.

suitable fluctuation spectrum to fit observations. Or one can bring in seed objects of ultrahigh density that are relics of an early phase transition that occurred after the inflationary epoch. Or one can even try to discard inflation entirely.

Hot and Cold

If hot dark matter produces too little small-scale structure and cold dark matter produces too much small-scale power, the logical compromise is a mixture. In fact, assigning the neutrino a mass of 8 eV provides one-third of the critical density in hot dark matter. With two-thirds cold, say, in the form of photinos, one can now construct a model that fits observations on both small and large scales. Hot dark matter guarantees large-scale power, and allows a match to the fluctuations measured by the COBE satellite. The dominance of cold dark matter guarantees that structure forms in the desired bottom-up sequence. The one-third contribution of hot dark matter suppresses sufficient small-scale structure that one can obtain an excellent match to our astronomical observations of galaxies and galaxy clusters. Inflation also works, since the universe is at the critical density.

What could be wrong with this scheme? The oldest stars in our galaxy are at least 14 billion years old: with the most optimistic uncertainties, one can push this age down by 2 billion years. Yet if the Hubble constant is 75 kilometers per second per megaparsec, as many astronomers believe it to be, the age of a universe at critical density is only 10 billion years. One has to hope that the Hubble constant will converge toward the lowest values being measured. Otherwise we would return to the depressing state of having stars older than the predicted age for the universe. An even more serious objection is that the admixture of hot dark matter delays galaxy and cluster formation until a relatively recent epoch of the universe. Whether galaxies and clusters can form sufficiently early is the darkest cloud on the horizon for mixed dark matter.

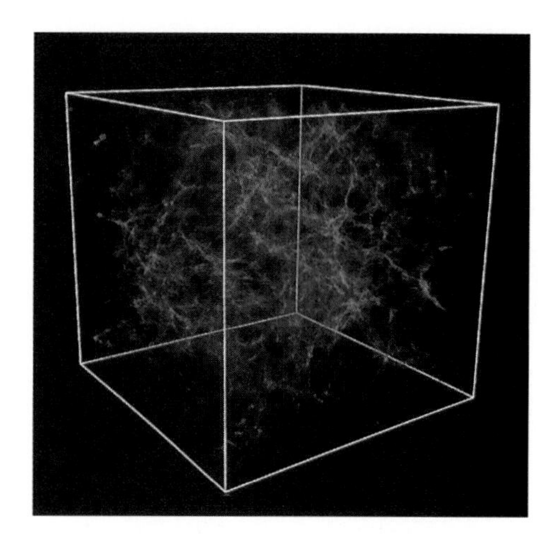

In this computer simulation, a universe of mixed cold and hot dark matter has evolved to create clusters of baryonic matter, whose distribution seems to match the observations of galaxy clusters by x-ray satellites. Dense regions are red and low-density regions blue in this cube measuring 200 megaparsecs on a side.

The Vacuum-Dominated Universe

The amount of matter observed on scales below about 10 megaparsecs unquestionably favors a low-density universe with $\Omega = 0.1$ or 0.2. On the largest scales, our techniques for measuring both matter and its fluctuations become progressively less reliable. If one takes the view that large-scale determinations of Ω should be distrusted, then an open universe is clearly preferred because it best matches observations on smaller scales. Since in an open universe, galaxies are decelerating more slowly than in a universe at critical density, the age of the universe is longer, by about 50 percent. Thus this also is the model that may be necessary if the Hubble constant settles down to about 75 kilometers per second per megaparsec.

But the open model leads to a serious quandary. Gravity has less time to operate in growing fluctuations, and hence fluctuations must have been larger at the epoch of last scattering, when the photons of the cosmic microwave background became free from matter. We should thus see correspondingly larger temperature fluctuations in the cosmic microwave background today. From the inflationary expectation of scale-invariant fluctuations, one can predict precisely what observers of the microwave background should see on various angular scales. The results are discouraging: the temperature fluctuations observed are weaker than those predicted. One solution discussed in the previous chapter was to abandon inflation, to desert cold dark matter, and to renege on primordial adiabatic fluctuations. A baryon-dominated open universe may then just work, at the cost of doing some injustice to the primordial nucleosynthesis prediction.

There is another resolution of this difficulty: bring back the vacuum density, as a reincarnation of Einstein's "greatest blunder," the cosmological constant. In 1916, Einstein originally envisioned a static, nonexpanding universe, yet he realized that a static universe was unstable and would collapse. To reconcile the model of a static universe with his theory, Einstein invented the idea of cosmic repulsion, enshrined as the cosmological constant term in his equations of general relativity. Of course, the universe turned out not to be static. However, one can try to revive the cosmological constant to save a low-density flat universe.

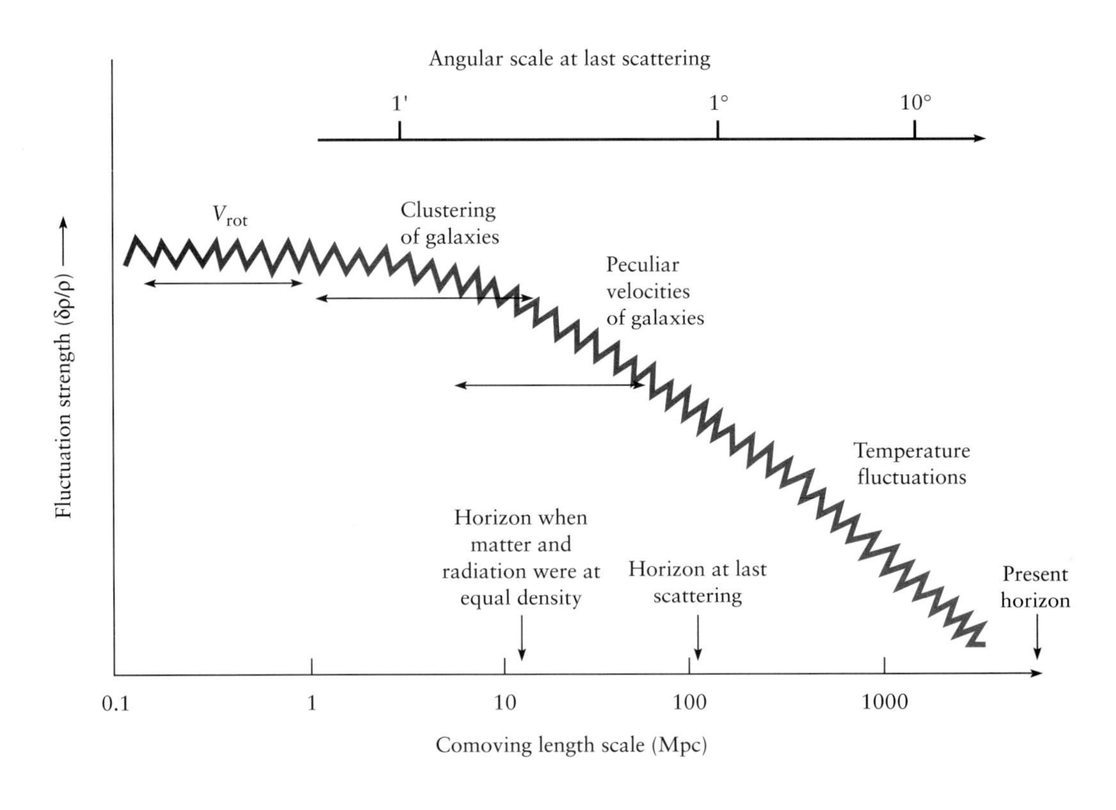

Density fluctuations are measured to be large on small scales and small on large scales. Measurements of the cosmic microwave background can be taken at a variety of scales. When such measurements are made on small scales of a degree or less, they probe similar scales to those studied by galaxy clustering and galaxy peculiar velocities.

A modern interpretation of the cosmological constant moves it from one side of Einstein's equations of general relativity, where it appears as a repulsive force, to the opposite side, where it is placed among the source terms for gravity. These terms are just the density of matter and energy in the universe. One can then reinterpret the cosmological constant as an energy density of the vacuum, with no associated matter attached to it. The constant simply adds to the energy density of the universe. One can still have a flat inflationary universe, since combining vacuum density, that is, the cosmological constant, with the appropriate fraction of the density in cold dark matter will add up to a critical density.

A vacuum-dominated universe of cold dark matter has one large advantage: what is assumed in the way of cold dark matter is what is actually observed by measuring the orbits of stars in galaxies and the motions of galaxies in clusters. One does not need to play with ideas such as the biasing of cold dark matter in order to hide it from view. However, there are two shadows threatening the vacuum-dominated universe. First, any large-scale flows, from 10 to 100 megaparsecs, predicted by the model do not seem able to achieve a sufficiently large velocity. Such flows typically travel at velocities of several hundreds of kilometers per second on scales

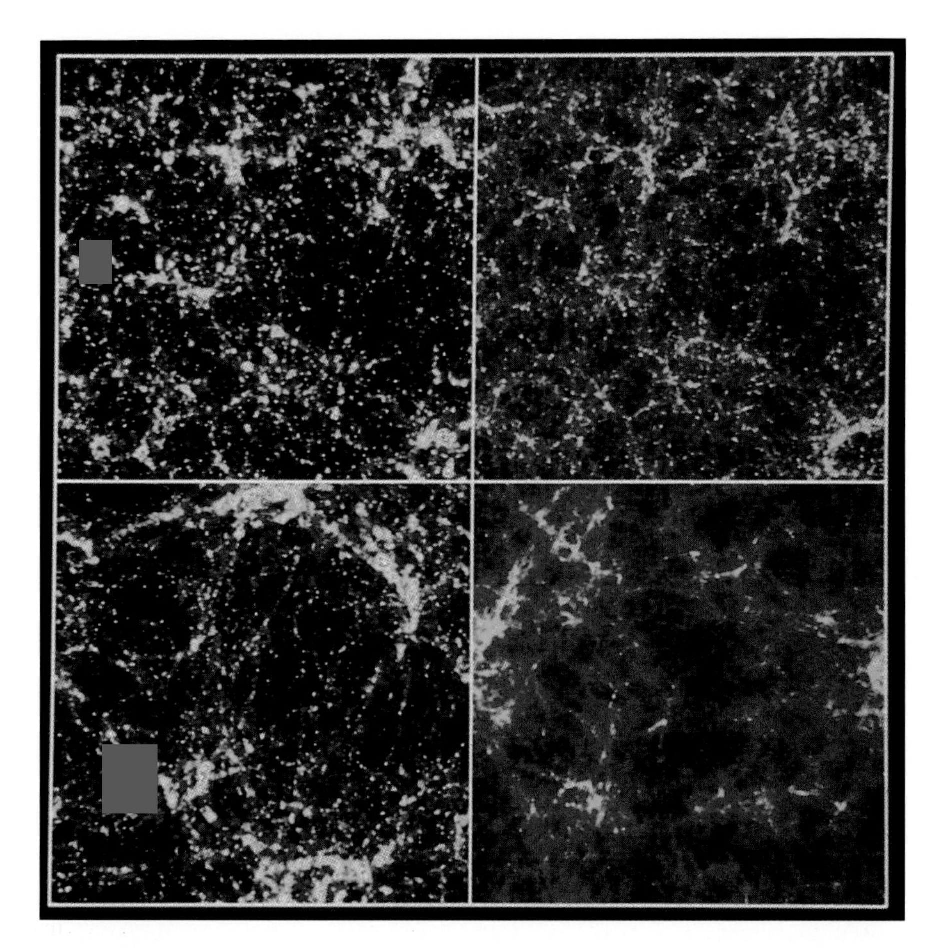

A computer simulation predicts the mass density in three different cosmological models. Each square is the two-dimensional projection of a cube of comoving size 100 megaparsecs, seen at a redshift of 1, when the universe was one-third its present age: top left is the cold dark matter universe, bottom left is the vacuum-dominated cold dark matter universe, and bottom right is a mixed model, one-third hot and two-thirds cold. The cold dark matter universe at a redshift of 4 is shown in the top right panel. The mixed model fails to produce enough galaxy clusters, while the vacuum-dominated model produces more large-scale structure than the cold dark matter model, in better accord with observations.

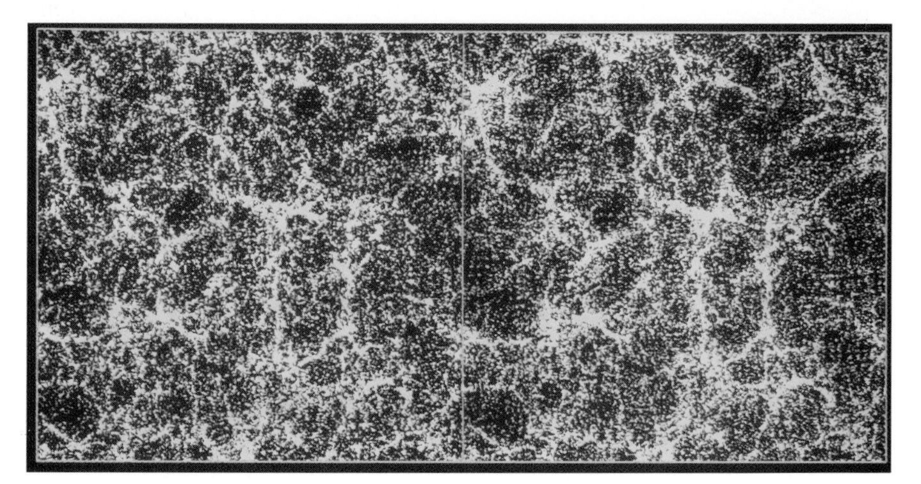

This computer simulation compares the present-day large-scale structure produced in a universe of cold dark matter (left) and that produced in a low-density, vacuum-dominated universe (right).

of tens of megaparsecs. These speeds are most easily generated in a universe at critical density, where the influence of density fluctuations is not expended in fighting the expansion energy, but helps drive large-scale flows.

The second problem is that galaxy clusters may form too early. In a low-density universe, the epoch when the universe first is matter-dominated is delayed by a factor of up to 10 in the expansion of the universe. The delay has an effect on the shape of the fluctuation spectrum, since these fluctuations grow only after the universe becomes matter-dominated. The horizon at the onset of matter domination marks a natural transition between the smaller scales that begin growth at that epoch, and the larger scales that enter the horizon later and have less time to grow. In a universe at critical density, the scale at the transition is roughly the scale from which a cluster forms. All smaller scales attain a similar amplitude and therefore form the first structures at more or less the same time. The galaxies have a slight headstart over clusters, perhaps by a factor of 3 in redshift, or expansion factor. However, in a low-density universe such as the vacuum-dominated one, the transition epoch is delayed and galaxies no longer have much of an advantage over clusters. Galaxies and galaxy clusters form early, and one would not observe many clusters forming at present, in apparent contradiction with what we see with x-ray telescopes. On large scales, density fluctuations are considerably stronger than in the cold dark matter universe. Indeed, the vacuum-dominated universe might even produce fluctuations too strong to jibe with those we observe in the cosmic microwave background.

Much more data, from both near and far, on the galaxy distribution will be needed before these issues can be properly resolved.

Primordial Defects

Cosmologists wanting to reconcile their theories of galaxy origins with the observed large-scale structure have been compelled to go to extreme measures. One school of thought argues that perhaps the very early universe is not as simple as inflationary theory predicts. According to inflation, any bizarre relics from the beginning of time, such as magnetic monopoles, are inflated away. There is expected to be at most one such relic per observable horizon, unless they are re-created after inflation. Some theorists have argued that such relics could have appeared during the phase transition at the end of the grand unification era, when the nuclear and electromagnetic forces behaved as different aspects of a single force. Much as cracks appear when ice is warmed, or as fracture stresses appear in brittle materials, the demise of grand unification may have left topological knots behind that contain trapped remnants of the earlier dense phase of the universe.

These so-called topological defects can take on various forms: they may be pointlike, stringlike, or sheetlike.

Surviving relics of the grand unification era include some rather exotic objects called "textures" and "cosmic strings." A texture is a "knot" of high energy density and a string is a linelike relic of high energy density, both in an otherwise empty space that contains an energy field, the Higgs field. This field is named after the physicist Peter Higgs, who first conjectured that such fields exist, at rather lower energies, in order to account for the masses of quarks and leptons. A Higgs field helps explain how the symmetry between the weak nuclear force and the electromagnetic forces, previously indistinguishable, is spontaneously broken at an energy of about 200 GeV, when the universe was one-hundredth of a nanosecond old. Quarks and leptons subsequently materialize as massive particles.

More generally, Higgs fields appear frequently in modern elementary particle theory. A Higgs field can interact with other particles and carry energy density, which is potential energy that depends only on the value of the field at each point. There is also kinetic energy associated with gradients in the field, due both to its non-uniform spatial distribution and to any time variation of the energy field.

A texture is a non-uniform Higgs distribution, for simplicity's sake assumed to be spherically symmetric, whose energy density arises solely from the field's kinetic energy. It is a three-dimensional topological knot. A string is a one-dimensional topological defect of leftover Higgs field energy. It is a defect in the sense that it is a topological relic of an earlier phase; although almost all of the volume of space has safely made the transition to the new, lower-energy phase of matter, trapped regions of the old phase are left behind. Both textures and strings are distinct from pointlike field defects (monopoles) or two-dimensional defects (walls), which may also be produced in phase transitions in the early universe. Of these, only strings and textures have been used to construct theories for large-scale structures.

A phase transition lays down the initial Higgs field randomly in causally disconnected regions, and with the Higgs field come the relic defects, also laid down randomly. Their randomness creates a non-uniformity that cannot be smoothed out, since the regions are not causally connected. In this manner, the fluctuations are generated that much later are required for the formation of structure in the universe. In order for such relic defects to be useful for forming galaxies, the phase transition must have occurred during the grand unified epoch when the temperature of the universe was 10^{15} GeV.

The case of a texture is particularly interesting. Once the distance light has traveled since creation becomes comparable to the radius of a texture, the texture collapses and drags in with it all of the associated energy density. When it reaches a size of about 10^{-30} centimeter, the knot unwinds

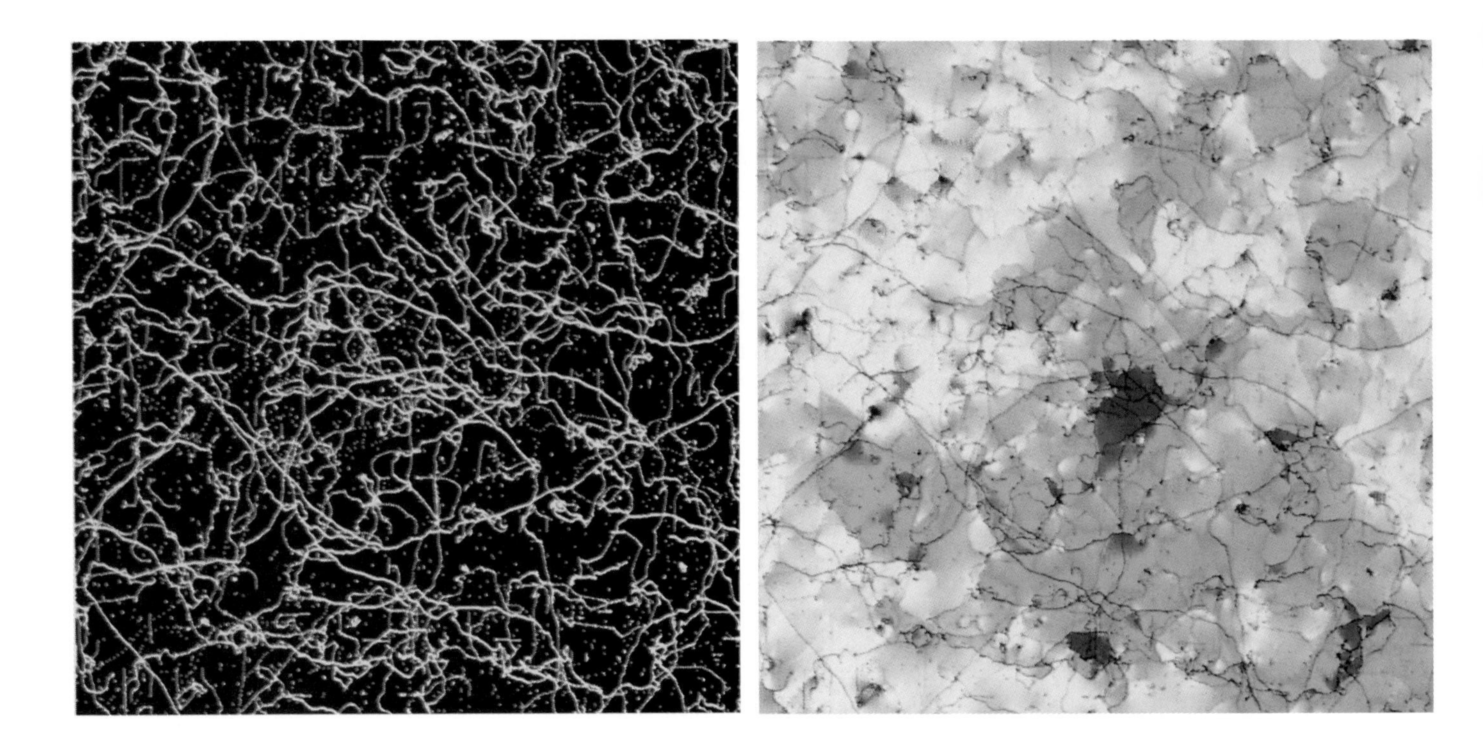

Left: A network of cosmic strings projected on the sky, at a redshift of 1000, by a computer simulation. The strings are left-over relics of high density from the very early universe; they serve as gravitational seeds for the growth of structure. Long strings intersect and form loops that eventually shrink and decay by emission of gravitational radiation. Right: A simulation of a 10-degree-square patch of the cosmic microwave background shows how cosmic strings could leave their imprint in the form of temperature fluctuations with telltale sharp edges.

itself. In this process, it emits a burst of weakly interacting massless particles known as goldstone bosons. The energy density associated with the texture, and also with the expanding spherical shell of bosons, pushes ordinary matter around, which then forms the structure we see today. The ambient matter responds by producing density fluctuations that eventually collapse to form galaxies, clusters of galaxies, and structures on even larger scales. An apparently undetectable background sea of goldstone bosons is left behind.

Strings, by contrast, are complex tangles of relic energy. The length of a string is infinite but its width is only 10^{-25} centimeter. As the universe expands, the strings, under immense tension, whiz around at near light speed. They intersect one another, forming loops that subsequently decay by emission of radiation in the form of gravitational waves. If strings exist, the sea of gravitational waves should be detectable by its effect on pulsar timing. The jostling around of the pulsar relative to the observer, tiny though the effect is, leaves a flaw in the accuracy of the pulsar clock, which manifests itself as a trace of irreducible noise. Before they decay, however, both the loops as well as the long strings have acted as gravitational seeds around which matter is accreted.

Do textures and strings leave any other signs of their presence? The gravitational potential wells of textures blueshift light falling in and redshift light climbing out, whereas light passing by a string is gravitationally lensed. Both these effects would produce temperature fluctuations in the cosmic microwave background toward the direction of the texture or the string. The models tell us that if the strings or textures created the observed large-scale structure, these temperature fluctuations should be detectable. In contrast to more standard theories of structure formation, the predicted distribution of amplitudes is nonrandom. Rather, a unique distribution of fluctuation amplitudes is predicted on the sky. By analyzing the temperature fluctuations in the cosmic microwave background, cosmologists should be able to fully test the defect theory and probe to a mere 10^{-36} second after the big bang.

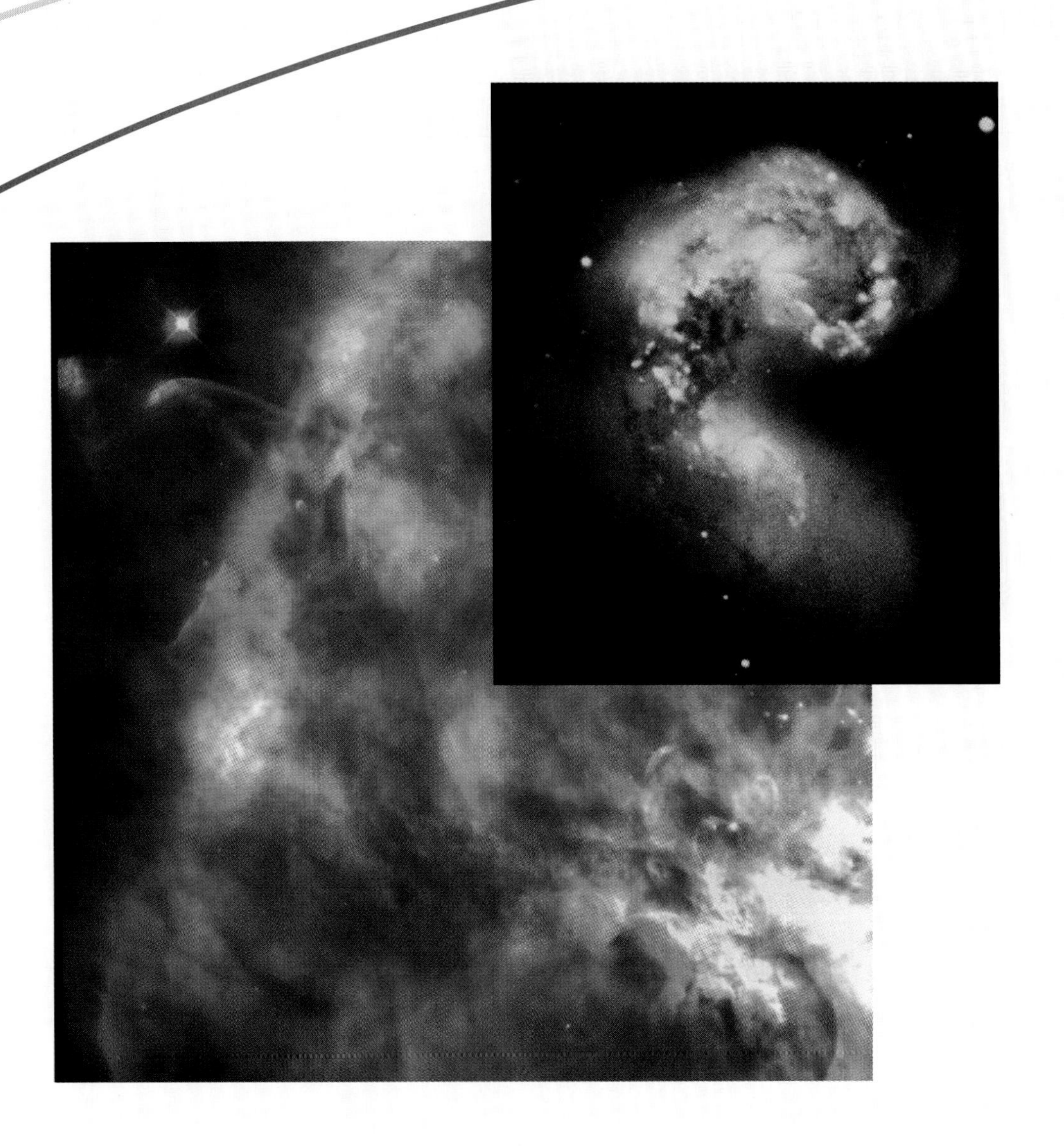

Right: *The Antennae, a pair of interacting galaxies otherwise known as NGC 4038/9.*
Left: *The Orion nebula, a region of vigorous star formation, as viewed by the Hubble Space Telescope.*

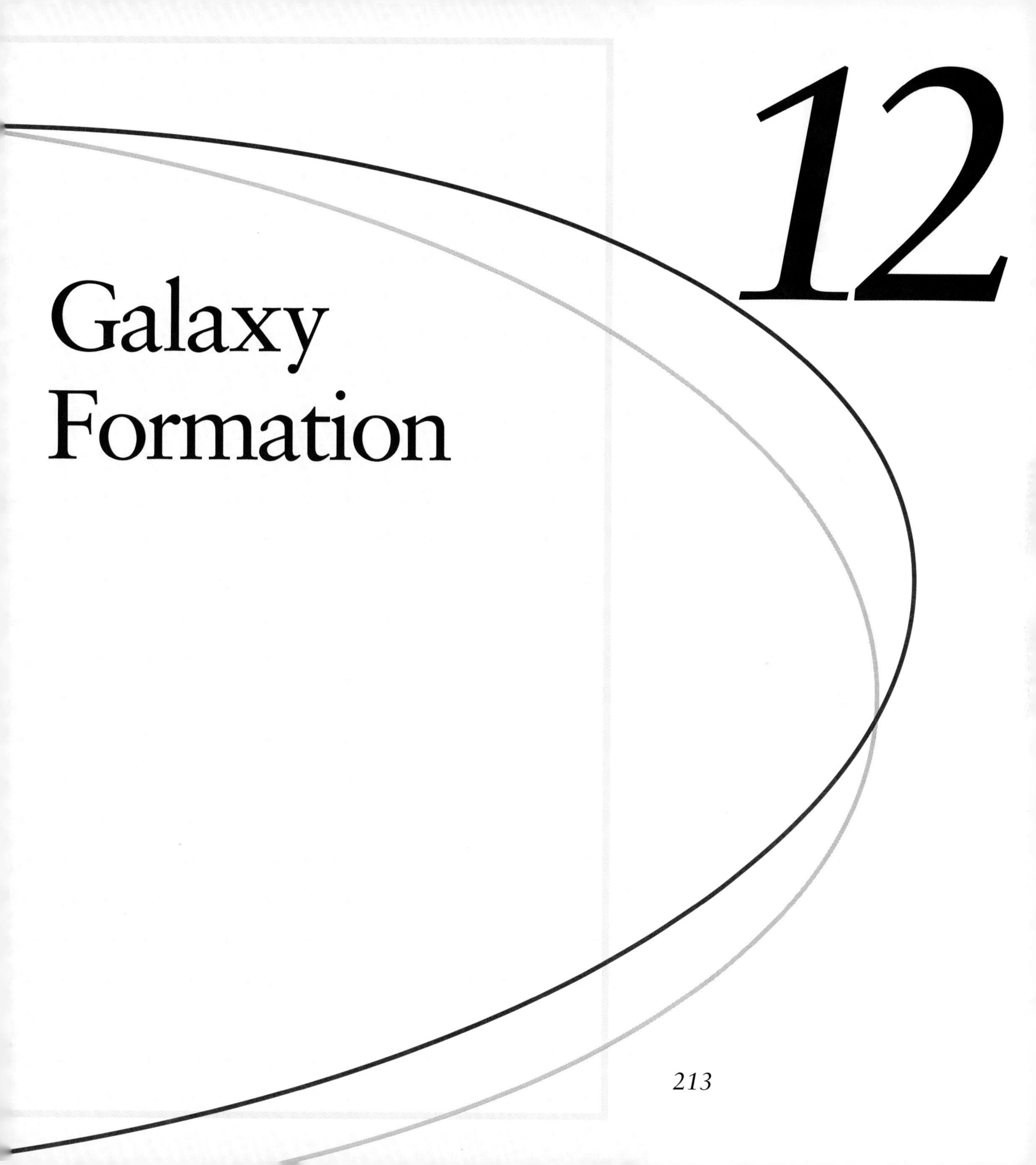

12

Galaxy Formation

Galaxies are the beacons that enable cosmologists to trace the universe back in time, toward a much earlier, less differentiated state. One finds that galaxies invariably are the tools or the stepping stones used to test any cosmological model. Indeed, almost all of our attempts to decipher the geometry of the universe and its fate have been frustrated by our lack of understanding of how galaxies evolve. Galaxies are the best-studied reservoirs of dark matter, and within them, they contain the fossilized clues to a youthful and more vigorous phase of the universe. We, after all, are firmly implanted within a galaxy, and understanding our origin on the cosmic scene is inseparable from deciphering the mysteries that surround the formation of the galaxies in the remote past.

Thus one of the ultimate goals of cosmology is to understand how galaxies formed. Unfortunately, this is where the physics becomes messy and poorly understood. Now we have to contend with dark matter that evolves collisionlessly, conserving its energy, and with gas that evolves by dissipating its energy. We have to incorporate star formation, a process that is not well understood in the vicinity of the sun, let alone in a remote, forming galaxy. We will need to extrapolate poorly understand processes over periods of billions of years and over distances of millions of parsecs. This chapter describes our best attempts at constructing a model for galaxy formation.

Formation of the Stars

We have learned certain facts about the continuing formation of stars. For example, stars are forming throughout the disk of the Milky Way, in a region that spans a thickness of about 200 parsecs and a diameter of 20 kiloparsecs. The sites of star formation in our galaxy are observed to be the densest, coldest, and darkest gas clouds in the interstellar medium. These clouds range in mass from 10,000 to 1 million $M_\odot$ and are formed predominantly of molecular hydrogen.

The short-lived massive stars illuminate the regions where stars are forming. The interstellar clouds, where star birth is occurring, as well as the sites of the young stars, are concentrated into *spiral arms*. Our local arm is called the Orion arm. Other spiral arms in the Milky Way are Centaurus, Sagittarius, and Perseus, as one moves out from the center of the galaxy. Each arm appears to wind around the galaxy only one time or less. This was initially a surprise to astronomers: one would expect the arms to wind up as the galaxy rotates if they were long-lasting phenomena, since the inner parts make more rotations.

The longevity of spiral arms is explained by *density waves*, regions of alternating compression and rarefaction. Clouds of interstellar gas swirl round the disk of a spiral galaxy in nearly circular orbits. A tidal force continuously acts on the stars and gas clouds in the disk and causes matter to be compressed in traveling density waves that create distorted orbits. In our galaxy, the Large Magellanic Cloud appears to be the source of the tidal force. The distortion it creates causes a bunching-up of orbits in some locations, much like a freeway traffic jam. The differential rotation of the galaxy draws the density waves into a spiral pattern. Thus the spiral arms are simply where the orbits have bunched up. These areas of congestion are seen as bright areas, and the pattern formed by this process gives us the familiar spiral structure of our own galaxy, as well as other galaxies.

As if in congested traffic, clouds occasionally collide in the zones of high density created by the tidal field. Stars, which are much smaller, will not collide. Interstellar clouds are highly inelastic: when they collide, they tend to coalesce, like making a snowball. The colliding clouds form complex aggregations that gradually build up in size as they orbit the galaxy. The clouds become denser as they grow more massive, and the gravitational force gets stronger. In this way, clouds grow to the point at which they are unstable. The more massive clouds in the spiral arms eventually fragment into clusters of many stellar-mass blobs. These are protostars, objects that are about to become stars.

Whether these clouds can collapse to form stars depends on the Jeans criterion, which we previously discussed in the context of galaxy formation. To see what characteristics tend to make a cloud collapse, one balances self-gravity in a cloud of mass M, radius R, and density ρ against the central pressure force. Self-gravity is proportional to the second power of cloud mass, but the opposing pressure force is proportional only to the cloud mass. If gravity dominates, as it must do in a sufficiently massive cloud, then a cloud can collapse.

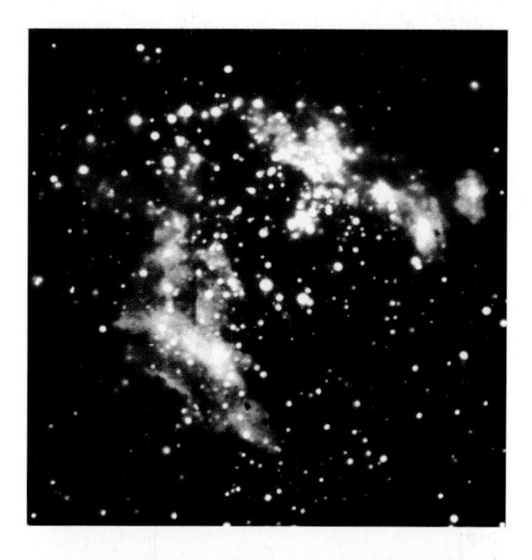

The star-forming region known as the Omega nebula, M17, seen at infrared (top) and optical (bottom) wavelengths. Many more stars are detected in the infrared because optical light is heavily obscured by dust in the molecular gas cloud where the stars are forming.

The critical scale above which collapse occurs is the Jeans length, R_{Jeans}, equal to $V_s/\sqrt{G\rho}$. Here G is Newton's constant of gravity and V_s is the sound speed, equal to $(kT/m)^{1/2}$, where T is temperature, k is Boltzmann's constant, and m is the proton mass. The Jeans length is the critical size of a cloud of specified density and temperature. If the cloud is larger than R_{Jeans}, it collapses. A more useful way to express the Jeans length is as $10T^{1/2}n^{-1/2}$ parsecs, where the temperature T is in degrees Kelvin, and n is the density in atoms per cubic centimeter. The corresponding Jeans mass is $20T^{3/2}n^{-1/2}$ $M_\odot$. For a typical molecular cloud having a temperature of about 30 degrees Kelvin and a density of about 1000 molecules per cubic centimeter, the Jeans length is 2 parsecs and the Jeans mass is 100 $M_\odot$. A cloud of this size will eventually produce many stars.

The spiral arms in nearby galaxies are demarcated by dense clouds of gas, dust clouds, and associations of luminous stars called O and B stars. These three stages in the evolution of a cloud are found in sequence across

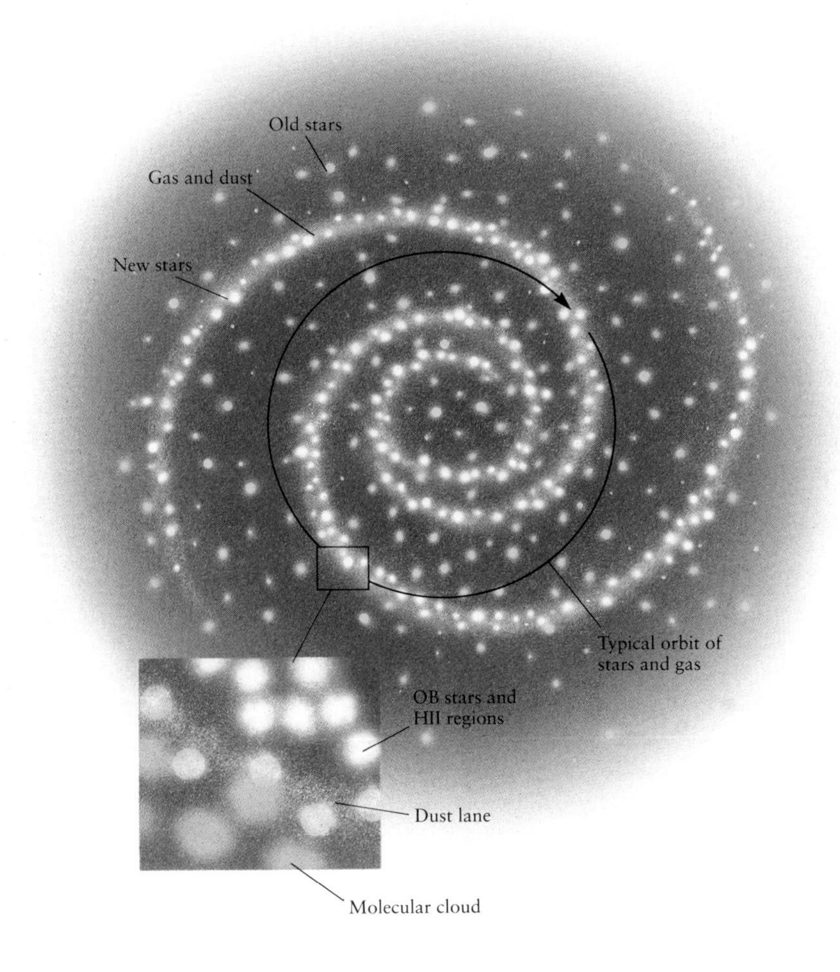

When interstellar gas clouds enter the spiral density wave, they are first compressed, then become optically visible as dark dust clouds, and finally fragment to form massive young stars.

the spiral arms from the inner to the outer edges. This arrangement supports the idea that clouds are compressed as they encounter density waves, with the densest clouds then becoming visible as dust concentrations and finally forming stars. The luminous O and B stars in these associations highlight spiral arms like beads on a string.

The onset of star formation as the orbiting clouds enter the spiral density wave has been likened to a contagious epidemic. The short-lived massive stars are concentrated in the compression region of the density wave and fade away within a few millions of years. Hence the spiral arms appear regular and well defined. Spiral arms are continuously being regenerated, and hence they appear to persist.

Because the parent cloud was rotating, so is each fragment after the cloud breaks up. Thus each fragment initially collapses to a disk of dust and gas that is also supported by rotational torques. The inner core contracts, heats up, and forms a star, really a protostar, that at first is powered by gravity alone. The disk may break up into planets that orbit the central star. The ultraviolet radiation and wind from the star soon ionizes and disperses the surrounding gas.

The star will remain gravitationally powered as long as it has enough gravitational potential energy available, which is released by the contracting protostar. The protostar's initial store of this energy amounts to GM^2/R, for a star of mass M and radius R. Consider a star that is similar to the present-day sun. Its gravitational potential energy is about 4×10^{48} ergs. If its protostellar luminosity is 10 $L_\odot$, or 4×10^{34} ergs per second, it can continue shining on gravitational energy alone for about 10^{14} seconds, or 3 million years. This is the typical duration of the protostellar phase. During this phase, a star is visible as an infrared source, since it is still surrounded by an extended cocoon of absorbing (and reradiating) gas. Soon the hydrogen in the stellar core ignites, and a stable hydrogen-burning star is born.

Dying Stars

Stars range in mass from 0.1 to 100 $M_\odot$. A massive star of 20 to 50 $M_\odot$ spends only a few million years on the main sequence, in its hydrogen-burning phase, whereas a star of a solar mass burns hydrogen quiescently for about 10 billion years. The low-mass stars are long-lived, destined to outlive the sun by far. Stellar deaths play a critical role in galaxy formation.

The supernovae that form neutron stars, called Type II supernovae, are produced by dying massive stars of between 8 and 50 $M_\odot$. These rapidly evolving massive stars are responsible for producing most of the heavy elements, such as carbon, oxygen, and silicon, from which we are made. Stars more massive than 8 $M_{\odot}$ evolve through a complete series of cycles of nuclear burning, as shells successively form of hydrogen, helium, carbon, nitrogen, oxygen, and silicon, all surrounding an iron core. At the end of the star's life, the iron implodes to form a neutron star, there being no further source of energy from nuclear fusion once iron is synthesized. The resulting energy release powers a supernova explosion. The explosion sends a blast wave propagating through the massive hydrogen-rich envelope of the precursor giant star, liberating substantial amounts of heavy elements into the interstellar medium.

A supernova of Type II characteristically has strong hydrogen and helium spectral lines. One expects the hydrogen-rich massive stars that are the precursors of such supernovae to be found in the star-forming, gas-rich regions for which the spiral arms are noted. And in fact Type II supernovae are found only in the disks of spiral galaxies today and presumably of protogalaxies in the past. The perversity of astronomical nomenclature requires that the characteristic supernovae of Population I be Type II.

But, as we may recall, there is another type of supernova, called Type I, which is found in all types of galaxies and, in particular, in regions where there are no massive stars. The Type I supernova is associated with the deaths of low-mass stars, usually in binary systems.

Stars of mass less than about 8 $M_{\odot}$ lose most of their mass while they are red giants and supergiants. These stars develop stellar winds at the onset of helium burning and eventually spew out their envelopes as planetary nebulae. Low-mass stars therefore end up as white dwarfs of around 0.6 $M_{\odot}$. These stars are mostly made of carbon or oxygen, since unlike in the early universe, stars succeed in converting helium into carbon.

Many stars are members of binary systems, and these will have a somewhat different fate. If a white dwarf has a more massive close companion, it will accrete hydrogen-rich gas. This is an explosive mixture that leads to a nova outburst as bright as a million suns. A nova fades after a month, but often repeats many years later.

If the white dwarf accretes substantial mass, the star is destabilized. This happens when two white dwarfs are present in a close binary system; they will eventually merge into one as the system evolves. The white dwarf accretes so much matter that it becomes more massive than its stability limit of 1.4 $M_{\odot}$ and collapses.

Such a star is an excellent bomb. Although it is initially supported by degeneracy pressure rather than pressure from thermal energy, it soon becomes disrupted by runaway thermonuclear reactions. During the collapse, the compression heats up the star's oxygen and helium, which provides an excellent source of nuclear fuel. The nuclear reactions commence, and because the star initially is still degenerate, the gas does not at first expand, so allowing a vastly more explosive situation to develop. As the temperature rises, ordinary thermal pressure eventually dominates over degeneracy pressure, and the pressure build-up makes an explosion inevitable. The white dwarf self-destructs completely in a supernova explosion, brighter than a billion suns. Heavy elements such as carbon, oxygen, and iron are explosively ejected in the rapidly expanding debris, but there is an almost complete absence of hydrogen. Indeed, supernovae of this type are distinguished from Type II supernovae by the absence of hydrogen lines in their spectra. These supernovae, of Type I, are characteristic of low-mass, long-lived stars in old stellar populations. They are observed in elliptical galaxies, which have exclusively old stars, as well as in the galactic spheroid population.

Supernovae Remnants and Enrichment

Supernovae of both types produce gaseous remnants: expanding shells of gas that are rich in heavy elements. The neutron-rich environment makes it possible for elements much heavier than iron to be synthesized during the explosion. All of this debris is spewed out into the interstellar medium. The debris has been observed to be radioactive; indeed, this radioactivity is known to be responsible for much of the light given off by the supernova. As the light fades away over about a year, it diminishes at a rate that resembles the decay of a radioactive element with a half-life of months. Theory suggests the following possibility. About 0.6 $M_\odot$ of radioactive nickel (Ni^{56}) is produced when the iron core forms. The nickel decays into cobalt (Co^{56}), with a half-life of 6 days, which in turn decays into iron (Fe^{56}); the half-life for the decay from cobalt to iron requires an additional 79 days. The radioactive energy is released in the form of gamma rays and electrons, and these are absorbed by the stellar envelope. The resulting glow accounts for the observed light decay.

Supernovae are rare: they occur at a rate of only about one per century per galaxy. The last one seen in the Milky Way galaxy was recorded by Kepler in 1604. The closest supernova in the last 400 years was SN 1987A, observed in the Large Magellanic Cloud in February of 1987 and well studied since. Astronomers observing the explosion at infrared wavelengths also detected emission lines of freshly synthesized cobalt and nickel. Moreover, the light from SN 1987A decayed with the characteristic half-life of radioactive cobalt, 79 days. This provided proof of the scenario for the radioactive powering of a supernova when the gamma rays from cobalt decay were detected in its radiation.

SN 1987A was close enough for the precursor star to have been photographed and catalogued. This event turned out to be quite unusual for a

During a supernova, a shock wave rips out through the exploding star's interior layers and heats them to more than 5 billion degrees Kelvin, triggering nucleosynthesis. Part of the silicon and sulfur fuses to form radioactive nickel-56 (stage 1); some of the oxygen at the bottom of the next shell burns to silicon and sulfur (stage 2), and neon and magnesium in the inner part of the shell burn to oxygen (stage 3). The shock propagates through the remaining material without triggering further nucleosynthesis (stage 4).

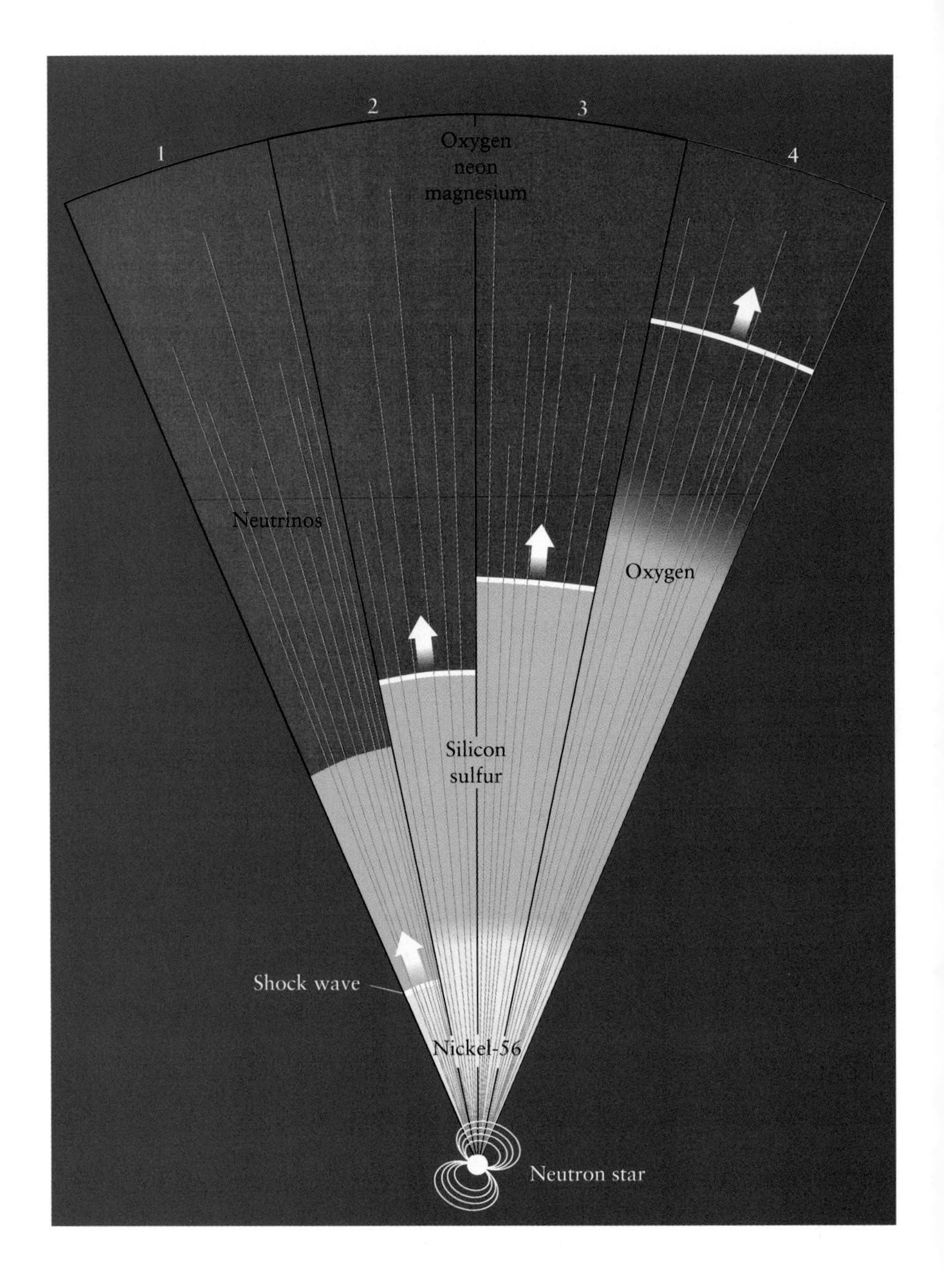

Type II supernova. For example, it brightened in only three hours, much more rapidly than expected for this type of supernova. Its precursor star was not a red supergiant as is usual, but a much smaller blue supergiant, with a radius equal to 50 solar radii rather than 500 solar radii. The star's small size helps explain the rapid rise of the supernova light and other peculiarities. For example, SN 1987A was 10 times less luminous than other Type II supernovae, reddened rapidly, and expanded very fast, at one-tenth the speed of light. All these characteristics can be understood as the result of energy dissipating more rapidly during the expansion of the compact precursor star. That star was probably a red supergiant about 10,000 years ago, but had since lost some mass and become much smaller.

Supernovae and their remnants are prolific sources of cosmic rays. These high-energy particles, mostly protons and electrons but including some heavier nuclei, are detected directly by observing the gamma rays emitted as cosmic ray particles interact with interstellar gas, or they can be detected by observing synchrotron radio emission produced by fast electrons accelerated in the interstellar magnetic field. The cosmic rays are also measured by experiments on high mountain sites, in balloons, and on satellites.

Type I supernovae produce most of the iron in the universe, while Type II supernovae produce most of the oxygen and other heavy elements. Altogether, they eject a hundred or more freshly synthesized isotopes into their environments. Type II supernovae would have also exploded during the billion-year-long era of galaxy formation if short-lived massive stars were present, as seems inevitable, whereas the Type II supernovae, evolving from long-lived low-mass stellar precursors, would have contributed their enriched debris to the interstellar medium over a longer timescale. The debris from exploding stars is recycled into interstellar clouds, which collapse to form new stars out of the old debris. Chemical evolution proceeds over the life cycles of many generations of stars. But how did everything begin? Were the first stars any different from stars forming today?

The First Stars

The stars that formed early in the galaxy seem indistinguishable in many respects from stars that form today. We know this because many of these stars are with us still. The Hertzsprung-Russell diagram, which plots stellar luminosity against surface temperature, provides a way to diagnose the ages of normal, hydrogen-burning stars. The position of the star when it turns off the main sequence indicates how long the observed era of hydrogen burning lasted, while its position on the horizontal branch indicates how long it has been burning helium. Globular clusters dated by this technique are inferred to have formed about 14 billion years ago, with an age spread of about 2 billion years. These are the oldest stars in the galaxy.

How do we know anything about the stars that were present at the beginning of the galaxy? True, some stars are very long-lived: a star of mass half that of the sun will continue to burn hydrogen for at least twice the age of the galaxy. However, there are many low-mass stars around, very few of which could have formed in the initial collapse of our galaxy. Despite the fact that the first stars are few in number, they have a unique attribute that enables astronomers to recognize them. At the beginning of the galaxy, very few heavy elements had been produced. Consequently, the first stars would have had essentially no oxygen or iron. As time went on, supernovae ejected enriched debris into the interstellar gas, and stars continued to form out of this polluted environment. Thus, the more recently a star formed, the higher is its content of metals relative to hydrogen.

In fact, stars that formed relatively recently, for example, 10 million years ago, are about twice as metal-rich as the sun. Conversely, stars that formed before the formation of the solar system, some 4.6 billion years ago, are metal-poor in comparison with the sun. The oldest stars in the galaxy have only one-hundredth or one-thousandth of a percent of the metal content of the sun, as measured spectroscopically. These metal-poor stars are all members of Population II, the spheroidal component of our galaxy that contains only old stars.

Such extremely metal-poor stars are found as well in some of the outlying globular star clusters. The globular clusters near the galactic center are not so metal-poor, although they are still deficient relative to the sun. In the galactic halo, most stars have about one-tenth of the solar metal abundance, whereas in the disk the average abundance is roughly one-half the solar metallicity. The halo therefore must have formed before the disk, or at least its typical stars did.

With so low a metal content, we believe that the oldest stars in the Milky Way galaxy must have formed during the galaxy's earliest phase of formation. The massive old metal-poor stars have long since died. The surviving metal-poor stars are very rare, for reasons I shall speculate on below. Some astronomers label this mostly long-dead population of stars Population III. More properly, it is just the extreme tail of the spheroid Population II, since its long-lived members, below the mass of the sun, are still around, although very hard to detect.

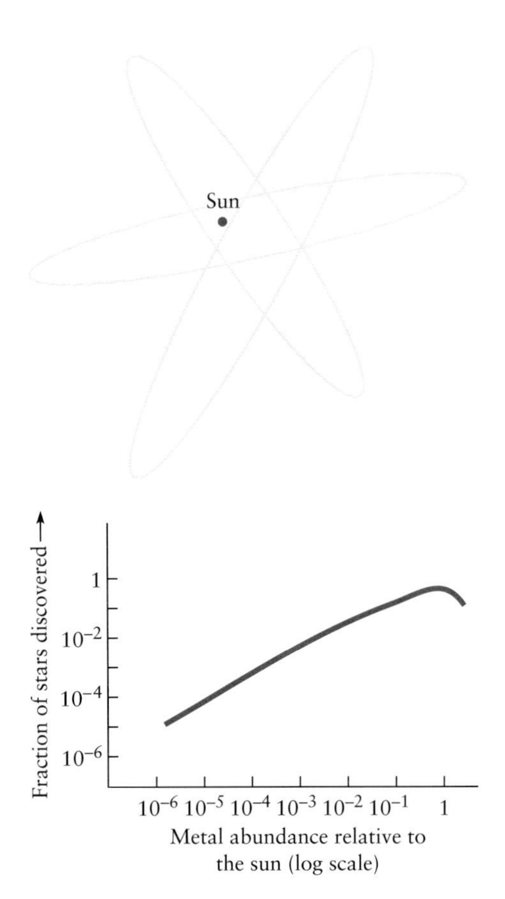

Top: The oldest stars, found in the halo, have eccentric orbits, inclined to the galactic plane at random angles. Bottom: Most stars in the galaxy are metal-poor with respect to the sun, although extremely metal-poor stars are quite rare.

Another clue to the nature of the first stars comes from studying the metals they produced and ejected. These ejecta were subsequently incorporated into the old galactic spheroid stars. The patterns of abundance that can be determined from stellar spectra, such as the ratio of oxygen to nitrogen or to iron, tell us something about the masses of the stars that produced these elements. We find that the ratio of oxygen to iron in the oldest metal-rich stars is enhanced by about a factor of 3 relative to younger metal-rich stars. A 10 $M_\odot$ star, for example, synthesizes much less oxygen than a 30 $M_\odot$ star, and produces far more carbon and nitrogen. Short-lived massive stars exploded as Type II supernovae to produce oxygen early in the galaxy, whereas less massive stars evolved more slowly to enrich later generations of stars and thereby account for the observed shift in abundance pattern with age. We infer that the first stars, which formed some 14 billion years ago, spanned essentially the same range of masses as stars that form today.

The fact that the first stars were similar to conventional stars is somewhat of a surprise, since conditions were very different in the early protogalaxy. There were no heavy elements, and no dust. The clouds from which the first stars formed consisted of hydrogen atoms, rather than hydrogen molecules. In the absence of the trace molecules and heavy atoms that act as coolants, the primordial clouds must have been much warmer than molecular clouds today.

Estimates of the Jeans mass tell us the size of the smallest clouds that can begin to collapse under their own self-gravity. From these estimates, we infer that the masses of primordial clouds were probably around 10^6 or 10^7 $M_\odot$. In addition, we know the gas temperature from the fundamental properties of atomic hydrogen, which has energy levels that do not allow any cooling below a few thousand degrees. But because molecular hydrogen was present in trace amounts, the clouds could cool to about 500 degrees Kelvin. Knowing the mass and the temperature is sufficient for us to calculate the density, which turns out to be a few thousand atoms per cubic centimeter, again rather similar to clouds today. The main difference, then, is that primordial clouds were warmer but rather more massive than present-day interstellar clouds. They also lacked heavy elements and dust grains.

There is a tendency for these characteristic differences to compensate for each other when a primordial cloud collapses and fragments to finally form stars. The higher temperature means that pressure support from the

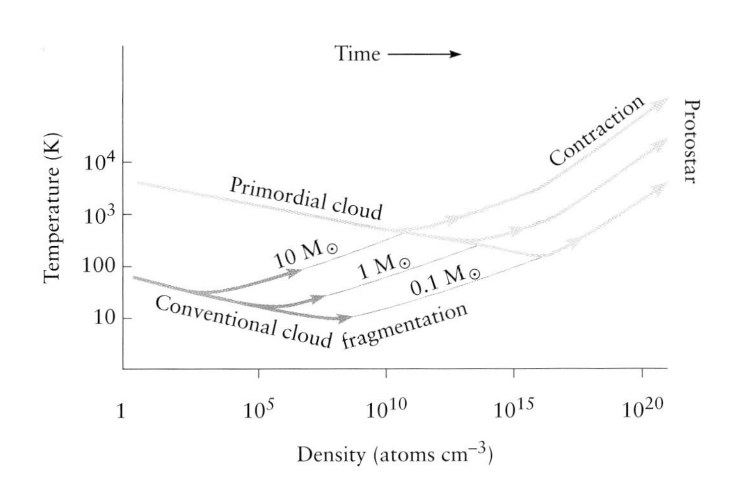

The fragmentation of a collapsing gas cloud leads to the formation of stars. The cloud cools as it collapses, becoming denser; and smaller and smaller fragments separate out eventually to contract to form protostars. Stars in the range between 0.1 and 100 $M_\odot$ are formed regardless of whether the cloud began contracting long ago, at a high temperature and mass, in an interstellar medium with no heavy elements or whether it began contracting more recently, at a lower temperature and mass, in an interstellar medium relatively abundant in heavy elements.

gas is more important, and initially allows more massive fragments. The lack of dust grains makes the cloud much more transparent to radiation, and so the cooling persists, as the cloud collapses, to a higher density than in conventional clouds. The continued cooling delays the rise of pressure forces in combating gravity as the cloud collapses. Fragments can therefore collapse to a higher density than would otherwise be possible, but they end up at a higher temperature. In other words, the fragments are the same mass, although at a higher density that is compensated for by a higher temperature. Clouds end up producing much the same final range of fragments irrespective of whether they collapse at the beginning of the galaxy or today.

Chemical Evolution

Stars that are formed at later stages in the lifetime of the galaxy are progressively more enriched with heavy elements. We refer to this progressive enrichment as chemical evolution. The solar system, which formed 4.6 billion years ago, reflects the chemical state of the galaxy at that time. About 2 percent of the mass of the sun consists of elements heavier than helium, as do the masses of the giant outer planets, notably Jupiter and Saturn. The inner planets are not a fair example of presolar chemical abundances, since they have not held on to any of the hydrogen and helium from the protosolar nebula. A region such as the Orion nebula, where stars are presently forming, contains about twice as many heavy elements, per unit of hydrogen, as the sun. This difference is only to be expected, since the interstellar medium has had an additional 4.6 billion years or so to accumulate ejecta from nearby supernovae.

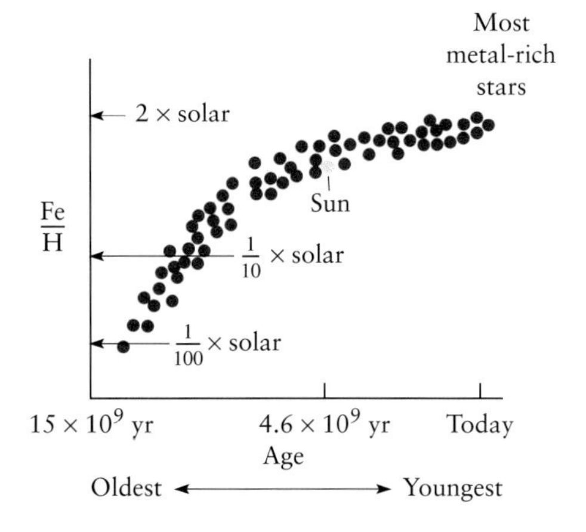

The abundance of iron, as more generally of the metals, relative to hydrogen is a measure of the average age of a star.

Practically all of the gas in the galaxy has settled into a thin disk, where the ejecta from stars and explosions accumulates with time. The galactic disk is where all of the new stars are now being formed, although most of its stars are somewhat older than the sun, and have, on the average, about one-half of the solar metallicity.

About half of the refractory (that is, nonvolatile) heavy elements in the interstellar gas are locked up in dust grains. These grains are small solid particles, such as silicates, that condensed in outflows from giant stars at late stages of their evolution. Refractory grains condense at high temperature in stellar outflows and volatile grains at low temperature in interstellar clouds. Consequently, coatings of volatile ices develop around the grains in cold molecular clouds. When these clouds become massive enough to break up into fragments, the rotation of the galaxy, and in turn of the clouds, causes disks to develop around each putative protostar. The ice-coated grains fall into a very thin mid-plane around the

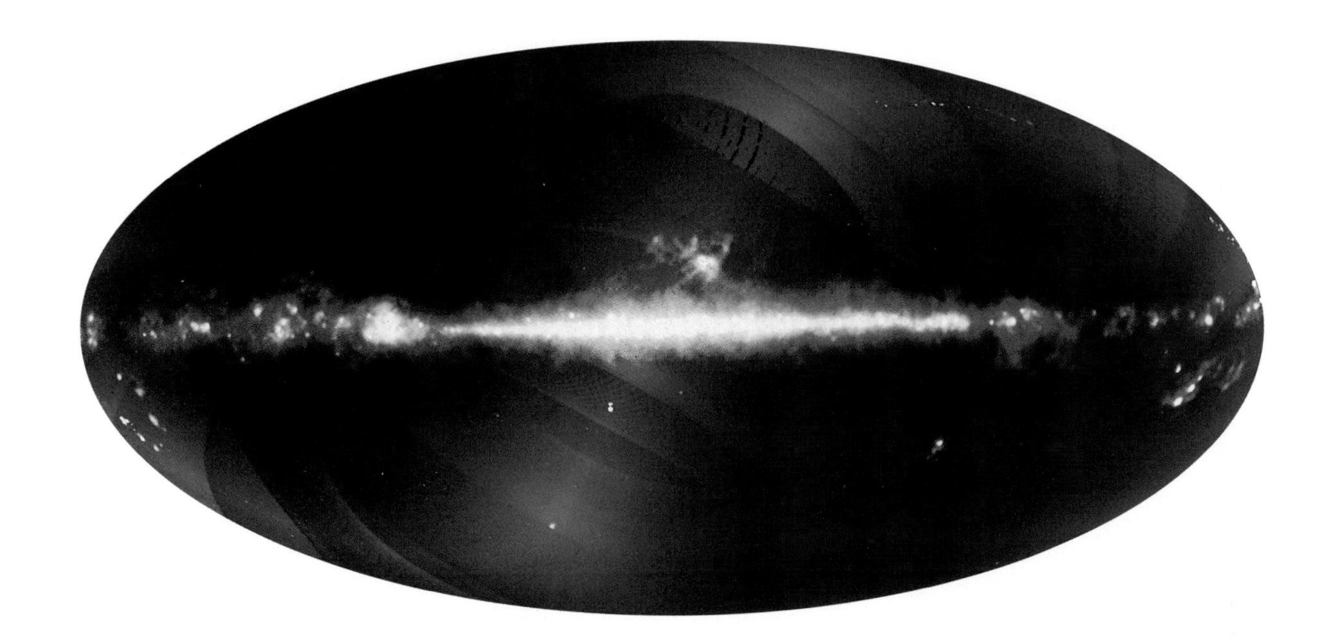

The plane of the Milky Way stretches horizontally across the center of this image, taken by the Diffuse Infrared Background Experiment on the Cosmic Background Explorer. This image records radiation detected at wavelengths of 25, 60, and 100 micrometers in blue, green, and red, respectively. At these wavelengths, the dominant feature in the sky is radiation from cold dust located in vast clouds of dust and gas in the disk of our galaxy.

emerging star. The disk itself breaks up to eventually form a planetary system.

Most stars in the galactic spheroid have less than about 3 percent of the solar metal abundance. We interpret this to mean that the spheroid formed stars before the gas settled into a disk. Why are the oldest stars in the spheroid outside of the disk? An explanation is the following. Once formed, stars retain the velocities with which they were born. Stars that formed during the initial collapse of the protogalaxy would acquire large orbital velocities, and thus spheroid stars are invariably high in velocity relative to the sun. Indeed, this is a defining characteristic of Population II. However, stars that form late, once the protogalactic gas has collapsed to a disk, have largely lost the "out-of-the-plane" component of velocities, because much of the kinetic energy of the gas is radiated away. Only the circular component of motion survives in the gas disk. Thus the stars that subsequently form are all moving in circular orbits and in the same direction. In other words, early-forming stars constitute the spheroid while late-forming stars make up the disk.

One problem remains: why are there so few stars remaining in the halo of really low, essentially zero, metallicity? The answer may be that during the formation of the galaxy there were relatively more short-lived, massive stars than are found today and relatively fewer long-lived, low-mass stars. If so, there would be few survivors of this phase.

The Formation of Ellipticals and Spirals

The scenario for the formation of spheroid and disk lends itself to the formation of elliptical and spiral galaxies as well. Ellipticals are round systems that contain only old, Population II stars, similar in mass to the sun. Spirals have both old spheroidal components and disks containing an admixture of old stars, young stars, and gas. The probable explanation for these differences goes back to each galaxy's initial collapse, which lasted less than 1 billion years. Ellipticals formed their stars very efficiently in a dramatic burst during the initial collapse a few billion years ago, whereas spirals formed only some stars, perhaps amounting to 10 percent of their mass, during this phase. Most of the stars in a spiral galaxy must have formed more gradually, over the past 10 billion years, after the gas had already formed a disk. Spiral galaxies form stars at a more or less steady rate, and continue to do so until the raw material, the interstellar gas and dust, is exhausted. They are only now, after some 10 billion years of star formation, running down their gas supply.

An interesting suggestion has been put forward to explain why stars formed so rapidly in elliptical galaxies. Circumstantial evidence points the way. Ellipticals are found mostly in dense clusters of galaxies, whereas spirals are present in loose groups and in the low-density outer regions of clusters. One presumes that the tidal jostling between gas-rich protogalaxies as galaxy clustering develops would trigger immense amounts of star formation. Ellipticals form in the dense cluster cores, where tidal effects are important. Outside the clusters, the gas collapses relatively quiescently, and takes 10 billion years or more to form the galactic disks.

Why are ellipticals round and spiral galaxies flat? The answer is simple: spiral galaxies rotate. Indeed, their shape is due to the rotational motions of stars, and results from centripetal forces balancing gravitational attraction. The rotation is a consequence of the fact that the galaxies formed out of turbulent, inhomogeneous gas clouds. There may have been several density peaks in reasonable proximity, each destined to form a galaxy. In the case of our own galaxy, for example, one major companion was forming at about the same time, the Andromeda galaxy. As the self-gravitating protogalactic clouds develop and become centrally concentrated, they inevitably exert tidal torques on one another via the effects of gravity. Even though the net angular momentum in the region is zero, and stays zero, individual protogalaxies acquire spin. As a protogalaxy collapses further, it spins more rapidly, since it conserves angular momentum. The contraction of the gas ceases only when the centripetal force balances gravity. Since the cloud can continue to collapse along the axis of rotation, it ends up as a flattened gas disk. A disk is a disk because gas clouds slowly spiraled together over billions of years as they lost orbital kinetic energy.

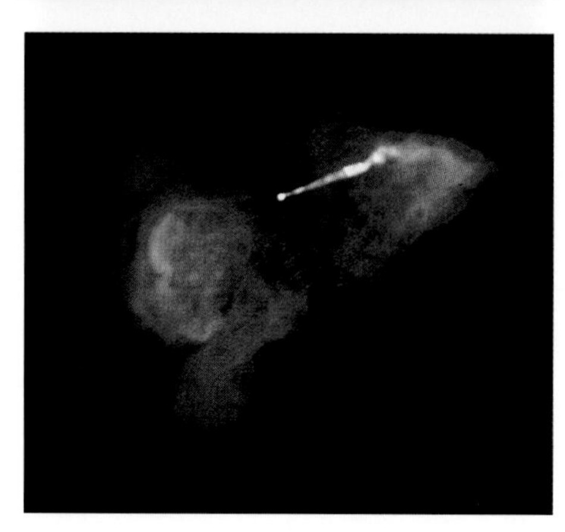

An optical image (top) and a radio image (bottom) of the giant elliptical galaxy Messier 87. Jets of synchrotron emission emanate from a massive black hole thought to lie at the galaxy center; they produce giant lobes of radio emission that extend beyond the scale of the luminous galaxy. The black hole is believed to have formed as gas clouds accreted during an early merger that also resulted in the formation of the galaxy itself.

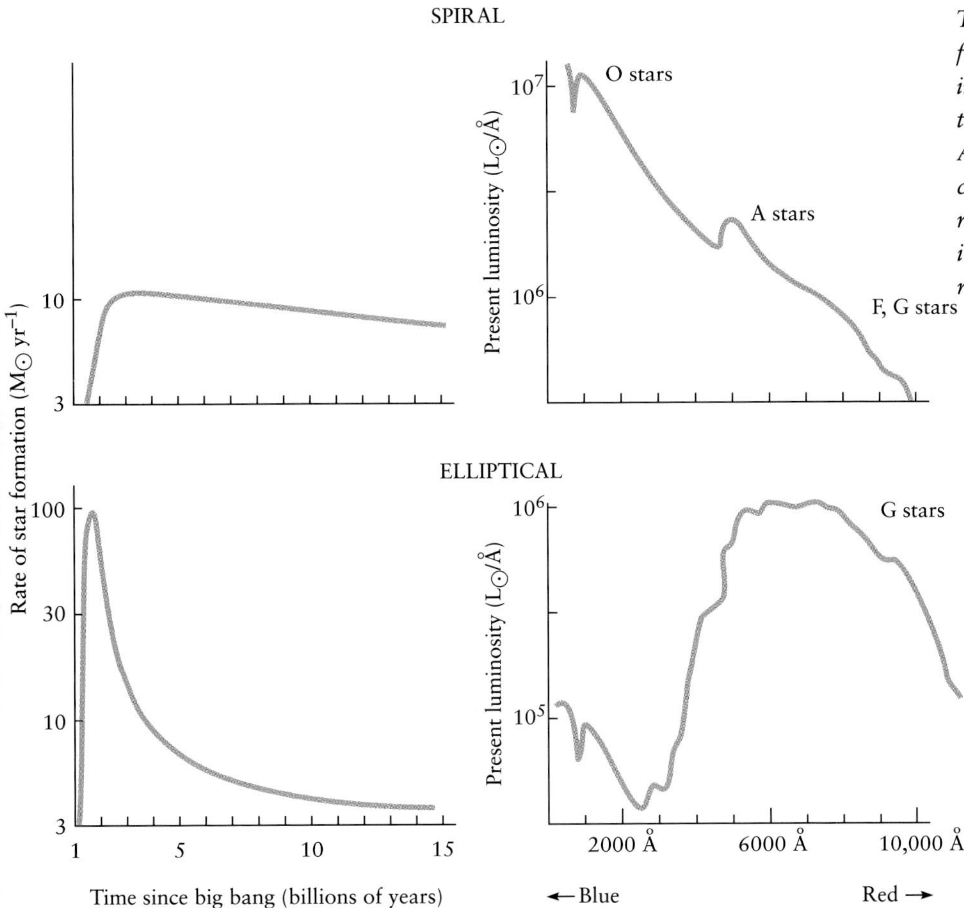

Top: Stars in spiral galaxies continue to form at a steady rate, whereas most stars in elliptical galaxies formed early, when the galaxies were first created. Bottom: As a consequence, younger, bluer stars contribute most to the luminosity of spiral galxies, whereas most of the luminosity of elliptical galaxies comes from older, redder stars.

In contrast, elliptical galaxies are not supported by rotational forces. An elliptical galaxy is round because the stars have motions that are not confined to a plane; their orbits are elliptical and randomly oriented. Their shapes are plausibly explained if pairs of disk galaxies, oriented at some arbitrary angle, merged.

Galaxy Mergers

Mergers, especially of protogalaxies, occur in dense regions of the universe where the relative velocities between galaxies are a few hundreds of kilometers per second or less. At higher velocities of encounter, galaxies fly through each other like ghosts passing through a wall. Thus most galaxy mergers occur in galaxy groups, where the relative motions of galaxies are slow, rather than between faster-moving galaxies in clusters. When two disks merge together, the final result is a round system, or elliptical galaxy. Much can be learned about how galaxies formed by studying nearby galaxies for the rare present-day examples of galaxies still in the process of merging. Processes that are uncommon today may have been common early in time when galaxies formed. These rare phenomena can give invaluable clues to cosmic evolution.

Galaxies are "sticky" systems: a close encounter between two galaxies will, if at sufficiently low speed, inevitably lead to a merger. Gas clouds within the galaxies undergo collisions, even though stars do not, and the colliding clouds lose orbital energy. Their orbits become more radial, and the frequency of cloud collisions increases still further because of the tidal forces exerted as the stellar systems merge together. The result is a massive, dense cloud of gas in the center of the merging galaxy system. Such a cloud is unstable: it must collapse and fragment into many stars. Thus mergers between galaxies appear to trigger a violent outburst of star formation, similar to the star bursts in which the stars of elliptical galaxies evidently were born. The star burst eventually fades away, and the energy released from exploding stars sweeps clean any remaining interstellar medium, to leave behind a galaxy of aging stars.

Such episodes of star formation are observed in the most luminous star-forming galaxies, which are inevitably in the final throes of a merger. Such objects are rare today, because the phenomenon is short-lived. However, the aftereffects of such mergers are relatively common. Many apparently normal, nearby elliptical galaxies are found on close inspection to reveal

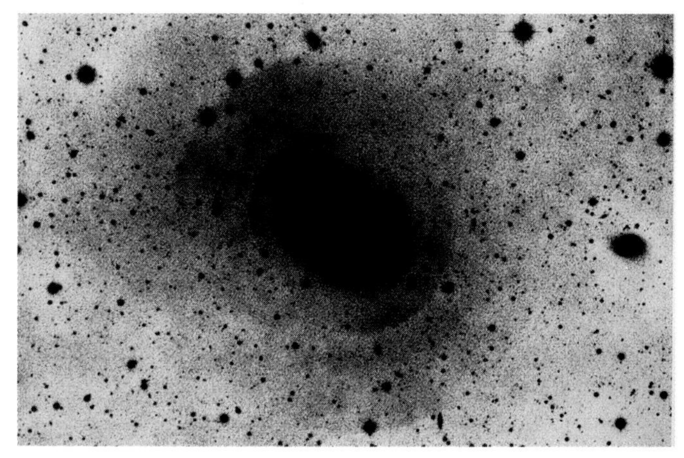

The elliptical galaxy NGC 3923 shows nothing unusual in a normally exposed negative photograph at the left. However, the heavily overexposed image at the right reveals several concentric shells that extend more than 100 kiloparsecs. The shells are relics of a galaxy merger that occurred about 3 billion years ago. Note that the two photographs are on the same scale.

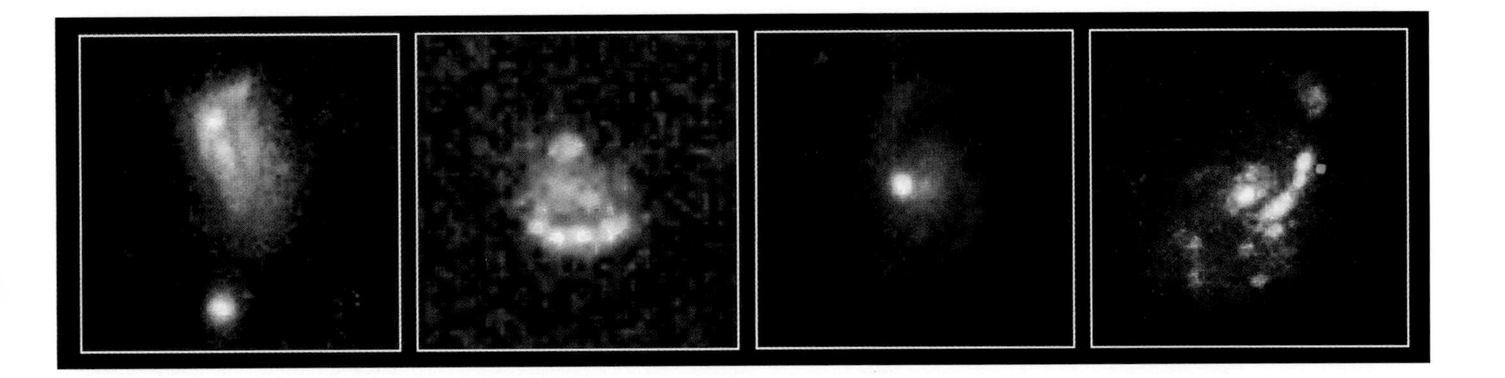

signs of a merger in their distant past. The evidence is in the form of very faint shells of stars seen at the outermost periphery of these galaxies. The long-ago mergers caused disturbances in the gravity field, analogous to the ripples in a pond that persist after a stone is thrown into the water; like the ripples in a pond, the disturbances slowly fade. Over billions of years, however, stars accumulate in the gravitational potential wells of these ripples. The arrangement of these stars marks the transient occurrence of a merger that, in the central regions at least, has produced what to casual inspection is a normal elliptical galaxy. Other features, such as dust lanes or knots of younger stars, are occasionally found that also bear witness to a more violent past.

Images of remote galaxies obtained with the refurbished Hubble Space Telescope. A far higher fraction of very distant galaxies like these, compared with the 1 or 2 percent of nearby galaxies, are extremely disturbed in shape. The distortions are attributed to strong interactions or mergings with companion galaxies, followed by outbursts of star formation.

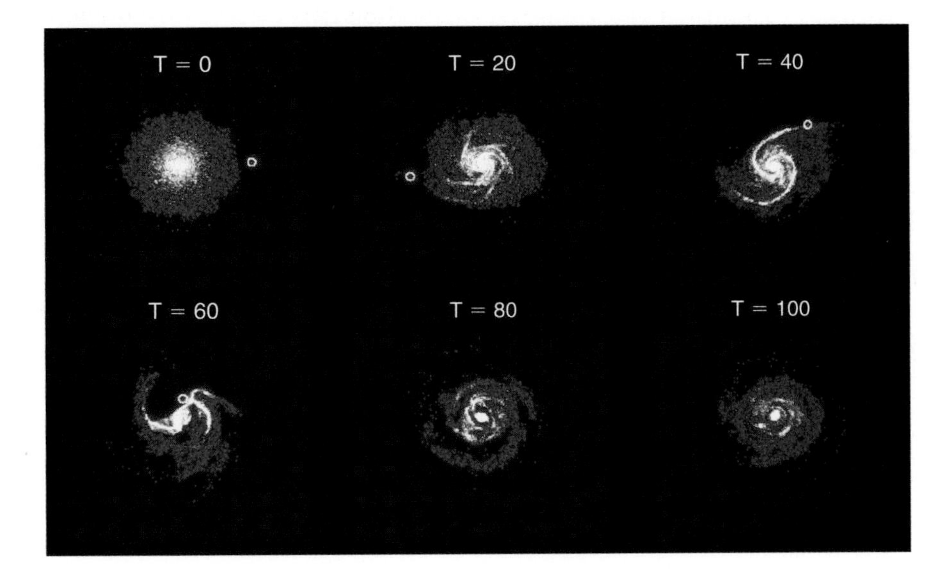

In this computer simulation, a spiral galaxy is shown accreting a dwarf satellite galaxy, whose center is denoted by the white circle. The interaction sets off a burst of star formation, which is most intense in the area colored white. The time for each successive snapshot is shown above each frame in units of 13 million years.

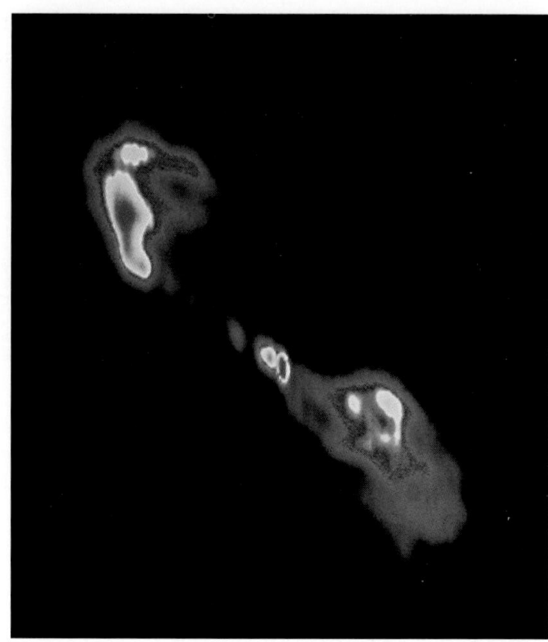

Left: The elliptical galaxy Centaurus A, seen here in an optical view, is undergoing a merger with a gas-rich companion galaxy. The nearest known violent galaxy, Centaurus A is a strong emitter of radio emission and x-rays. Right: Giant lobes of radio emission emerge from a jet originating in the nucleus of the optical part of Centaurus A. The hypothesized supermassive black hole in the nucleus is being fed by gas supplied by the merger and is fueling the radio outburst.

Computer simulations have succeeded in demonstrating that a merger between two gas-rich spirals, the most common galactic systems in the universe, results in the formation of an elliptical galaxy. The elliptical is characterized by a smooth, centrally concentrated, spheroidal distribution of old red stars, whereas its precursors are disk galaxies with patchy dust lanes and many clusters of young, blue stars. The merger greatly accelerates the rate at which clouds form, aggregate, and collapse. Hence it exhausts almost all of the gas reservoir in a violent outburst of star formation. One observes a similar phenomenon for the most luminous infrared galaxies, which are identified as galaxies undergoing violent bursts of star formation. In many instances, one can, with the most sensitive images, actually identify the merging systems.

The Cosmological Hierarchy

Mergers are rare today. Perhaps one galaxy in a hundred undergoes such strong tidal interactions. However, cosmological theory strongly suggests that mergers were a common phenomenon in the past. Since structure developed in a hierarchy, from small to large scales, we expect most galaxies to have formed before the galaxy groups that we see in place today. The most massive objects in the universe, clusters and superclusters of galaxies, are just forming at the present epoch. Studies, especially maps of the hot intracluster gas at x-ray wavelengths, have revealed that many clusters of galaxies are morphologically young systems. The gas has a clumpy distribution, as if it has not had time to become completely mixed. Eventually, the gas will appear uniform, spherically symmetric, and centrally concentrated. This indeed is the distribution of gas observed in mature galaxy clusters.

On megaparsec scales today we are witnessing the gravitationally driven aggregation and merging of matter into giant galaxy clusters. On smaller scales, individual galaxies must have formed by a similar process in the remote past. When the universe was perhaps a billion years old, a tenth or less of its present age, clouds with the mass of dwarf galaxies must have merged frequently to eventually form individual massive galaxies, surrounded by relic dwarf satellites. Our own Local Group is such an aggregation. The Local Group is dominated by the Milky Way and the Andromeda galaxy, which are accompanied by a handful of dim dwarfs.

This reasoning suggests that dwarfs are the building blocks of giant galaxies. Indeed, the abundant population of faint blue galaxies in the early universe is almost certainly composed of dwarfs undergoing a transient burst of star formation. Redshift surveys of the faintest galaxies have revealed that at great distances, corresponding to a redshift of 0.5 or so, there is an abundant population of star-forming galaxies with no similar local counterparts. The nearby universe may be littered with the fossilized relics of such early episodes of star formation. Exceedingly dim nearby galaxies are indeed found that are barely discernible above the night sky. These are probably the failed counterparts of today's bright galaxies. Nature most likely fails many times before it succeeds, and galaxies, both dim dwarfs and giants, may be testimony to this process.

Indeed, giant gas clouds are found in the distant universe, detectable because they absorb light, in the spectral lines of a few abundant elements, from an even more distant quasar. Such absorbing gas clouds cover about 20 percent of the sky, as a transparent cosmic wallpaper. These clouds most likely contracted and formed stars, to evolve into galaxies similar to the Milky Way. Occasionally though, they may have formed giant, dim galaxies of low surface brightness where gravity has never been sufficiently strong to provoke vigorous star formation. Only by studying all of the denizens of the cosmic zoo can we hope to ever understand the complex processes that led to galaxy formation.

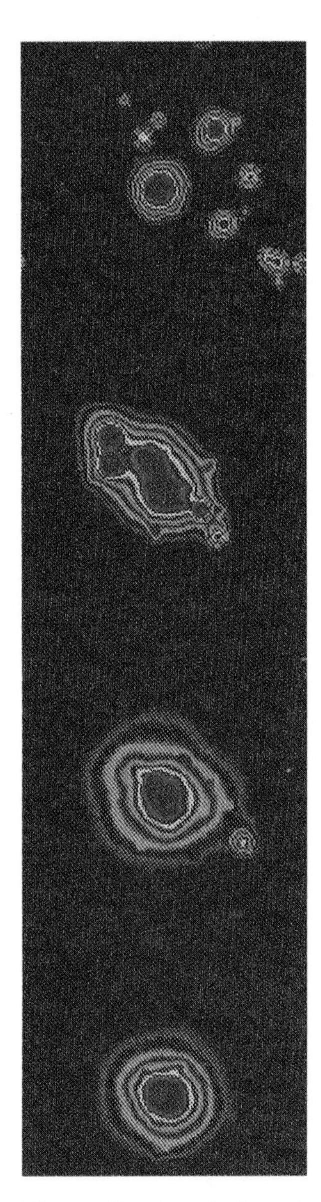

The predicted evolution of a cluster of galaxies, as mapped by an x-ray telescope. The hot x-ray–emitting gas traces the dark matter and galaxy distribution, and is shown (from top to bottom) at redshifts of 0.7, 0.3, 0.1, and today. The present-day galaxy cluster is atypical in that it has the uniform appearance of a mature cluster.

Closing the Circle

perfectly uniform big bang would be unacceptable: fluctuations are an ssential ingredient. These density fluctuations were created at the thresh-ld of cosmic time. There must have been deviations early on to make, over ne subsequent billions of years, the obvious structure of the modern uni-erse. It is thanks to such irregularities that the universe was able to evolve contain the complex array of structures that we see.

The cosmic microwave background, with its near perfect blackbody pectrum at the observed temperature of 2.73 degrees Kelvin, bears wit-ess to the early hot phase of the big bang. Its existence encourages cos-nologists to extrapolate the theory of the big bang back in time to within 0^{-43} second of the initial singularity, the instant at which classical physics, s opposed to the as yet poorly understood theory of quantum gravity, first ecomes applicable. Quantum fluctuations were inevitably present at this poch, when the fundamental scale of elementary particles was of the same rder as their Schwarzschild radius. These fluctuations were boosted in cale, from microscopic to macroscopic dimensions, by an inflationary hase of the expansion of the universe that occurred about 10^{-35} second fter the big bang. From these fluctuations, all structure subsequently prang.

A million years after the big bang, the primeval fog lifted, as the radi-tion field let the primeval plasma free from its previously unrelenting grip. he fluctuations then began to grow, to eventually evolve into proto-alaxies. Ultimately, the neutralization of the primordial plasma allowed ydrogen atoms to accumulate into the great clouds that were the precur-ors of galaxies.

The ultimate test of such a model is how well it predicts temperature uctuations in the cosmic microwave background. The model described nakes predictions that are intriguingly close to the latest measurements of

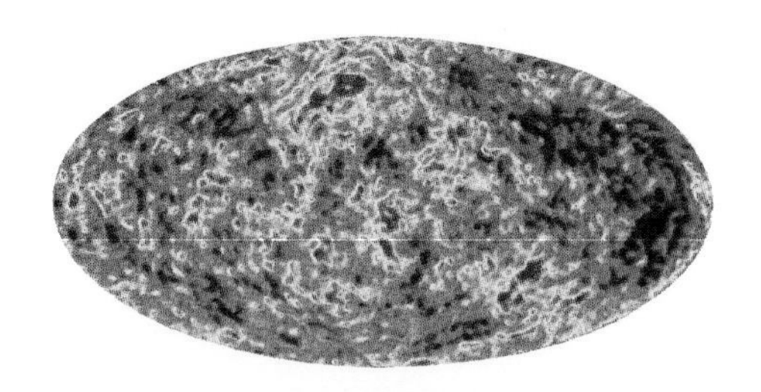

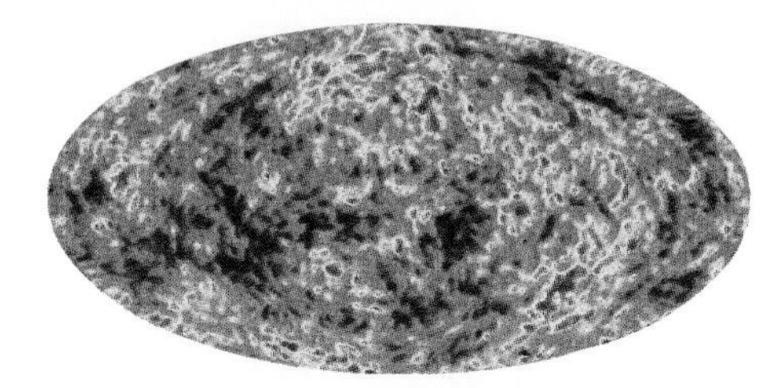

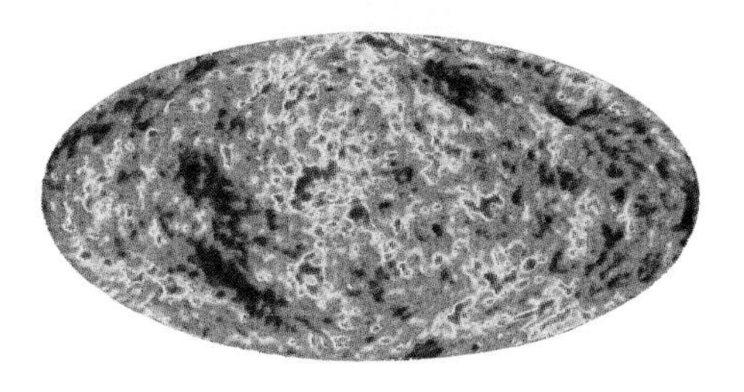

The three ovals show the large-scale fluctuations in the cosmic microwave background predicted by three different models of large-scale structure. All are cold dark matter models, but the initial distribution of primordial fluctuation strengths has been varied to give progressively more small-scale power to form galaxy clusters, as one goes from top to bottom. When we have better measurements of the temperature fluctuations, we will eventually be able to reconstruct the initial conditions from which large-scale structure originated.

that background's slight irregularities. The remarkable degree of uniformity observed in the cosmic microwave background radiation testifies to the near homogeneity of the very early universe. What is even more remarkable is that the small fluctuations that exist were found at more or less the expected level. Blotches, both hot and cold spots, were detected by the COBE satellite over angular scales of 10 degrees at a level of 30 microkelvins, or about one-thousandth of a percent of the average level of the cosmic microwave background itself. The universe, a million years after the big bang, is inferred to have been almost completely, but not quite, homogeneous. Infinitesimal density fluctuations were present in the primordial universe with an amplitude of about 10^{-5}, precisely the amplitude required (to within a factor of 2!) for galaxies to have formed by the present epoch as a consequence of growth through the action of gravitational instability. This is the current paradigm to which cosmologists subscribe.

Astronomers are now mapping these primordial fluctuations in the cosmic fireball on smaller angular scales. Only over scales of a few degrees or less would one hope to directly detect the precursor seeds from which the observed large-scale structure in the galaxy distribution arose. We should eventually, over the next decade, be able to explore how the distribution of primordial fluctuation amplitudes varies with scale. In principle, one could hope to distinguish cold dark matter from hot dark matter, primordial adiabatic from primordial entropy fluctuations, and even an open universe from one at the critical density. All of these cosmological alternatives leave an ineradicable signature in the sky.

The fingerprint is unmistakable, given data of high enough quality. The solution, however, is not going to be immediately forthcoming; our efforts are hindered by the "dirty windshield" effect. Our galaxy is not perfectly transparent at microwave frequencies. We have to contend with a noisy, uncertain foreground of dust emission, synchrotron radiation, and bremsstrahlung from interstellar matter at a level of 10 percent or more of the signal we are seeking. Improved sensitivity, broader frequency coverage, and more complete sky coverage will help penetrate the galactic murk.

We are on the threshold of confirming whether our present ideas about the evolution of the universe and the origin of large-scale structure have any validity. This is a golden age in cosmology, which is rapidly becoming as respectable a science as, say, archaeology, and the field promises to provide an equally reliable probe of a far more remote past. Once we map the distribution of primordial fluctuation amplitudes with scale, perhaps within a decade, our future evolution will be deciphered.

It has been said that our destiny is bleak, to end up pulverized into a fiery mush were the universe to recontract, or else abysmally cold and dark were it to expand forever. "The more the universe seems comprehensible, the more it also seems pointless" was the gloomy reaction of cosmologist Steven Weinberg. Not all cosmologists share this state of terminal depression. For there is the prospect of infinite knowledge even in the infinite future. Freeman Dyson has argued that sentient beings may even survive the proton decays that will occur some 10^{32} years hence. There is cause for optimism. The painstakingly accumulated information content of the universe may even survive a future big crunch, to re-emerge, phoenixlike, in another cycle of the big bang.

We simply do not know enough at present to offer intelligent critiques of these speculations. It is a time for hope, for marveling at the beauty and simplicity of the big bang theory that is gradually falling into place, and for anticipating the future discoveries that will fill the remaining gaps in our knowledge.

For Further Reading

Popular books on modern cosmology

Barrow, J. D. *Theories of Everything.* New York: Ballantine, 1992.

Barrow, J. D., and J. Silk. *The Left Hand of Creation.* New York: Basic Books, 1983; Oxford University Press, 1993.

Bartuski, M. *Through a Universe Darkly.* New York: HarperCollins, 1993.

Carrigan, R. A., and W. P. Trower. *Particle Physics in the Cosmos.* Readings from Scientific American. New York: W. H. Freeman, 1989.

———. *Particles and Forces.* Readings from Scientific American. New York: W. H. Freeman, 1990.

Davies, P. C. W. *The Edge of Infinity.* London: Dent, 1981.

Davies, P. C. W., and J. Brown. *Superstrings: A Theory of Everything?* Cambridge: Cambridge University Press, 1992.

Ferris, T. *Coming of Age in the Milky Way.* New York: Morrow, 1988.

———. *The Red Limit: The Search for the Edge of the Universe.* New York: Morrow, 1983.

Gribbin, J. *In Search of the Big Bang.* New York: Bantam, 1986.

———. *In the Beginning.* Boston: Little, Brown, 1993.

Gribbin, J., and M. J. Rees. *Cosmic Coincidences: Dark Matter, Mankind, and Anthropic Cosmology.* New York: Bantam, 1989.

Hawking, S. W. *A Brief History of Time.* New York: Bantam, 1988.

Kaufmann, W. *The Cosmic Frontiers of General Relativity.* Boston: Little, Brown, 1977.

Krauss, L. *The Fifth Essence: The Search for Dark Matter in the Universe.* New York: Basic Books, 1989.

Lederman, L., and D. N. Schramm. *From Quarks to the Cosmos.* Scientific American Library. New York: W. H. Freeman, 1989.

Lightman, A. *Ancient Light.* Cambridge, Mass.: Harvard University Press, 1991.

Riordan, M., and D. Schramm. *The Shadows of Creation.* New York: W. H. Freeman, 1991.

Pagels, H. *Perfect Symmetry.* London: M. Joseph, 1985.

Silk, J. *The Big Bang,* 2nd ed. New York: W. H. Freeman, 1989.

———. *Cosmic Enigmas.* New York: American Institute of Physics, 1994.

Weinberg, S. *Dreams of an Ultimate Theory.* New York: Basic Books, 1993.

Zee, A. *Fearful Symmetry.* New York: Macmillan, 1986.

Popular books on the history of cosmology

Bertotti, B., et al. *Modern Cosmology in Retrospect.* Cambridge: Cambridge University Press, 1990.

Charon, J. *Cosmology.* New York: McGraw-Hill, 1977.

North, J. *The Measure of the Universe.* New York: Dover, 1965.

Popular books on the philosophy and theology of cosmology

Jaki, S. *Cosmos and Creator.* Edinburgh: Scottish Academic Press, 1980.

Davies, P. *God and the New Physics.* New York: Simon and Schuster, 1985.

———. *The Mind of God.* New York: Simon and Schuster, 1992.

Peters, T. *Cosmos as Creation.* Nashville: Abingdon Press, 1989.

Books on cosmology on a slightly higher but nonmathematical level

Barrow, J. D. *The World Within the World.* Oxford: Oxford University Press, 1988.

Berendzen, R., R. Hart, and D. Seeley. *Man Discovers the Galaxies.* New York: Science History Publications, 1976.

Close, F. *The Cosmic Onion.* London: Heinemann, 1983.

Cornell, J., ed. *Bubbles, Voids and Bumps in the New Cosmology.* Cambridge: Cambridge University Press, 1989.

Davies, P. C. W., ed. *The New Physics.* Cambridge: Cambridge University Press, 1989.

Drees, W. *Beyond the Big Bang: Quantum Cosmology and God.* La Salle: Open Court, 1990.

Harrison, E. *Cosmology: The Science of the Universe.* Cambridge: Cambridge University Press, 1981.

Longair, M. *The Origins of the Universe.* Cambridge: Cambridge University Press, 1990.

Luminet, J-P. *Black Holes.* Cambridge: Cambridge University Press, 1993.

Peat, F. D. *Superstrings and the Search for a Theory of Everything.* Chicago: Contemporary Books, 1988.

Tayler, R. J. *Hidden Matter.* Chichester, England: Ellis Harwood, 1991.

Weinberg, S. *The First Three Minutes.* New York: Basic Books, 1977.

Cosmology books on a more technical level

Barrow, J. D., and F. J. Tipler. *The Anthropic Cosmological Principle.* Oxford: Oxford University Press, 1986.

Kolb, E., and M. S. Turner. *The Early Universe.* Redwood City, Calif.: Addison-Wesley, 1990.

Linde, A. *Particle Physics and Inflationary Cosmology.* New York: Harwood, 1990.

Narlikar, J. V. *Introduction to Cosmology,* 2nd ed. Cambridge: Cambridge University Press, 1993.

Padmanabhan, T. *Structure Formation in the Universe.* Cambridge: Cambridge University Press, 1993.

Peebles, P. J. *Principles of Physical Cosmology.* Princeton, N.J.: Princeton University Press, 1993.

Shu, F. *The Physical Universe.* Mill Valley, Calif.: University Science Books, 1982.

Tryon, E. P. "Cosmic inflation," in *The Encyclopedia of Physical Science and Technology,* Vol. 3. New York: Academic Press, 1987, pp. 709–43.

Vilenkin, A., and E. P. S. Shellard, *Cosmic Strings and Other Topological Defects.* Cambridge: Cambridge University Press, 1993.

Sources of Illustrations

Background photograph for all chapter openers © 1994 Roger Ressmeyer/Starlight. Line illustrations are by Rolin Graphics and George Retseck.

Chapter 1
page 4: Ian Gatley and Michael Merril/NOAO. Starlight Photography. *page 7:* Tony Hallas. *page 9:* J. Trümper, Max-Planck-Institut für Extraterrestrische Physik, Garching. *page 13:* (top left and right) TERSCH. *page 14:* David Malin/Anglo-Australian Observatory. *page 18:* Los Alamos National Laboratory. *page 23:* David Malin/Anglo-Australian Observatory.

Chapter 2
page 26: (left) Vela: David Malin/Australian Observatory; (right) European Southern Observatory. *page 33:* (top) Harvard University Archives; (bottom) Hale Observatories. Courtesy AIP Emilio Segrè Visual Archives. *page 37:* (left and inset) Space Telescope Science Institute/NASA. *page 38:* David Malin/Anglo-Australian Observatory. *page 41:* (bottom) National Optical Astronomy Observatories. *page 43:* Space Telescope Science Institute/NASA. *page 46:* European Southern Observatory. *page 48:* NRAO.

Chapter 3
page 50: NASA Cosmic Background Explorer Satellite (COBE). Differential Microwave Radiometer (DMR). Goddard Space Flight Center, Greenbelt, Maryland. DMR Principal Investigator: George F. Smoot, Space Sciences Laboratory, Lawrence Berkeley Laboratory, Physics Department, University of California, Berkeley. *page 53:* Cramponi Brussels. *page 54:* Roger Ressmeyer/Starlight. *page 55:* (top) Physics Today Collection. AIP Emilo Segrè Visual Archives. *page 56:* Adapted from J. C. Mathei et al. *Astrophysical Journal,* vol. 420, p. 439 (1994). *page 57:* NASA Cosmic Background Explorer Satellite (COBE). Differential Microwave Radiometer (DMR). Goddard Space Flight Center, Greenbelt, Maryland. DMR Principal Investigator: George F. Smoot, Space Sciences Laboratory, Lawrence Berkeley Laboratory, Physics Department, University of California, Berkeley. *pages 62 and 63:* (top left and top right) E. Bunn, University of California, Berkeley.

Chapter 4
pages 64 and 71: CERN. page 77: Fermi.

Chapter 5
page 84: David Malin/Anglo-Australian Observatory. *page 93:* CERN. *page 96:* Dennis di Cicco.

Chapter 6
page 98: George Efstathiou, Oxford University. *page 110:* Joseph Silk, European Southern Observatory.

Chapter 7
page 114: Yannick Mellier, Observatoire de Toulouse. *page 118:* David Malin/Anglo-Australian Observatory. *page 122:* Yannick Mellier, Observatoire de Toulouse. *page 123:* Nick Kaiser, CITA, University of Toronto. *page 124:* Avishai Dekel, Hebrew University, Jerusalem. *page 126:* Adapted from J. Linsky et al. *Astrophysical Journal,* vol. 407, p. 694 (1994). *pages 127 and 128:* Adapted from T. Walker et al. *Astrophysical Journal,* vol. 376, p. 51 (1991).

Chapter 8
page 132: Lawrence Livermore National Laboratory. From Charles Alcock et al. "Possible gravitational

microlensing of a star in the Large Magellanic Cloud," *Nature,* vol. 365, no. 6447, (Oct. 14, 1993). *page 141:* David Malin/Anglo-Australian Observatory. page 140: IMB/Astronomical Society of the Pacific. *page 141:* Malin/Pasachoff/Caltech, 1992. *page 143:* Energetic Gamma Ray Experiment Telescope Science Team. *page 144:* Bildarchiv Preussischer Kulturbesitz, Berlin. *page 147:* Space Telescope Science Institute/NASA. *page 151:* Space Telescope Science Institute/NASA. *page 152:* Space Telescope Science Institute/NASA. *page 153:* A. Readhead, California Institute of Technology. *page 154:* Stephen C. Unwin, California Institute of Technology.

Chapter 9

page 158: NASA. *page 164:* (bottom) Adapted from A. Wolfe et al. *Astrophysical Journal,* vol. 404, p. 480 (1993). *page 167:* S. D. M. White, University of Cambridge/ROSAT. *page 169:* (bottom) Anthony Lasenby, MRAO.

Chapter 10

page 170: (left) S. Djorgovski, University of California, Berkeley. From Scientific American, p. 74 (Oct. 1983); (right) Ed Bertschinger, MIT. *page 176:* (bottom) Alex Szalay. *page 177:* Carlos Frenk, University of Durham. *page 181:* F. Bouchet, Institut d'Astrophysique, and L. Hernquist, University of California, Santa Cruz. *page 190:* NASA Cosmic Background Explorer Satellite (COBE). Differential Microwave Radiometer (DMR). Goddard Space Flight Center, Greenbelt, Maryland. DMR Principal Investigator: George F. Smoot, Space Sciences Laboratory, Lawrence Berkeley Laboratory, Physics Department, University of California, Berkeley.

Chapter 11

page 192: (left) Margaret Geller, Luis da Costa, John Huchra, and Emilio Falco, Smithsonian Astrophyscial Observatory; (right) Valerie de Lapparent, Margaret Geller, and John Huchra, Smithsonian Astrophysical Observatory. *page 195:* (top) Bob Williams, Space Telescope Science Institute, and Alan Dressler, Carnegie Institute; (bottom left) Carlos Frenk, University of Durham; (bottom right) M. Seldner, B. L. Siebers, E. J. Groth, and P. J. E. Peebles, *Astronomical Journal,* vol. 82, p. 249 (1977). *page 198:* Valerie de Lapparent, Margaret Geller, and John Huchra, Smithsonian Astrophysical Observatory. *page 199:* (top) Frank Summer, Princeton University Observatory. *page 201:* D. Mathewson et al. *Astrophysical Journal,* vol. 389, p. L5 (1992). *page 204:* Frank Summer, Princeton University Observatory. *page 205:* Greg Bryan and Michael Norman/NCSA. *page 207:* (top) Ed Bertschinger, MIT; (bottom) George Efstathiou, Oxford University. *page 210:* Albert Stebbins, University of Chicago.

Chapter 12

page 212: (left) C. Robert O'Dell, William Marsh Rice University, Houston; (right) David Malin, Anglo-Australian Observatory. *page 215:* National Optical Astronomy Observatories. *page 225:* NASA. *page 226:* (top) David Malin/Anglo-Australian Observatory; (bottom) NRAO. *page 228:* D. Malin and D. Carter, *Astrophysical Journal,* p. 274–534 (1983). *page 229:* (top) Lars Hernquist, University of California, Santa Cruz; (bottom) Richard Griffiths, Johns Hopkins University. *page 230:* (left) NOAO; (right) NRAO. *page 231:* A. Evrard, University of Michigan. *page 234:* E. Bunn, University of California, Berkeley.

Index

Other books in the Scientific American Library Series

SUN AND EARTH
by Herbert Friedman

ISLANDS
by H. William Menard

DRUGS AND THE BRAIN
by Solomon H. Snyder

EXTINCTION
by Steven M. Stanley

EYE, BRAIN AND VISION
by David H. Hubel

SAND
by Raymond Siever

THE HONEY BEE
by James L. Gould and Carol Grant Gould

ANIMAL NAVIGATION
by Talbot H. Waterman

SLEEP
by J. Allan Hobson

FROM QUARKS TO THE COSMOS
by Leon M. Lederman and David N. Schramm

SEXUAL SELECTION
by James L. Gould and Carol Grant Gould

THE NEW ARCHAEOLOGY AND THE ANCIENT MAYA
by Jeremy A. Sabloff

A JOURNEY INTO GRAVITY AND SPACE-TIME
by John Archibald Wheeler

SIGNALS
by John R. Pierce and A. Michael Noll

BEYOND THE THIRD DIMENSION
by Thomas F. Banchoff

DISCOVERING ENZYMES
by David Dressler and Huntington Potter

THE SCIENCE OF WORDS
by George A. Miller

ATOMS, ELECTRONS, AND CHANGE
by P. W. Atkins

VIRUSES
by Arnold J. Levine

DIVERSITY AND THE TROPICAL RAINFOREST
by John Terborgh

STARS
by James B. Kaler

EXPLORING BIOMECHANICS
by R. McNeill Alexander

CHEMICAL COMMUNICATION
by William C. Agosta

GENES AND THE BIOLOGY OF CANCER
by Harold Varmus and Robert A. Weinberg

SUPERCOMPUTING AND THE TRANSFORMATION OF SCIENCE
by William J. Kaufmann III and Larry L. Smarr

MOLECULES AND MENTAL ILLNESS
by Samuel H. Barondes

EXPLORING PLANETARY WORLDS
by David Morrison

EARTHQUAKES AND GEOLOGICAL DISCOVERY
by Bruce A. Bolt

THE ORIGIN OF MODERN HUMANS
by Roger Lewin

THE EVOLVING COAST
by Richard A. Davis, Jr.

THE LIFE PROCESSES OF PLANTS
by Arthur W. Galston

IMAGES OF MIND
by Michael I. Posner and Marcus E. Raichle

THE ANIMAL MIND
by James L. Gould and Carol Grant Gould

MATHEMATICS: THE SCIENCE OF PATTERNS
by Keith Devlin